Molecular Pathology: Predictive, Prognostic, and Diagnostic Markers in Tumors

Editor

LYNETTE M. SHOLL

SURGICAL PATHOLOGY CLINICS

www.surgpath.theclinics.com

Consulting Editor
JASON L. HORNICK

September 2016 • Volume 9 • Number 3

ELSEVIER

1600 John F. Kennedy Boulevard • Suite 1800 • Philadelphia, Pennsylvania, 19103-2899

http://www.theclinics.com

SURGICAL PATHOLOGY CLINICS Volume 9, Number 3
September 2016 ISSN 1875-9181, ISBN-13: 978-0-323-46268-6

Editor: Lauren Boyle
Developmental Editor: Donald Mumford

Surgical Pathology Clinics (ISSN 1875-9181) is published quarterly by Elsevier Inc., 360 Park Avenue South, New York, NY 10010. Months of issue are March, June, September, and December. Business and Editorial Office: Elsevier Inc., 1600 John F. Kennedy Blvd., Ste. 1800, Philadelphia, PA 19103-2899. Accounting and Circulation Offices: Elsevier Inc., 3251 Riverport Lane, Maryland Heights, MO 63043. Periodicals postage paid at New York, NY and at additional mailing offices. Subscription prices are $200.00 per year (US individuals), $263.00 per year (US institutions), $100.00 per year (US students/residents), $250.00 per year (Canadian individuals), $300.00 per year (Canadian Institutions), $250.00 per year (foreign individuals), $300.00 per year (foreign institutions), and $120.00 per year (international & Canadian students/residents). Foreign air speed delivery is included in all Clinics' subscription prices. All prices are subject to change without notice. **POSTMASTER:** Send address changes to Surgical Pathology Clinics, Elsevier, 3251 Riverport Lane, Maryland Heights, MO 63043. **Customer Service: 1-800-654-2452 (US). From outside the United States, call 1-314-447-8871. Fax: 1-314-447-8029. E-mail: JournalsCustomerServiceusa@elsevier.com (for print support)** and JournalsOnlineSupport-usa@elsevier.com **(for online support)**.

Reprints. For copies of 100 or more, of articles in this publication, please contact the Commercial Reprints Department, Elsevier Inc., 360 Park Avenue South, New York, NY 10010-1710. Tel. 212-633-3874; Fax: 212-633-3820; E-mail: reprints@elsevier.com.

Surgical Pathology Clinics of North America is covered in MEDLINE/PubMed (Index Medicus).

Contributors

CONSULTING EDITOR

JASON L. HORNICK, MD, PhD
Director of Surgical Pathology, Director, Immunohistochemistry Laboratory, Brigham and Women's Hospital, Associate Professor of Pathology, Harvard Medical School, Boston, Massachusetts

EDITOR

LYNETTE M. SHOLL, MD
Department of Pathology, Brigham and Women's Hospital, Assistant Professor, Harvard Medical School, Boston, Massachusetts

AUTHORS

JUDITH V.M.G. BOVÉE, MD, PhD
Department of Pathology, Leiden University Medical Center, Leiden, The Netherlands

CHRISTOPHER C. GRIFFITH, MD, PhD
Department of Pathology and Laboratory Medicine, Emory University, Atlanta, Georgia

DONNA E. HANSEL, MD, PhD
Chief, Division of Anatomic Pathology; Professor, Department of Pathology, University of California, San Diego, La Jolla, California

CALEB HO, MD
Department of Pathology, Memorial Sloan Kettering Cancer Center, New York, New York

JASON L. HORNICK, MD, PhD
Director of Surgical Pathology, Director, Immunohistochemistry Laboratory, Brigham and Women's Hospital, Associate Professor of Pathology, Harvard Medical School, Boston, Massachusetts

BROOKE E. HOWITT, MD
Associate Pathologist, Department of Pathology, Brigham and Women's Hospital, Instructor in Pathology, Harvard Medical School, Boston, Massachusetts

JASON T. HUSE, MD, PhD
Department of Pathology, Memorial Sloan-Kettering Cancer Center, New York, New York

MICHAEL J. KLUK, MD, PhD
Department of Pathology, Weill Cornell Medical College, New York, New York

FRANK C. KUO, MD, PhD
Associate Professor of Pathology, Director for Assay Development, Center for Advanced Molecular Diagnostics, Brigham and Women's Hospital, Harvard Medical School, Boston, Massachusetts

ADRIÁN MARIÑO-ENRÍQUEZ, MD, PhD
Department of Pathology, Brigham and Women's Hospital, Harvard Medical School, Boston, Massachusetts

JONATHAN A. NOWAK, MD, PhD
Associate Pathologist, Department of
Pathology, Brigham and Women's Hospital,
Instructor of Pathology, Harvard Medical
School, Boston, Massachusetts

LAUREN L. RITTERHOUSE, MD, PhD
Department of Pathology, Brigham and
Women's Hospital, Clinical Fellow, Harvard
Medical School, Boston, Massachusetts

RAJA R. SEETHALA, MD
Department of Pathology and Laboratory
Medicine, Presbyterian University Hospital,
University of Pittsburgh, Pittsburgh,
Pennsylvania

LYNETTE M. SHOLL, MD
Department of Pathology, Brigham and
Women's Hospital, Assistant Professor,
Harvard Medical School, Boston,
Massachusetts

AATUR D. SINGHI, MD, PhD
Assistant Professor, Department of Pathology,
University of Pittsburgh Medical Center,
Pittsburgh, Pennsylvania

JAMES P. SOLOMON, MD, PhD
Pathology Resident, Department of Pathology,
University of California, San Diego, La Jolla,
California

BRIAN K. THEISEN, MD
Department of Pathology, University of
Pittsburgh Medical Center, Pittsburgh,
Pennsylvania

ABIGAIL I. WALD, PhD
Department of Pathology, University of
Pittsburgh Medical Center, Pittsburgh,
Pennsylvania

Contents

> Although initial attempts at using ancillary studies in salivary gland tumor classification were viewed with skepticism, numerous advances over the past decade have established a role for assessment of molecular alterations in the diagnosis and potential prognosis and treatment of salivary gland tumors. Many monomorphic salivary tumors are now known to harbor defining molecular alterations, usually translocations. Pleomorphic, high-grade carcinomas tend to have complex alterations that are often further limited by inaccuracy of initial classification by morphologic and immunophenotypic features. Next-generation sequencing techniques have great potential in many aspects of salivary gland tumor classification and biomarker discovery.

> Advances in lung cancer genomics have revolutionized the diagnosis and treatment of this heterogeneous and clinically significant group of tumors. This article provides a broad overview of the most clinically relevant oncogenic alterations in common and rare lung tumors, with an emphasis on the pathologic correlates of the major oncogenic drivers, including *EGFR*, *KRAS*, *ALK*, and *MET*. Illustrations emphasize the morphologic diversity of lung adenocarcinoma, including genotype-phenotype correlations of genomic evolution in tumorigenesis. Molecular diagnostic approaches, including PCR-based testing, massively parallel sequencing, fluorescence in situ hybridization, and immunohistochemistry are reviewed.

> The comprehensive molecular profiling of cancer has dramatically altered conceptions of numerous tumor types, particularly with regard to their fundamental classification. In the case of primary brain tumors, the widespread use of disease-defining biomarker sets is profoundly reshaping existing diagnostic entities that had been designated solely by histopathological criteria for decades. This review describes recent progress for diffusely infiltrating gliomas of adulthood, the most common primary brain tumor variants. More specifically, it details how routine incorporation of a handful of highly prevalent molecular alterations robustly designates refined subclasses of glioma that transcend conventional histopathological designations.

> Although there have been many recent discoveries in the molecular alterations associated with urothelial carcinoma, current understanding of this disease lags behind

many other malignancies. Historically, a two-pathway model had been applied to distinguish low- and high-grade urothelial carcinoma, although significant overlap and increasing complexity of molecular alterations has been recently described. In many cases, mutations in *HRAS* and *FGFR3* that affect the MAPK and PI3K pathways seem to be associated with noninvasive low-grade papillary tumors, whereas mutations in *TP53* and *RB* that affect the G1-S transition of the cell cycle are associated with high-grade in situ and invasive carcinoma. However, recent large-scale analyses have identified overlap in these pathways relative to morphology, and in addition, many other variants in a wide variety of oncogenes and tumor-suppressor genes have been identified. New technologies including next-generation sequencing have enabled more detailed analysis of urothelial carcinoma, and several groups have proposed molecular classification systems based on these data, although consensus is elusive. This article reviews the current understanding of alterations affecting oncogenes and tumor-suppressor genes associated with urothelial carcinoma, and their application in the context of morphology and classification schema.

This article focuses on the diagnostic, prognostic, and predictive molecular biomarkers in uterine malignancies, in the context of morphologic diagnoses. The histologic classification of endometrial carcinomas is reviewed first, followed by the description and molecular classification of endometrial epithelial malignancies in the context of histologic classification. Taken together, the molecular and histologic classifications help clinicians to approach troublesome areas encountered in clinical practice and evaluate the utility of molecular alterations in the diagnosis and subclassification of endometrial carcinomas. Putative prognostic markers are reviewed. The use of molecular alterations and surrogate immunohistochemistry as prognostic and predictive markers is also discussed.

Molecular testing in colorectal cancer helps to address multiple clinical needs. Evaluating the mismatch repair pathway status is the most common use for molecular diagnostics and this testing provides prognostic information, guides therapeutic decisions and helps identify Lynch syndrome patients. For patients with metastatic colorectal cancer, testing for activating mutations in downstream components of the EGFR signaling pathway can identify patients who will benefit from anti-EGFR therapy. Emerging molecular tests for colorectal cancer will help further refine patient selection for targeted therapies and may provide new options for monitoring disease recurrence and the development of treatment resistance.

Within the past few decades, there has been a dramatic increase in the detection of incidental pancreatic cysts. It is reported a pancreatic cyst is identified in up to 2.6% of abdominal scans. Many of these cysts, including serous cystadenomas and pseudocysts, are benign and can be monitored clinically. In contrast, mucinous cysts, which include intraductal papillary mucinous neoplasms and mucinous cystic neoplasms, have the potential to progress to pancreatic adenocarcinoma. In this review, we discuss the current management guidelines for pancreatic cysts, their underlying genetics, and the integration of molecular testing in cyst classification and prognostication.

Sarcomas are infrequent mesenchymal neoplasms characterized by notable morphological and molecular heterogeneity. Molecular studies in sarcoma provide refinements to morphologic classification, and contribute diagnostic information (frequently), prognostic stratification (rarely) and predict therapeutic response (occasionally). Herein, we summarize the major molecular mechanisms underlying sarcoma pathogenesis and present clinically useful diagnostic, prognostic and predictive molecular markers for sarcoma. Five major molecular alterations are discussed, illustrated with representative sarcoma types, including 1. the presence of chimeric transcription factors, in vascular tumors; 2. abnormal kinase signaling, in gastrointestinal stromal tumor; 3. epigenetic deregulation, in chondrosarcoma, chondroblastoma, and other tumors; 4. deregulated cell survival and proliferation, due to focal copy number alterations, in dedifferentiated liposarcoma; 5. extreme genomic instability, in conventional osteosarcoma as a representative example of sarcomas with highly complex karyotype.

Application of next-generation sequencing (NGS) on myeloid neoplasms has expanded our knowledge of genomic alterations in this group of diseases. Genomic alterations in myeloid neoplasms are complex, heterogeneous, and not specific to a disease entity. NGS-based panel testing of myeloid neoplasms can complement existing diagnostic modalities and is gaining acceptance in the clinics and diagnostic laboratories. Prospective, randomized trials to evaluate the prognostic significance of genomic markers in myeloid neoplasms are under way in academic medical centers.

Lymphoid neoplasms show great diversity in morphology, immunophenotypic profile, and postulated cells of origin, which also reflects the variety of genetic alterations within this group of tumors. This review discusses many of the currently known genetic alterations in selected mature B-cell and T-cell lymphoid neoplasms, and their significance as diagnostic, prognostic, and therapeutic markers. Given the rapidly increasing number of genetic alterations that have been described in this group of tumors, and that the clinical significance of many is still being studied, this is not an entirely exhaustive review of all of the genetic alterations that have been reported.

SURGICAL PATHOLOGY CLINICS

Preface

CrossMark

Lynette M. Sholl, MD
Editor

Advances in sequencing and cytogenetic technology have facilitated widespread adoption of molecular diagnostics in clinical management. Implications for the practicing pathologist are broad. The fundamental role of molecular markers is perhaps best established in sarcomas, where specific mutational, amplification, and fusion events are central to the diagnosis, particularly in the setting of tumors with poorly differentiated or variant morphology (Mariño-Enríquez and Bovée). The routine use of cytogenetic studies in salivary gland tumors, where oncogenic fusion events are common, has aided in more specific characterization of tumors with overlapping morphologic features and may be used to confirm the presence of potentially targetable alterations such as NTRK gene fusions in mammary analogue secretory carcinoma (Seethala and Griffith). The recognition of distinct pathways driving tumorigenesis has transformed previously heterogeneous diagnostic entities into more specific categories with predictive and prognostic implications. Distinctive molecular pathways define subtypes of endometrial carcinomas: defective DNA repair pathways in mismatch repair and POLE-defective tumors, PI3K pathway activation in common low-grade tumors, and *TP53* mutations in high-grade and/or serous-type tumors (Ritterhouse and Howitt). Similarly, comprehensive genomic studies of bladder cancer have expanded our understanding of the biologic diversity across the tumor type, with some notable genotype-phenotype correlations, including between micropapillary growth patterns and the presence of *ERBB2* activating mutations (Soloman and Hansel). Prognostically, molecular classification has been shown to outperform morphologic classification of diffuse gliomas, where dysregulated metabolic pathways driven by IDH mutations define most low-grade gliomas as distinctive entities from more aggressive IDH wild-type gliomas/glioblastomas (Huse).

In other settings, molecular markers may facilitate a more definitive diagnosis when morphologic clues are indeterminate or lacking. In myeloid disorders, including bone marrow cytopenias and cytosis, the presence of characteristic mutations in oncogenes, splicing factors, epigenetic regulators, and cohesins can confirm the presence of clonal hematopoiesis when clinical or pathologic features are equivocal (Kuo). In pancreatic cystic neoplasms, the incorporation of molecular analysis significantly improves on existing clinical, radiographic, and cytologic testing algorithms for the detection of advanced neoplasia (Theisen, Wald, and Singhi).

In the clinic, knowledge of specific molecular markers is central to therapeutic planning for some tumor types. Practice guidelines recommend testing for *EGFR* mutation and *ALK* and *ROS1* rearrangements in all patients with advanced stage lung adenocarcinoma in order to select appropriate therapy (Sholl). The identification of recurrent *BRAF* V600E mutations in hairy cell leukemia has opened up new opportunities for targeted therapy in patients with refractory disease (Ho and Kluk). In colon adenocarcinoma, RAS gene testing should be performed prior to initiating EGFR-blocking therapy (Nowak and Hornick). Familial cancer risk is informed by molecular analysis of colon and endometrial tumors in particular, and hereditary factors in other tumor types are likely to become clear as genomic analyses in patients with cancer are routinely incorporated into practice.

Surgical Pathology 9 (2016) ix–x
http://dx.doi.org/10.1016/j.path.2016.06.001

Diverse approaches may be applied to genomic and epigenomic analysis, including through in situ cytogenetic techniques, molecular methods, or protein surrogates via immunohistochemistry. This issue emphasizes the incorporation of these methods into clinical practice and emphasizes the diagnostic, predictive, and prognostic role of molecular markers across a wide range of tumor types.

Lynette M. Sholl, MD
Department of Pathology
Brigham and Women's Hospital
Harvard Medical School
75 Francis Street
Boston, MA 02115, USA

E-mail address:
lmsholl@partners.org

Molecular Pathology
Predictive, Prognostic, and Diagnostic Markers in Salivary Gland Tumors

Raja R. Seethala, MD[a],*, Christopher C. Griffith, MD, PhD[b]

KEYWORDS

• Salivary • Molecular • Translocations • Mutations • Classification

---

Key points

- Recognition of defining molecular alterations, usually translocations, in monomorphic salivary gland tumors has emerged as a paradigm for tumor diagnosis.

- Proper front-end traditional morphologic and immunophenotypic characterization improves the impact of the complex alterations seen in pleomorphic high-grade salivary carcinomas.

- Mucoepidermoid carcinoma, adenoid cystic carcinoma, mammary analog secretory carcinoma, and hyalinizing clear cell carcinomas harbor translocations that are readily testable by fluorescence in situ hybridization.

- Salivary duct carcinoma is defined by an apocrine phenotype.

- The taxonomy of polymorphous low-grade adenocarcinoma and cribriform adenocarcinoma of salivary gland is currently debated and has recently incorporated the findings of next-generation sequencing identifying reproducible mutations and translocations in the *PRKD* family of genes.

ABSTRACT

Although initial attempts at using ancillary studies in salivary gland tumor classification were viewed with skepticism, numerous advances over the past decade have established a role for assessment of molecular alterations in the diagnosis and potential prognosis and treatment of salivary gland tumors. Many monomorphic salivary tumors are now known to harbor defining molecular alterations, usually translocations. Pleomorphic, high-grade carcinomas tend to have complex alterations that are often further limited by inaccuracy of initial classification by morphologic and immunophenotypic features. Next-generation sequencing techniques have great potential in many aspects of salivary gland tumor classification and biomarker discovery.

OVERVIEW

Salivary gland tumors are rare but morphologically diverse. The significant histologic overlap between tumor types with different biological behavior not only poses diagnostic challenges but also colored early attempts at using ancillary studies to refine diagnoses with a good deal of skepticism. However, the ensuing decades bore witness to significant refinements of morphologic criteria, developments of newer immunohistochemical markers, and the discovery of key molecular alterations in a variety of different tumor types. The initial defeatist attitude towards ancillary testing in salivary gland tumors has been slowly replaced with a more optimistic, integrative mindset. Although morphologic features remain the cornerstone of salivary gland tumor assessment, ancillary testing can now help refine diagnosis, and in

---

Disclosures: None.
[a] Department of Pathology and Laboratory Medicine, Presbyterian University Hospital, University of Pittsburgh, 200 Lothrop Street, Pittsburgh, PA 15213, USA; [b] Department of Pathology and Laboratory Medicine, Emory University Hospital, Midtown 550 Peachtree Street, Atlanta, GA 30308, USA
* Corresponding author.
E-mail address: seethalarr@upmc.edu

surgpath.theclinics.com

some instances provide prognostic and predictive value.

ASSOCIATED GENETIC CHANGES/ ALTERATIONS

Given the rarity of salivary gland tumors, the molecular understanding of these tumors was initially slow to evolve. Limitations have included a scarcity of in vitro and animal models and paucity of high-quality clinical annotation, limiting the relevance of any molecular alterations. Hereditary associations and precursor lesions are rare. Importantly, many tumor types are consistently misdiagnosed, which affects the accuracy of the available molecular findings. **Table 1** summarizes current key molecular alterations in salivary gland tumors.

DIAGNOSIS-DEFINING MOLECULAR ALTERATIONS: A NEW PARADIGM IN MONOMORPHIC SALIVARY GLAND TUMORS

One of the earliest reproducible translocations in salivary gland tumors, the t(11;19) in mucoepidermoid carcinoma, was discovered in 1994 by conventional karyotyping,[1] but it was nearly a decade later that the translocation partners *CRTC1* (*MECT1*) and *MAML2* were resolved.[2] In the following 5 to 10 years the diagnostic and prognostic uses and limitations of this translocation were established.[3–6] Now, mucoepidermoid carcinoma is regarded as one of the first to exemplify a new paradigm in monomorphic salivary gland tumors; namely that there is a high likelihood of a reproducible genetic alteration defining these tumors. Adenoid cystic carcinoma shares a similar story with mucoepidermoid carcinoma and is often characterized by an *MYB-NFIB* translocation.[7] Some consistent alterations were already established, notably *PLAG1* and *HMGA2* translocations in pleomorphic adenomas,[8,9] but largely ignored given that they characterized a benign tumor.

Other more recent diagnosis-defining translocations were established with a mixture of serendipity and astute correlation with the morphologic features of entities at other sites. For instance, *ETV6-NTRK3*–translocated

---

Table 1
Key molecular alterations in salivary gland tumors

Tumor	Chromosomal Alteration	Gene	Prevalence (%)
Pleomorphic adenoma	8q12 12q13–15 rearrangements	*PLAG1* *HMGA2*	25–30 10–15
Epithelial-myoepithelial carcinoma	11p15.5	*HRAS*	25
Tubulotrabecular basal cell adenoma/adenocarcinoma	3p22.1	*CTTNB1* mutation	60–70
Membranous basal cell adenoma/adenocarcinoma	16q12–13	*CYLD1* loss of heterozygosity/mutation	75–80
Mucoepidermoid carcinoma	t(11;19)(q21;p13) t(11;15)(q21;q26)	*CRTC1-MAML2* *CRTC3-MAML2*	40–80 ~5
Salivary duct carcinoma	17q21.1 3q26.32	*ERBB2* amplification *PIK3CA* mutation	~40 ~20
Adenoid cystic carcinoma	t(6;9)(q22–23;p23–24)	*MYB-NFIB*	25–50
Mammary analogue secretory carcinoma	t(12;15)(p13;q25) t(12;XXX)	*ETV6-NTRK3* *ETV6-XXX*	~95–98 (defining) ~2–5 (defining)
Hyalinizing clear cell carcinoma	t(12;22)(q21;q12)	*EWSR1-ATF1*	~80–90 (defining)
Myoepithelial carcinoma	t(22;XXX)	*EWSR1-XXX*	~40 clear cell morphology 0 non–clear cell morphology
Polymorphous low-grade adenocarcinoma/Cribriform adenocarcinoma of minor salivary origin	14q12 t(1;14)(p36.11;q12) t(X;14)(p11.4;q12)	*PRKD1* mutation *ARID1A-PRKD1* *DDX3X-PRKD1* *PRKD2* rearrangement *PRKD3* rearrangement	~20 ~24 ~13 ~16

mammary analog secretory carcinoma was discovered based on the morphologic similarities to another *ETV6-NTRK3*–translocated tumor, namely secretory carcinoma of breast.[10] Similarly, a translocation involving *EWSR1* was initially conjectured in hyalinizing clear cell carcinoma in part based on morphologic similarities to *EWSR1*-translocated cutaneous/soft tissue–type myoepitheliomas.[11] The resemblance of hyalinizing clear cell carcinoma to clear cell mucoepidermoid carcinoma also led investigators to speculate on the existence of a defining translocation. As this concept further crystallized, advances in high-throughput next-generation sequencing identified driving alterations in other rare tumors, including PRKD translocations and mutations in the polymorphous low-grade adenocarcinoma-cribriform adenocarcinoma of salivary gland spectrum of tumors.

HIGH-GRADE PLEOMORPHIC SALIVARY GLAND MALIGNANCIES: THE IMPORTANCE OF WALKING BEFORE RUNNING IN APPLYING ANCILLARY STUDIES

Clinically, it is especially important to advance the understanding of pleomorphic high-grade primary salivary gland carcinomas. Although the monomorphic salivary tumors tend to be indolent even if malignant and are typically amenable to surgical resection, the pleomorphic high-grade malignancies often present at advanced stage necessitating adjuvant treatment, for which conventional approaches have limitations. Thus although diagnostic markers are still of interest, therapeutically targetable molecular alterations are especially desirable. Tumors in this category may either represent de novo carcinomas or high-grade or transformed versions of monomorphic tumors outlined earlier that have accumulated additional molecular alterations.[12] It is logical to assume that the molecular profiles of pleomorphic salivary gland tumors show more complexity reflecting a higher level of genomic instability, making them difficult to characterize as cleanly as described earlier. However, the most significant impediment remains fundamental: these entities are consistently misdiagnosed, and in some instances the current taxonomic scheme is flawed.

The prototypical tumor type in this group is salivary duct carcinoma. This tumor has been popularly defined as an adenocarcinoma resembling ductal carcinoma of breast. As such, an analogous phenotype is assumed in salivary duct carcinoma. However, a molecular classification similar to that of breast carcinoma has been

attempted.[13] Markers such as HER2 were studied as a potential therapeutic target akin to ductal carcinoma of breast, but with less success when applied to patients.[14,15] On further scrutiny, the perceived heterogeneity of salivary duct carcinoma is rooted in its nonspecific definition. Simply put, any high-grade salivary carcinoma with a ductal phenotype may resemble ductal carcinoma of breast. Without more specific criteria, this entity would be prone to misdiagnosis from the start. However, with the discovery of androgen receptor (AR) reactivity in salivary duct carcinoma, the possibility of restricting salivary duct carcinoma to one with an apocrine phenotype became plausible.[16] A recent study shows that most tumors labeled as AR-negative salivary duct carcinoma are other entities.[17] Thus, before a particular molecular phenotype and associated therapeutic implications are assigned to a carcinoma type, it is critical that supporting studies show a more rigorous pathologic classification.

MOLECULAR PATHOLOGY OF CLASSIC/ COMMON TUMORS

MUCOEPIDERMOID CARCINOMA

Mucoepidermoid carcinoma is the most common salivary gland carcinoma in both adults and children. Most occur in the parotid, although they have been described in all upper aerodigestive tract sites. This tumor is defined by a population of uncommitted intermediate-type cells that transition toward mucinous and glandular differentiation, recapitulating normal salivary excretory duct phenotype.[18] Variant morphologies arise as a result of stromal features (ie, sclerosing) or modifications of various cell populations (ie, oncocytic, clear cell).[4,19] This tumor is arguably the only salivary gland tumor in which grade is both prognostically and therapeutically significant. Low-grade tumors have an indolent behavior and are typically cured by surgery alone, whereas high-grade tumors may be locoregionally aggressive and occasionally lethal, and often require neck dissection and adjuvant therapy. Intermediate-grade tumors vary in their behavior depending on the grading system used, making standardization of treatment difficult.[20,21]

Roughly 40% to 80% of tumors are characterized by a *CRTC1-MAML2* translocation, t(11;19)(q21;p13),[6] whereas another ~5% are characterized by a *CRTC3-MAML2* translocation, t(11;15)(q21;q26).[22,23] These translocations were initially thought to favor low-grade and intermediate-grade tumors and to represent independent favorable prognosticators.[3,5] However, the prevalence of this translocation in high-grade tumors is greater in more recent studies, considerably diminishing the prognostic value.[6,19,24,25] Furthermore, even positive translocation status does not guarantee indolent behavior because a subset of tumors with concurrent *CDKN2A* deletion show lethal behavior.[26,27] Thus the main utility of translocation testing in a clinical setting is to establish the diagnosis on biopsy, confirm diagnosis with variant morphologies, and to distinguish high-grade mucoepidermoid carcinoma (Fig. 1) from more aggressive entities such as salivary duct carcinoma and adenosquamous carcinoma.[4,24]

High-grade *MAML2* rearrangement–negative mucoepidermoid carcinomas have been reported to show a greater number of chromosomal gains and losses compared with low-grade tumors or rearrangement-positive high-grade tumors.[26]

ADENOID CYSTIC CARCINOMA

Adenoid cystic carcinoma is a common salivary gland carcinoma at both major and minor salivary sites and is fairly common in the sinonasal tract. Adenoid cystic carcinoma is a highly infiltrative biphasic proliferation of abluminal myoepithelial and luminal ductal cells arranged in a tubular, cribriform, or solid growth pattern. It is a prototypical blue or basaloid salivary tumor because all cell types tend to show scant cytoplasm and angulated, hyperchromatic nuclei. This tumor is characterized clinically by its slow but relentless progression. Local recurrence and distant metastases are common, but nodal disease is rare. Tumors are typically graded based on solid tumor composition.[20] Although conventional adenoid cystic carcinoma of all growth patterns is isomorphic, high-grade transformation (or dedifferentiation) has been described and consists of pleomorphic cells with necrosis, loss of biphasic arrangement of cells, and exuberant mitotic activity.[28]

The main molecular alteration in adenoid cystic carcinoma is the *MYB-NFIB* translocation t(6;9)(q22–23;p23–24) noted in 30% to 85% of cases (Fig. 2).[29–32] However, it is noted that *MYB* messenger RNA (mRNA) and MYB protein overexpression are noted in almost all adenoid cystic carcinomas, indicating that *MYB* alterations other than the aforementioned translocation exist. There is some suggestion that MYB mRNA levels may be linked to solid growth and aggressive behavior[33]; to date, *MYB* status has not been shown to consistently correlate with prognosis or other clinicopathologic features.[29,31] *MYB-NFIB* translocations and MYB protein expression have been noted in adenoid cystic carcinomas with high-grade transformation, although it is not necessarily required for transformation.[34]

Other genetic alterations in adenoid cystic carcinoma are varied, with solid tumors showing a high number of copy number alterations.[32] Consistent chromosomal loss at 1p36.33-p35.3 implicates several candidate tumor suppressor genes. Recurrent mutations in the fibroblast growth factor–insulinlike growth factor –phosphatidylinositol 3 kinase pathway in about 30% of cases, and

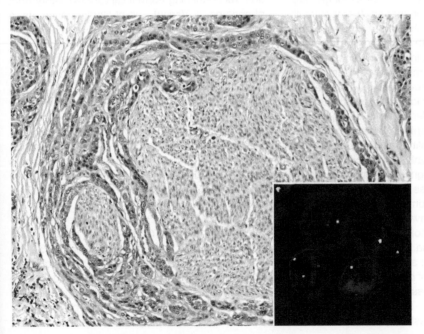

Fig. 1. High-grade mucoepidermoid carcinoma with extensive large nerve perineural invasion (hematoxylin-eosin [H&E], original magnification ×100). The aggressiveness should raise consideration for other entities. (*Inset*) *MAML2* break-apart fluorescence in situ hybridization (FISH) showing a split orange and green signal in several of the cells, indicating a translocation, thus confirming the diagnosis.

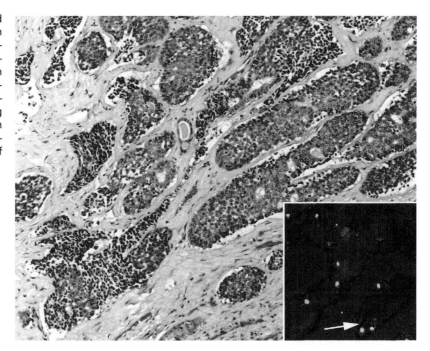

Fig. 2. Solid adenoid cystic carcinoma with some atypia raising consideration for incipient high-grade transformation (H&E, original magnification ×100). (*Inset*) *MYB-NFIB* fusion FISH showing a fused yellow signal in many cells (*arrow*), supporting the presence of this translocation.

the Notch signaling pathway in about 12% to 13% of cases may yield candidates for targeted therapy.[35–37] Recently, a subset (35%) of *MYB-NFIB*–negative cases have been described to harbor a t(8;9) translocation resulting in a novel *MYBL1-NFIB* translocation.

SALIVARY DUCT CARCINOMA

Salivary duct carcinoma is the prototypical primary salivary gland high-grade adenocarcinoma, and although generally described as having morphologic features akin to high-grade breast ductal carcinoma, it is more properly classified as having morphologic and molecular features of breast carcinoma with luminal AR-positive, apocrine phenotype.[17,38] Salivary duct carcinomas are most commonly encountered in older men and often arise from a preexisting pleomorphic adenoma. There is no role for grading in salivary duct carcinoma because aggressive behavior is seen in most cases. Variant morphologies occur rarely, including sarcomatoid, basaloid, papillary, and micropapillary. Given the aggressive nature, advanced stage at presentation, and poor prognosis associated with salivary duct carcinoma, it is hoped that identification of alternative molecular targets for therapy can improve outcomes compared with standard radiation and chemotherapy, which have had limited success.

When defined as a high-grade adenocarcinoma with apocrine phenotype, salivary duct carcinoma almost uniformly expresses AR (**Fig. 3**A, B).[17] As such, detection of AR expression by immunohistochemistry is a useful diagnostic aid in resolving the differential diagnosis with other high-grade carcinomas but also presents a potential therapeutic target. Several studies have shown benefit with androgen-deprivation therapy alone or in combination with more conventional radiotherapy in some patients.[39,40] Because of the rarity of this carcinoma and the limited experience with androgen-deprivation therapy, correlation between the intensity of AR expression and response has not been determined and there is currently no accepted threshold to define AR positivity.

The similarities between salivary duct carcinoma and breast carcinoma also include *ERBB2* (HER2) amplification as determined by fluorescence in situ hybridization (FISH), which is reported in 20% to 30% of cases.[14,38,41] HER2 overexpression can also be detected by immunohistochemistry and 20% to 40% of salivary duct carcinomas show 3+ membranous reactivity when defined using breast scoring criteria (**Fig. 3**C). Specific definitions of HER2 overexpression and amplification for salivary duct carcinoma have not been determined and currently these are extrapolated from the more robust experience with breast carcinoma. Treatment with anti-HER2 therapy in combination with bevacizumab[15] and

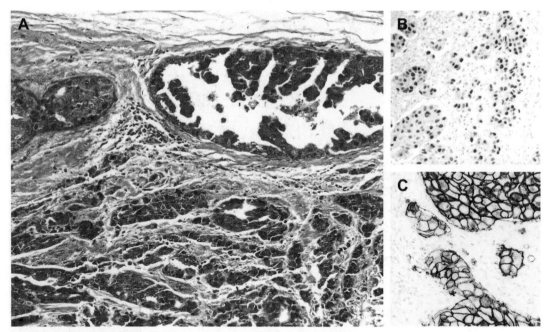

Fig. 3. Salivary duct carcinoma. (*A*) The tumor resembles a breast carcinoma (H&E, original magnification ×100) with micropapillary and apocrine morphology. (*B*) AR is strongly expressed in almost all salivary duct carcinomas (original magnification ×200). (*C*) A subset also shows 3+ HER2/neu expression by immunohistochemistry (original magnification ×400).

chemoradiation[42] has resulted in objective tumor response in some patients. However, complete response to anti-HER2 therapy is rare in salivary duct carcinoma and there is evidence that additional mutations involving *TP53* or *HRAS* or loss of *PTEN* may decrease the efficacy of HER2 blockade.[38]

Additional common molecular alterations in salivary duct carcinoma include mutations of *TP53*, *PIK3CA*, and *HRAS* and loss of *PTEN*.[38,43,44] *BRAF* mutations, *FGFR1* amplification, *CDKN2A/ p16* loss, and *MDM2* amplification are seen in small numbers of cases.

MAMMARY ANALOG SECRETORY CARCINOMA

Mammary analog secretory carcinoma is one of the most recently recognized and described salivary gland tumors and shows striking morphologic and molecular similarity to secretory carcinoma of the breast. This low-grade adenocarcinoma is characterized by tumor cells with fairly abundant highly vacuolated cytoplasm, usually low to moderate cytonuclear grade, and a variety of growth patterns, including cystic, papillary, and follicular.[10] Before recognition as a distinct entity, these tumors were often classified as zymogen granule–poor acinic cell carcinoma or adenocarcinoma not otherwise specified.[10,45]

Like its breast counterpart, mammary analog secretory carcinoma also shares a defining translocation involving *ETV6* most commonly fused to *NTRK3* as a result of a t(12;15)(p13;q25) translocation (Fig. 4).[10] Less commonly, translocations between *ETV6* and an as-yet unidentified partner gene also occur (ie, *ETV6*-X) and early evidence suggests a more aggressive phenotype in these cases.[46,47] In hopes of capitalizing on this molecular alteration, clinical trials using selective tropomyosin receptor kinase (TRK) inhibitors have been initiated.

POLYMORPHOUS LOW-GRADE ADENOCARCINOMA/CRIBRIFORM ADENOCARCINOMA OF SALIVARY GLAND

Polymorphous low-grade adenocarcinoma is a low-grade salivary carcinoma most commonly arising from palate and showing a tubulofascicular growth pattern (Fig. 5A). Recently, a subgroup of these tumors with more papillary glomeruloid or cribriform growth have been reclassified as cribriform adenocarcinoma of salivary glands given the frequent extrapalatal location (usually base of tongue) and higher propensity for lymph node metastases (Fig. 5B).[48] Whether cribriform adenocarcinoma of salivary gland is a distinct entity or simply a more aggressive form of polymorphous low-grade adenocarcinoma is a current taxonomic debate. Tumors histologically classified as

Fig. 4. Mammary analog secretory carcinoma showing the characteristic vacuolated cytoplasm and follicular growth pattern (H&E, original magnification ×100). (*Inset*) *ETV6* break-apart FISH showing several cells with gene rearrangement.

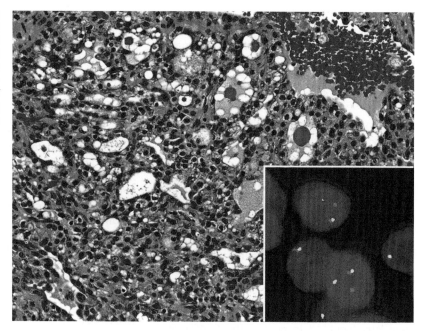

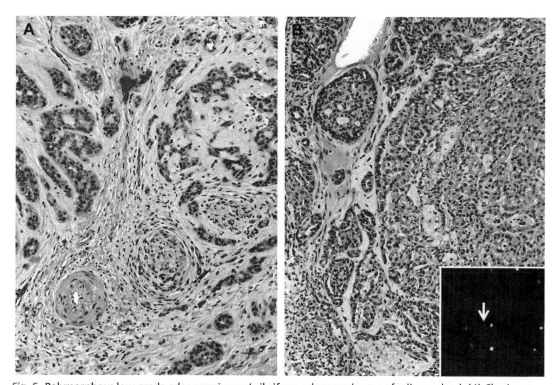

Fig. 5. Polymorphous low-grade adenocarcinoma/cribriform adenocarcinoma of salivary gland. (*A*) Classic neurotropic tubulofascicular growth in a palatal polymorphous low-grade adenocarcinoma (H&E, original magnification ×100). (*B*) This solid, cribriform to papillary base-of-tongue tumor has similar nuclear features but a distinctive growth pattern that some would categorize as cribriform adenocarcinoma of salivary gland (H&E, original magnification ×100). (*Inset*) Translocations in the *PRKD* family. Break-apart FISH showing a cell (*arrow*) with a *PRKD1* rearrangement. (*Courtesy of* Dr Ilan Weinreb, University Health Network, Toronto, ON, Canada.)

polymorphous low-grade adenocarcinoma frequently showed *PRKD-1* gene mutations,[49] whereas those classified as cribriform adenocarcinoma of minor salivary glands more frequently had chromosomal rearrangements (see **Fig. 5B**, inset) of *PRKD* genes (*PRKD-1*, *PRKD-2*, and *PRKD-3*).[48]

PLEOMORPHIC ADENOMA AND CARCINOMA EX PLEOMORPHIC ADENOMA

Pleomorphic adenoma is the most common salivary gland tumor and, although there is a characteristic morphology with ductal cells surrounded by myoepithelial cells and a myoepithelial derived chondromyxoid stroma, can occasional represent a diagnostic challenge, especially on small biopsies. Although benign, there is the risk of malignant transformation with pleomorphic adenoma (ie, carcinoma ex pleomorphic adenoma).

Early conventional karyotyping studies showed translocations at 8q12 and 12q13-15, which were later identified to contain the *PLAG1* and *HMGA2* genes, respectively.[50,51] Detection of these chromosomal rearrangements has the potential to be diagnostically useful in resolving the differential diagnosis of pleomorphic adenoma and its mimickers, especially on biopsy and fine-needle aspiration material. Overexpression of PLAG1 by immunohistochemistry is also possible and is more sensitive than FISH (positive in >90% of pleomorphic adenomas) but less specific (**Fig. 6**).[52–54] In the small number of cases tested to date, polymorphous low-grade adenocarcinoma, which often enters the differential diagnosis with pleomorphic adenoma, can show PLAG1 overexpression in a quarter of cases.

Importantly, detection of *PLAG1* or *HMGA2* rearrangements does not equate to a benign diagnosis in all cases because these are also detected in many cases of carcinoma ex pleomorphic adenoma.[55,56] Additional molecular alterations, such as amplification of 12q (especially *MDM2*), deletions of 5q23.2-q31.2, gains of 8q12.1 (*PLAG1*), gains of 8q22.1-q24.1 (*MYC*), and amplification of *ERBB2* (HER2), are likely important for progression to carcinoma ex pleomorphic adenoma.[55]

MOLECULAR PATHOLOGY OF RARE TUMORS

HYALINIZING CLEAR CELL CARCINOMA

Hyalinizing clear cell carcinoma is often a low-grade salivary gland tumor composed of clear cells with a squamous phenotype arranged in

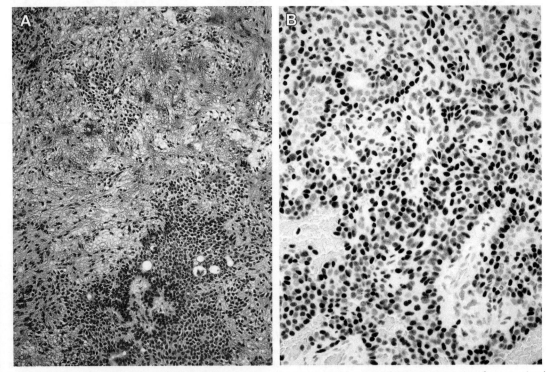

Fig. 6. Pleomorphic adenoma. (*A*) Typical pleomorphic adenoma chondromyxoid stroma streaming from a mixed ductal and myoepithelial tumor nest (H&E, original magnification ×200). (*B*) PLAG1 immunostain shows diffuse strong positivity (original magnification ×400). (*Courtesy of* Dr Vickie Y. Jo, Brigham and Women's Hospital, Boston, MA.)

nests and cords within a distinct fibrocellular hyalinized stroma. This tumor shares considerable morphologic, immunophenotypic, and molecular features with clear cell odontogenic carcinoma.[57]

Almost all tested cases of hyalinizing clear cell carcinoma show the *EWSR1-ATF* translocation [t(12;22)(q13;12)].[11] Detection of *EWSR1* rearrangements by break-apart FISH is readily available to aid in the diagnosis of hyalinizing clear cell carcinoma (Fig. 7) in challenging cases. However, taken out of context, *EWSR1* rearrangement is fairly promiscuous because it occurs in a wide variety of tumor types (eg, Ewing sarcoma, extraskeletal myxoid chondrosarcoma, desmoplastic small round cells tumor, soft tissue–type myoepithelioma), many of which can occur in the head and neck. More recently, a subset of salivary gland myoepithelial carcinomas have been reported to contain *EWSR1* rearrangements, further emphasizing the necessity of interpreting this translocation status in the right morphologic and immunophenotypic context.[58]

BASAL CELL ADENOMA AND ADENOCARCINOMA

Basal cell adenoma is a benign salivary gland tumor composed of predominantly basaloid cells arranged in tubulotrabecular, cribriform, membranous, or solid patterns. This tumor is biphasic with peripheral palisading of the outermost cell layer (Fig. 8A). More eosinophilic epithelial cells can also often be appreciated forming ductal structures with lumens. One of the more striking features in basal cell neoplasms is the abundant production of hyalinized basement membrane material in cylinders or membranes. An underappreciated characteristic of basal cell adenomas is an intervening myoepithelial derived stroma that is present to varying degrees, particularly in tubulotrabecular basal cell adenomas. Compared with basal cell adenoma, basal cell adenocarcinoma shows infiltration and may show higher cytonuclear grade and proliferation. In contrast with basal cell adenoma, membranous and solid patterns predominate, and basal cell adenocarcinoma has a higher propensity for recurrence and a potential for metastasis.

Membranous-type basal cell adenomas and adenocarcinoma (so-called dermal analog tumors) specifically show extensive morphologic similarity to dermal eccrine cylindromas and both are common manifestations in the rare autosomal dominant disease dermal cylindromatosis (Brooke-Spiegler syndrome). Loss of heterozygosity for the *CYLD* gene at 16q12-13, the same gene involved in heritable cylindromas, has been shown to occur in sporadic cases of membranous-type basal cell adenomas as well as in some basal cell adenocarcinomas.[59]

Basal cell adenomas, especially the tubulotrabecular types, are also characterized by nuclear expression of beta-catenin in the outermost basal layer in approximately 80% of cases (Fig. 8B) and

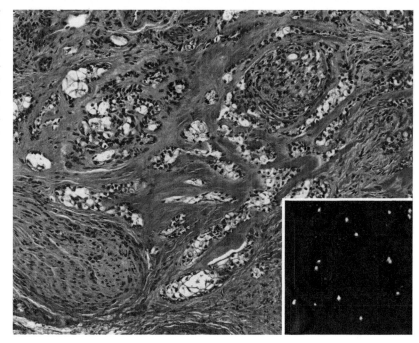

Fig. 7. Hyalinizing clear cell carcinoma showing a monomorphic but infiltrative tumor with perineural invasion, squamoid appearance, and distinctive hyaline stroma (H&E, original magnification ×100). (*Inset*) *EWSR1* break-apart FISH showing a rearrangement in the tumor.

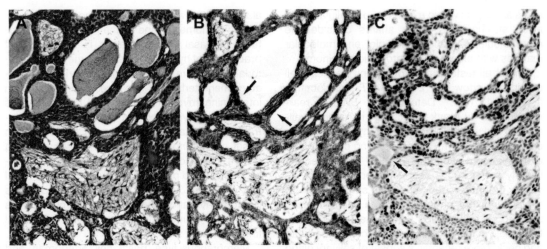

Fig. 8. Basal cell adenoma. (*A*) This tumor shows a tubulotrabecular and cribriform pattern with slight peripheral palisading of the outermost cell layer. These nests are distinct from the intervening myoepithelial derived stroma (H&E, original magnification ×200). (*B*) A subset of peripheral abluminal cells (*arrows*), and the myoepithelial derived stroma show nuclear accumulation of beta-catenin (original magnification ×200). (*C*) LEF-1 shows strong reactivity in the basal/myoepithelial cells and stroma, but is negative in the scattered ducts present (*arrow*) (original magnification ×200).

approximately 50% of cases show mutations of the *CTNNB1* gene.[60] Most basal cell adenocarcinomas also show nuclear localization of beta-catenin[61] but *CTNNB1* mutation testing has not been reported in this tumor type. Lymphoid enhancer-binding factor 1 (LEF-1), a transcription factor that interacts with beta-catenin in the Wnt pathway, also shows nuclear reactivity in approximately 60% to 70% of basal cell neoplasms (**Fig. 8**C).[62] Nuclear beta-catenin and LEF-1 reactivity is not seen with any frequency in other salivary gland tumor and is therefore useful as a diagnostic marker.

EPITHELIAL-MYOEPITHELIAL CARCINOMA

Epithelial-myoepithelial carcinoma is the prototypical biphasic salivary gland carcinoma composed of abluminal myoepithelial cells usually with abundant clear cytoplasm surrounding ductal structures. This low-grade malignancy often has a lobular invasive front. Recurrence is seen in fewer than half of patients and metastatic disease is even less common.

Epithelial-myoepithelial carcinoma is a rare tumor type and there are limited studies examining the molecular features. Some of these have identified *HRAS* codon 61 mutations in 26.7% to 33% of cases.[63,64] An isolated case has been noted to harbor a *EWSR1* rearrangement.[58]

MYOEPITHELIAL CARCINOMA

Myoepithelial carcinoma is a malignant salivary gland neoplasm composed almost entirely of

myoepithelial cells. Although some cases show similar morphology to benign myoepitheliomas with the only difference being invasive growth, most show additional features of malignancy with increased mitotic activity and myoepithelial cell atypia.

A recent report by Skalova and colleagues[58] identified *EWSR1* rearrangements in a subset of clear cell myoepithelial carcinomas. These cases shared some common histologic features, including necrosis, foci of squamous metaplasia, and hyalinization, and had more aggressive behavior compared with tumors without rearrangement.

TECHNIQUES

Molecular techniques used to detect alterations in a clinical setting are not different from those used at other organ sites.

Salivary tumor translocations have typically been detected in routine practice using break-apart FISH methodology with probe sets that flank the range of breakpoints for one of the translocation partners. This methodology can be applied readily to paraffin-embedded tissues of varying ages, and is even feasible on previously stained slides (ie, negative immunohistochemical control slides).[6,11,45] Fusion rather than break-apart probes can also be used, although these tend to be less sensitive and only detect 1 set of partners. Reverse transcription polymerase chain reaction for the fusion transcripts for a given tumor type is

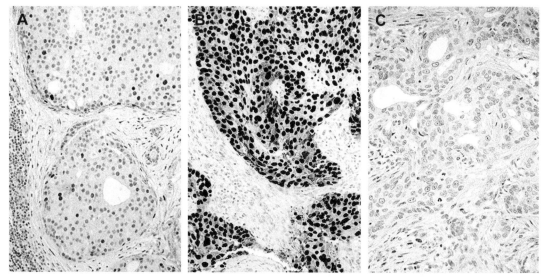

Fig. 9. Immunohistochemical staining for p53 in salivary duct carcinoma showing the spectrum of molecular correlates. (*A*) Wild-type *TP53* characteristically shows heterogeneous staining intensity and distribution for p53 in line with adjacent stroma and lymphocytes (original magnification ×200). (*B*) *TP53* missense mutations are characterized by extreme or strong positivity (original magnification ×200). (*C*) *TP53* deletions are characterized by extreme or complete negativity; note the wild-type expression in the desmoplastic stroma as an internal control (original magnification ×200). (*Courtesy of* Dr Simion Chiosea, University of Pittsburgh, Pittsburgh, PA.)

also feasible but requires good-quality mRNA and thus is not always applicable on paraffin tissues.[6,10] In addition, for tumors like adenoid cystic carcinoma, for which there may be a wide range of breakpoints and occasional insertions/deletions, testing for fusion transcripts may be complex.[32]

Gene amplifications (ie, HER-2) can be detected using FISH methodology with probes targeting the gene of interest as well as a chromosome enumeration probe. Criteria for defining therapeutically actionable amplification are not usually well established even for tumor types such as salivary duct carcinoma for various reasons noted earlier.[14,17] Point mutations, insertions, and deletions are typically detected by standard sequencing methods with amplification of the region of interest.[63] However, when potential mutations span a large region in a gene (ie, *TP53*), this adds complexity to sequencing.

The role of immunohistochemistry as a surrogate marker of a variety of molecular alterations is often underappreciated (**Fig. 9**). Immunohistochemical stains are readily available in a broad range of practice settings and easily adapted into a surgical pathology workflow. Although advantages and limitations depend on the marker of interest, immunostains in general are sensitive, but tend to be less specific markers of a molecular alteration, whether it is an amplification, mutation, or translocation.[29,38]

The high-throughput nature of next-generation sequencing methodologies in salivary gland tumors has allowed its application in discovery of novel genetic alterations in salivary gland tumors.[48,49] Clinical use is limited at this point, but in principle a panel with broad coverage of oncogenes and fusions can be applied and is feasible on paraffin-embedded tissues.[38]

REFERENCES

1. Nordkvist A, Gustafsson H, Juberg-Ode M, et al. Recurrent rearrangements of 11q14-22 in mucoepidermoid carcinoma. Cancer Genet Cytogenet 1994;74(2):77–83.

2. Tonon G, Modi S, Wu L, et al. t(11;19)(q21;p13) translocation in mucoepidermoid carcinoma creates a novel fusion product that disrupts a Notch signaling pathway. Nat Genet 2003;33(2):208–13.

3. Behboudi A, Enlund F, Winnes M, et al. Molecular classification of mucoepidermoid carcinomas-prognostic significance of the MECT1-MAML2 fusion oncogene. Genes Chromosomes Cancer 2006; 45(5):470–81.

4. Garcia JJ, Hunt JL, Weinreb I, et al. Fluorescence in situ hybridization for detection of MAML2 rearrangements in oncocytic mucoepidermoid carcinomas: utility as a diagnostic test. Hum Pathol 2011; 42(12):2001–9.

5. Okabe M, Miyabe S, Nagatsuka H, et al. MECT1-MAML2 fusion transcript defines a favorable subset

of mucoepidermoid carcinoma. Clin Cancer Res 2006;12(13):3902–7.

6. Seethala RR, Dacic S, Cieply K, et al. A reappraisal of the MECT1/MAML2 translocation in salivary mucoepidermoid carcinomas. Am J Surg Pathol 2010; 34(8):1106–21.

7. Nordkvist A, Mark J, Gustafsson H, et al. Nonrandom chromosome rearrangements in adenoid cystic carcinoma of the salivary glands. Genes Chromosomes Cancer 1994;10(2):115–21.

8. Mark J, Dahlenfors R, Ekedahl C, et al. Chromosomal patterns in a benign human neoplasm, the mixed salivary gland tumour. Hereditas 1982;96(1): 141–8.

9. Stenman G, Mark J, Ekedhal C. Relationships between chromosomal patterns and protooncogenes in human benign salivary gland tumors. Tumour Biol 1984;5(2):103–17.

10. Skalova A, Vanecek T, Sima R, et al. Mammary analogue secretory carcinoma of salivary glands, containing the ETV6-NTRK3 fusion gene: a hitherto undescribed salivary gland tumor entity. Am J Surg Pathol 2010;34(5):599–608.

11. Antonescu CR, Katabi N, Zhang L, et al. EWSR1-ATF1 fusion is a novel and consistent finding in hyalinizing clear-cell carcinoma of salivary gland. Genes Chromosomes Cancer 2011;50(7):559–70.

12. Nagao T. "Dedifferentiation" and high-grade transformation in salivary gland carcinomas. Head Neck Pathol 2013;7(Suppl 1):S37–47.

13. Di Palma S, Simpson RH, Marchio C, et al. Salivary duct carcinomas can be classified into luminal androgen receptor-positive, HER2 and basal-like phenotypes. Histopathology 2012;61(4): 629–43.

14. Clauditz TS, Reiff M, Gravert L, et al. Human epidermal growth factor receptor 2 (HER2) in salivary gland carcinomas. Pathology 2011;43(5): 459–64.

15. Falchook GS, Lippman SM, Bastida CC, et al. Human epidermal receptor 2-amplified salivary duct carcinoma: regression with dual human epidermal receptor 2 inhibition and anti-vascular endothelial growth factor combination treatment. Head Neck 2014;36(3):E25–7.

16. Fan CY, Wang J, Barnes EL. Expression of androgen receptor and prostatic specific markers in salivary duct carcinoma: an immunohistochemical analysis of 13 cases and review of the literature. Am J Surg Pathol 2000;24(4):579–86.

17. Williams L, Thompson LD, Seethala RR, et al. Salivary duct carcinoma: the predominance of apocrine morphology, prevalence of histologic variants, and androgen receptor expression. Am J Surg Pathol 2015;39(5):705–13.

18. Luna MA. Salivary mucoepidermoid carcinoma: revisited. Adv Anat Pathol 2006;13(6):293–307.

19. Schwarz S, Stiegler C, Muller M, et al. Salivary gland mucoepidermoid carcinoma is a clinically, morphologically and genetically heterogeneous entity: a clinicopathological study of 40 cases with emphasis on grading, histological variants and presence of the t(11;19) translocation. Histopathology 2011;58(4): 557–70.

20. Seethala RR. An update on grading of salivary gland carcinomas. Head Neck Pathol 2009;3(1):69–77.

21. Katabi N, Ghossein R, Ali S, et al. Prognostic features in mucoepidermoid carcinoma of major salivary glands with emphasis on tumour histologic grading. Histopathology 2014;65(6):793–804.

22. Fehr A, Roser K, Heidorn K, et al. A new type of MAML2 fusion in mucoepidermoid carcinoma. Genes Chromosomes Cancer 2008;47(3):203–6.

23. Nakayama T, Miyabe S, Okabe M, et al. Clinicopathological significance of the CRTC3-MAML2 fusion transcript in mucoepidermoid carcinoma. Mod Pathol 2009;22(12):1575–81.

24. Chiosea SI, Dacic S, Nikiforova MN, et al. Prospective testing of mucoepidermoid carcinoma for the MAML2 translocation: clinical implications. Laryngoscope 2012;122(8):1690–4.

25. Saade RE, Bell D, Garcia J, et al. Role of CRTC1/MAML2 translocation in the prognosis and clinical outcomes of mucoepidermoid carcinoma. JAMA Otolaryngol Head Neck Surg 2016;142(3): 234–40.

26. Jee KJ, Persson M, Heikinheimo K, et al. Genomic profiles and CRTC1-MAML2 fusion distinguish different subtypes of mucoepidermoid carcinoma. Mod Pathol 2013;26(2):213–22.

27. Anzick SL, Chen WD, Park Y, et al. Unfavorable prognosis of CRTC1-MAML2 positive mucoepidermoid tumors with CDKN2A deletions. Genes Chromosomes Cancer 2010;49(1):59–69.

28. Seethala RR, Hunt JL, Baloch ZW, et al. Adenoid cystic carcinoma with high-grade transformation: a report of 11 cases and a review of the literature. Am J Surg Pathol 2007;31(11):1683–94.

29. Brill LB 2nd, Kanner WA, Fehr A, et al. Analysis of MYB expression and MYB-NFIB gene fusions in adenoid cystic carcinoma and other salivary neoplasms. Mod Pathol 2011;24(9):1169–76.

30. Mitani Y, Li J, Rao PH, et al. Comprehensive analysis of the MYB-NFIB gene fusion in salivary adenoid cystic carcinoma: incidence, variability, and clinicopathologic significance. Clin Cancer Res 2010; 16(19):4722–31.

31. West RB, Kong C, Clarke N, et al. MYB expression and translocation in adenoid cystic carcinomas and other salivary gland tumors with clinicopathologic correlation. Am J Surg Pathol 2011; 35(1):92–9.

32. Persson M, Andren Y, Moskaluk CA, et al. Clinically significant copy number alterations and complex

rearrangements of MYB and NFIB in head and neck adenoid cystic carcinoma. Genes Chromosomes Cancer 2012;51(8):805–17.

33. Mitani Y, Rao PH, Futreal PA, et al. Novel chromosomal rearrangements and break points at the t(6;9) in salivary adenoid cystic carcinoma: association with MYB-NFIB chimeric fusion, MYB expression, and clinical outcome. Clin Cancer Res 2011; 17(22):7003–14.

34. Costa AF, Altemani A, Garcia-Inclan C, et al. Analysis of MYB oncogene in transformed adenoid cystic carcinomas reveals distinct pathways of tumor progression. Lab Invest 2014; 94(6):692–702.

35. Ho AS, Kannan K, Roy DM, et al. The mutational landscape of adenoid cystic carcinoma. Nat Genet 2013;45(7):791–8.

36. Keam B, Kim SB, Shin SH, et al. Phase 2 study of dovitinib in patients with metastatic or unresectable adenoid cystic carcinoma. Cancer 2015;121(15): 2612–7.

37. Stephens PJ, Davies HR, Mitani Y, et al. Whole exome sequencing of adenoid cystic carcinoma. J Clin Invest 2013;123(7):2965–8.

38. Chiosea SI, Williams L, Griffith CC, et al. Molecular characterization of apocrine salivary duct carcinoma. Am J Surg Pathol 2015;39(6):744–52.

39. Soper MS, Iganej S, Thompson LD. Definitive treatment of androgen receptor-positive salivary duct carcinoma with androgen deprivation therapy and external beam radiotherapy. Head Neck 2014; 36(1):E4–7.

40. Yamamoto N, Minami S, Fujii M. Clinicopathologic study of salivary duct carcinoma and the efficacy of androgen deprivation therapy. Am J Otolaryngol 2014;35(6):731–5.

41. Masubuchi T, Tada Y, Maruya S, et al. Clinicopathological significance of androgen receptor, HER2, Ki-67 and EGFR expressions in salivary duct carcinoma. Int J Clin Oncol 2015;20(1):35–44.

42. Limaye SA, Posner MR, Krane JF, et al. Trastuzumab for the treatment of salivary duct carcinoma. Oncologist 2013;18(3):294–300.

43. Griffith CC, Seethala RR, Luvison A, et al. PIK3CA mutations and PTEN loss in salivary duct carcinomas. Am J Surg Pathol 2013;37(8):1201–7.

44. Nardi V, Sadow PM, Juric D, et al. Detection of novel actionable genetic changes in salivary duct carcinoma helps direct patient treatment. Clin Cancer Res 2013;19(2):480–90.

45. Chiosea SI, Griffith C, Assaad A, et al. Clinicopathological characterization of mammary analogue secretory carcinoma of salivary glands. Histopathology 2012;61(3):387–94.

46. Ito Y, Ishibashi K, Masaki A, et al. Mammary analogue secretory carcinoma of salivary glands: a clinicopathologic and molecular study

including 2 cases harboring ETV6-X fusion. Am J Surg Pathol 2015;39(5):602–10.

47. Skalova A, Vanecek T, Simpson RH, et al. Mammary analogue secretory carcinoma of salivary glands: molecular analysis of 25 ETV6 gene rearranged tumors with lack of detection of classical ETV6-NTRK3 fusion transcript by standard RT-PCR: report of 4 cases harboring ETV6-X Gene fusion. Am J Surg Pathol 2016;40(1):3–13.

48. Weinreb I, Zhang L, Tirunagari LM, et al. Novel PRKD gene rearrangements and variant fusions in cribriform adenocarcinoma of salivary gland origin. Genes Chromosomes Cancer 2014;53(10):845–56.

49. Weinreb I, Piscuoglio S, Martelotto LG, et al. Hotspot activating PRKD1 somatic mutations in polymorphous low-grade adenocarcinomas of the salivary glands. Nat Genet 2014;46(11):1166–9.

50. Bullerdiek J, Hutter KJ, Brandt G, et al. Cytogenetic investigations on a cell line derived from a carcinoma arising in a salivary gland pleomorphic adenoma. Cancer Genet Cytogenet 1990;44(2):253–62.

51. Bullerdiek J, Wobst G, Meyer-Bolte K, et al. Cytogenetic subtyping of 220 salivary gland pleomorphic adenomas: correlation to occurrence, histological subtype, and in vitro cellular behavior. Cancer Genet Cytogenet 1993;65(1):27–31.

52. Matsuyama A, Hisaoka M, Nagao Y, et al. Aberrant PLAG1 expression in pleomorphic adenomas of the salivary gland: a molecular genetic and immunohistochemical study. Virchows Arch 2011;458(5): 583–92.

53. Rotellini M, Palomba A, Baroni G, et al. Diagnostic utility of PLAG1 immunohistochemical determination in salivary gland tumors. Appl Immunohistochem Mol Morphol 2014;22(5):390–4.

54. Bahrami A, Dalton JD, Shivakumar B, et al. PLAG1 alteration in carcinoma ex pleomorphic adenoma: immunohistochemical and fluorescence in situ hybridization studies of 22 cases. Head Neck Pathol 2012;6(3):328–35.

55. Persson F, Andren Y, Winnes M, et al. High-resolution genomic profiling of adenomas and carcinomas of the salivary glands reveals amplification, rearrangement, and fusion of HMGA2. Genes Chromosomes Cancer 2009;48(1):69–82.

56. Katabi N, Ghossein R, Ho A, et al. Consistent PLAG1 and HMGA2 abnormalities distinguish carcinoma ex-pleomorphic adenoma from its de novo counterparts. Hum Pathol 2015;46(1):26–33.

57. Bilodeau EA, Weinreb I, Antonescu CR, et al. Clear cell odontogenic carcinomas show EWSR1 rearrangements: a novel finding and a biological link to salivary clear cell carcinomas. Am J Surg Pathol 2013;37(7):1001–5.

58. Skalova A, Weinreb I, Hyrcza M, et al. Clear cell myoepithelial carcinoma of salivary glands showing EWSR1 rearrangement: molecular analysis of

94 salivary gland carcinomas with prominent clear cell component. Am J Surg Pathol 2015;39(3): 338–48.

59. Choi HR, Batsakis JG, Callender DL, et al. Molecular analysis of chromosome 16q regions in dermal analogue tumors of salivary glands: a genetic link to dermal cylindroma? Am J Surg Pathol 2002; 26(6):778–83.

60. Kawahara A, Harada H, Abe H, et al. Nuclear beta-catenin expression in basal cell adenomas of salivary gland. J Oral Pathol Med 2011;40(6):460–6.

61. Jung MJ, Roh JL, Choi SH, et al. Basal cell adenocarcinoma of the salivary gland: a morphological and immunohistochemical comparison with basal cell adenoma with and without capsular invasion. Diagn Pathol 2013;8:171.

62. Bilodeau EA, Acquafondata M, Barnes EL, et al. A comparative analysis of LEF-1 in odontogenic and salivary tumors. Hum Pathol 2015;46(2):255–9.

63. Chiosea SI, Miller M, Seethala RR. HRAS mutations in epithelial-myoepithelial carcinoma. Head Neck Pathol 2014;8(2):146–50.

64. Cros J, Sbidian E, Hans S, et al. Expression and mutational status of treatment-relevant targets and key oncogenes in 123 malignant salivary gland tumours. Ann Oncol 2013;24(10):2624–9.

The Molecular Pathology of Lung Cancer

Lynette M. Sholl, MD

KEYWORDS

- Lung carcinoma • Lung adenocarcinoma • EGFR • KRAS • ALK • MET • Histology

Key points

- Molecular analysis of lung adenocarcinoma permits targeted therapeutic approaches and is associated with improved outcomes.
- Epidermal growth factor receptor mutation correlates with lepidic/papillary/acinar/micropapillary adenocarcinoma subtypes and Kirsten rat sarcoma viral oncogene homolog mutation with solid-subtype and invasive mucinous adenocarcinoma; however, individual cases may show significant variation from published correlations.
- Intratumoral genomic heterogeneity contributes to the morphologic heterogeneity seen on pathology.
- A variety of molecular testing methods may be used for detection of common genomic alterations in lung cancer; massively parallel sequence may supersede many polymerase chain reaction–based targeted approaches, but use of a combination of methods is likely to provide the greatest clinical and technical sensitivity.

ABSTRACT

Advances in lung cancer genomics have revolutionized the diagnosis and treatment of this heterogeneous and clinically significant group of tumors. This article provides a broad overview of the most clinically relevant oncogenic alterations in common and rare lung tumors, with an emphasis on the pathologic correlates of the major oncogenic drivers, including *EGFR*, *KRAS*, *ALK*, and *MET*. Illustrations emphasize the morphologic diversity of lung adenocarcinoma, including genotype-phenotype correlations of genomic evolution in tumorigenesis. Molecular diagnostic approaches, including PCR-based testing, massively parallel sequencing, fluorescence in situ hybridization, and immunohistochemistry are reviewed.

OVERVIEW

Lung cancer ranks as the second most common form of cancer across all races and ethnicities in the United States.[1] The main types of lung cancer are adenocarcinoma, squamous cell carcinoma (SQC), large cell carcinoma, and high-grade neuroendocrine tumors, including small cell lung carcinoma (SCLC) and large cell neuroendocrine carcinoma.[2] Each lung cancer type has relatively unique clinicopathologic and molecular associations. Epidemiology studies have shown that the incidence of lung SQC has trended down and adenocarcinoma has trended up over the past few decades,[3] presumably reflecting shifts in smoking patterns and cigarette design in the form of changing chemical components and more effective filters.[4]

Despite an overall annual decrease in incidence, lung cancer remains a top cause of cancer-related deaths, affecting both men and women, a large age range, and smokers and nonsmokers. However, lung cancer–specific mortality has begun to decline in the past decade. This change is attributable both to earlier tumor detection using high-resolution chest computed tomography (CT) screening[5,6] and to routine use of therapies targeting specific tumor genomic alterations.[7]

Department of Pathology, Brigham and Women's Hospital, Harvard Medical School, 75 Francis Street, Boston, MA 02115, USA
E-mail address: lmsholl@partners.org

Surgical Pathology 9 (2016) 353–378
http://dx.doi.org/10.1016/j.path.2016.04.003
1875-9181/16/$ – see front matter © 2016 Elsevier Inc. All rights reserved.

Lung adenocarcinoma in particular is defined by a variety of mutually exclusive oncogenic driver alterations, the most common of which occur in the *Kirsten rat sarcoma viral oncogene homolog* (*KRAS*), *epidermal growth factor receptor* (*EGFR*), and *anaplastic lymphoma kinase* (*ALK*) genes, with the latter 2 concentrated in never smokers. Tumors harboring activating alterations in *EGFR* and *ALK*, as well as in other oncogenes, including *ROS1*, *RET*, *BRAF*, *ERBB2*, and *MET*, may respond to targeted inhibitors; the evidence for their association with a survival benefit has led to national guidelines recommending testing for *EGFR* and *ALK* alterations in all patients with metastatic lung adenocarcinoma before initiating first-line therapy.[8,9] Other lung carcinomas, including SQC, SCLC, and large cell carcinoma, tend to occur in heavy smokers and often lack a clear oncogenic driver. These tumors characteristically have high rates of *TP53* mutations and overall high mutational frequencies reflecting smoking-related mutagenesis.[10,11]

ASSOCIATED GENETIC ALTERATIONS

ADENOCARCINOMA

Our understanding of the genomic underpinnings of lung cancers has been advanced significantly by numerous large-scale sequencing efforts.[10–12] In adenocarcinoma, such efforts have confirmed the prevalence of established oncogenic driver alterations (Table 1) and have uncovered additional layers of genomic complexity, with probable implications for response to targeted therapies. *TP53* mutations are reported in 50% of lung adenocarcinoma, frequently overlapping with a variety of oncogenic driver alterations. Other commonly mutated genes in lung adenocarcinoma, such as *STK11*, *KEAP1*, *NF1*, *PIK3CA*, and *PTEN*, likely facilitate, rather than drive, tumorigenesis.[12] These alterations may nonetheless have prognostic and predictive implications. *STK11* is significantly co-mutated with *KRAS* in smokers and associated with worse outcomes.[13] Animal studies suggest that the presence of *STK11* mutations confer relative resistance to MEK inhibitors in *KRAS*-mutated lung adenocarcinomas.[14] Mutations in *PIK3CA*, *PTEN*, and *AKT1* are detected in a minority of *EGFR*-mutated lung adenocarcinoma and appear to confer relative resistance to EGFR tyrosine kinase inhibitors (TKIs).[15,16]

SQUAMOUS CELL CARCINOMA

Despite advances in genomics and precision medicine, efforts to apply targeted therapies outside of lung adenocarcinoma have largely been disappointing. A subset of SQC harbors oncogenic alterations, including in fibroblast growth factor receptor (FGFR) family members, PI3K/PTEN/AKT pathway, and *discoidin domain receptor tyrosine kinase 2* (*DDR2*) (Table 2); however, overall it is unclear that these alterations, existing as they do on a background of high mutational burden, near universal TP53 dysregulation, and multiple concomitant oncogenic mutations, will be readily targetable using a single inhibitor.[10] *FGFR1* amplification in particular has been examined as a biomarker for FGFR inhibitor therapy; in vitro studies suggest that *FGFR1*-amplified lung cancers are sensitive to a variety of targeted inhibitors,[17,18] but there are few data on the efficacy of these drugs in practice. Further, some studies suggest that FGFR1 mRNA and protein expression are better predictors of sensitivity to FGFR inhibitors than copy number.[19] Fortunately, the efficacy of immune checkpoint inhibitors, such as pembrolizumab and nivolumab, both programmed death-1 (PD-1) inhibitors, offers new therapeutic promise for patients with SQC.[20,21]

SMALL CELL LUNG CARCINOMA/LARGE CELL NEUROENDOCRINE CARCINOMA

Small cell lung carcinoma is typically centrally located and highly associated with a heavy smoking history, with only 2% of SCLC occurring in never smokers.[22] *TP53* and retinoblastoma 1(*RB1*) are inactivated in essentially all cases, on a background of a high somatic mutation rate consistent with smoking-related mutagenesis.[11] Amplification of MYC family oncogenes occurs in 3% to 7% of SCLC; in vitro studies suggest this is a targetable alteration[23] but there is limited evidence for clinical efficacy of agents targeting MYC. The profound cell cycle dysregulation seen in SCLC contributes to the distinct clinical presentation of rapid tumor growth, high proliferative index, and exquisite sensitivity to chemoradiotherapy, with inevitable relapse. In contrast to SCLC, large cell neuroendocrine carcinoma shows large cells with ample cytoplasm growing in nests with central comedo-type necrosis. This tumor type arises predominantly in the peripheral lung and has historically been considered a form of non–small cell carcinoma, with a prognosis intermediate between non–small cell and SCLC.[2] Genomic profiling studies have revealed that it is most closely related to SCLC, with loss of *TP53* and *RB1*, as well as a similar copy number profiles.[24] However, in practice this is a relatively rare diagnosis and systematic data are lacking to guide clinical therapy; current practice guidelines

Table 1
Oncogenes in lung adenocarcinoma with therapeutic implications

Gene	Alteration Type	Frequency	Clinical Characteristics	Pathologic Characteristics	Therapies
KRAS	Missense mutations at hotspots: Codons 12, 13, 61	25%	Smokers >> nonsmokers	Diverse pathologic correlates; enriched in invasive mucinous adenocarcinoma, large cell carcinoma	Chemotherapy; Clinical trials of MEK inhibitors[97]
EGFR	Missense and insertion-deletion mutations at hotspots in exons 18–21 >90% occur as Ex19del, L858R mutations	15%	Nonsmokers >> smokers Women > men Enriched in Asian populations	Historically associated with "bronchioloalveolar carcinoma"; however, these alterations are seen in diverse adenocarcinoma histotypes	First-line EGFR TKIs[9]: Erlotinib, Gefitinib, Afatanib
ALK	Rearrangements: EML4-ALK, most common	5%	Nonsmokers >> smokers Women ≈ men Younger patients	Correlates with solid and signet ring cell tumors with extracellular mucin, may be seen in other histotypes	Multikinase TKIs[98–102]: Crizotinib, Ceritinib, Alectinib
ROS1	Rearrangements: CD74-ROS1, SCL34A2-ROS1, EZR-ROS1, FIG1-ROS1	1%–2%	Nonsmokers >> smokers Women ≈ men Younger patients	Diverse pathologic features; some reports of correlation with psammomatous calcifications	Multikinase TKIs[103]: Crizotinib
RET	Rearrangements: KIF5B-RET, CCDC6-RET, NCOA4-RET	1%–2%	Nonsmokers >> smokers Women ≈ men Younger patients	Diverse pathologic features	Multikinase TKI trials[104–106]: Sunitinib, Cabozantinib, Vandetanib
BRAF	V600E and other missense mutations at hotspots in exons 11 and 15	4%	Smokers > nonsmokers	Limited data; some correlation with poorly differentiated histology	RAF-MEK inhibitor trials[38,107]: Dabrafenib + trametinib
ERBB2	Small exon 20 insertions Rarely may have concomitant amplification	2%	Nonsmokers >> smokers	Limited data; mixed patterns, correlation with moderately to poorly differentiated histology	HER inhibitor trials[108–110]: Trastuzumab, Neratinib, Afatinib, Lapatinib
MET	Exon 14 skipping mutations Amplification	3%	Smokers > nonsmokers Women ≈ men Older patients	Diverse pathologic features; early disease shows lepidic, acinar growth; advanced disease associated with pleomorphic carcinoma	Multikinase TKI trials[53–56]: Crizotinib, MGCD-265

Abbreviations: ALK, anaplastic lymphoma kinase; EGFR, epidermal growth factor receptor; KRAS, Kirsten rat sarcoma viral oncogene homolog; TKI, tyrosine kinase inhibitor.

Table 2
Recurrent alterations in lung squamous cell carcinoma with potential for therapeutic targeting

Gene	Alteration Type	Frequency	Molecular Details	Therapies
FGFR1	Amplification	17%	Gene copy number of ≥4 or FGFR1:CEP8 ratio of ≥2	Ponatinib[18,19,111,112] Pazopanib
FGFR2/3	Extracellular domain missense mutations, rearrangements (FGFR3-TACC3)	6%	Other alterations common including FGFR1 or CCND1 amplification, PIK3CA and RAS pathway activation	
DDR2	Missense mutations	4%	Mutations occur throughout the gene, including in the kinase domain and discoidin domain	Dasatinib[113,114] Rare reports of response; use limited by toxicity
PIK3CA	Missense mutations	16%	Other activating alterations common in receptor tyrosine kinases, RAS pathway	AKT. mTOR inhibitor trials[115]: MK-2206, Ridaforolimus
PTEN	Missense, loss-of-function mutations, gene deletion	15%	Other activating alterations common in receptor tyrosine kinases, RAS pathway	

Abbreviations: ALK, anaplastic lymphoma kinase; mTOR, mammalian target of rapamycin.

recommend treating large cell neuroendocrine carcinoma in the same manner as non-squamous non–small cell carcinoma.[25]

MOLECULAR PATHOLOGY OF CLASSIC TUMORS

INVASIVE LUNG ADENOCARCINOMA

Invasive adenocarcinoma of the lung comprises a genetically, clinically, and morphologically diverse collection of tumors. Invasive adenocarcinomas are currently subtyped according to their predominant histologic pattern (lepidic, acinar, papillary, micropapillary, or solid).[2] In practice, there are inherent challenges to identifying distinct morphologic subclasses of invasive adenocarcinoma, as most tumors come to clinical attention at a late stage and are diagnosed on small biopsies, often from metastatic sites. Even large multi-institutional genotyping studies have been underpowered to detect a correlation between genetic profile and morphologic subtype, in large part because surgical resections are rare in such studies.[26] However, there is substantial evidence to suggest that EGFR mutations are more common among tumors with predominant acinar, lepidic, micropapillary, and papillary morphology and are uncommon in solid-predominant tumors[27–29] (Fig. 1). In contrast, solid-predominant adenocarcinomas are significantly more likely to harbor KRAS mutations and arise in smokers[30] (Fig. 2).

In patients with stage I, resected adenocarcinoma, solid-subtype morphology predicts early recurrence and poor survival.[31] In other studies of surgically resected tumors, micropapillary subtype predicts for occult lymph node metastases and is associated with worse lung cancer–specific outcomes.[32,33] Within acinar-predominant tumors, the presence of a cribriform pattern correlates with an increased risk of recurrence, similar to that seen for solid and micropapillary predominant tumors.[34] Tumors with ALK, ROS1, and RET fusions have some unique histologic features, including extracellular mucin and prominent signet ring cells; interestingly, this group of fusion-positive tumors also shows a significant association with aggressive morphologic patterns including solid, cribriform, and micropapillary growth[35–37] (Fig. 3). although these findings have been more reproducible in Asian populations than in Western populations.[34]

BRAF and ERBB2 driver alterations are relatively rare in lung adenocarcinoma and are not universally interrogated in clinical practice. As a result, there are limited data on morphologic correlates for tumors with mutations in these genes. Given promising phase II trial results in patients with BRAF V600E mutated tumors,[38] it is likely that BRAF testing in lung cancer will become more routine and clinicopathologic correlates may become apparent.

Despite the statistical correlations between genotype and morphology, unusual genotype-phenotype matches are common in clinical practice and distinct morphologic patterns, including adenocarcinoma with fetal-lung–like morphology, may be seen in KRAS-mutated or EGFR-mutated

Fig. 1. Morphologic patterns of *EGFR*-mutated lung adenocarcinoma. (*A*) *EGFR* L858R-mutated lepidic pattern adenocarcinoma (H&E, 100×). (*B*) *EGFR* exon 20,773_774insH acinar pattern adenocarcinoma (H&E, 200×).

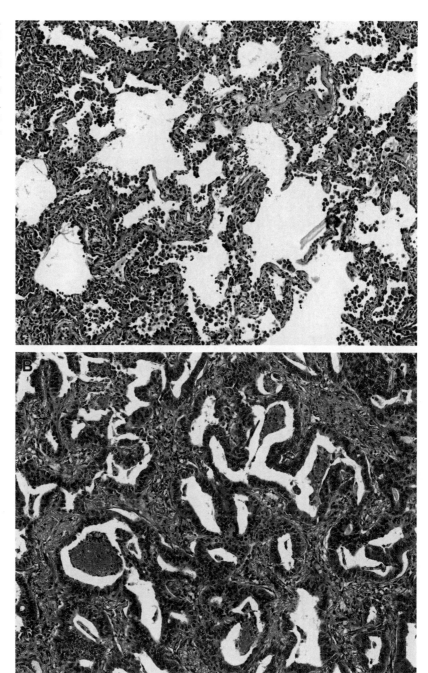

tumors[39] (**Fig. 4**). As a result, adenocarcinoma subtyping should generally not be used to select patients for molecular testing.[8] The wide variation in the histologic appearances of lung adenocarcinoma does, however, permit the pathologist to differentiate independent primary tumors from intrapulmonary metastases, a common challenge in clinical practice.[40–42] The addition of genetic analysis to morphologic subtyping substantially

improves our ability to determine relatedness of multiple lung tumors and outperforms the classic Martini-Melamed criteria.[42]

ADENOCARCINOMA IN SITU/MINIMALLY INVASIVE ADENOCARCINOMA

The use of high-resolution chest CT screening in at-risk populations, as well as increased use of

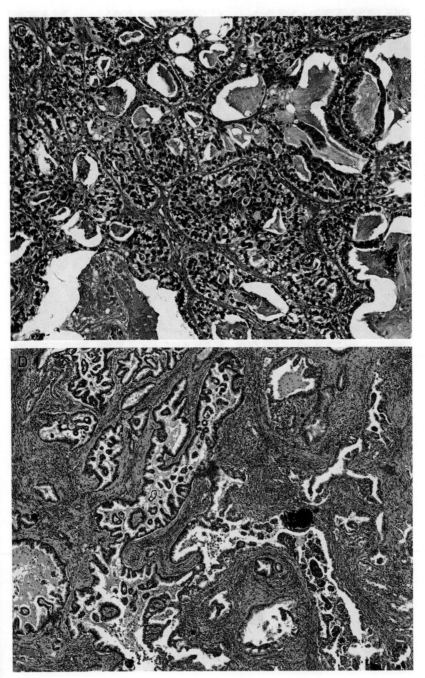

Fig. 1. (continued). (*C*) *EGFR* exon 19 ELREA746_750del, acinar pattern adenocarcinoma with an unusual degree of cytoplasmic glycogenation leading to a "fetal-adenocarcinoma appearance." (H&E, 200×) (*D*) *EGFR* exon 19 ELREA746_750del, micropapillary predominant adenocarcinoma with scattered psammomatous calcifications (H&E, 100×).

CT scanning for extrapulmonary indications, has led to increased detection of early-stage, asymptomatic lung tumors. Adenocarcinoma in situ (AIS) is a proliferation of malignant-appearing alveolar epithelial cells with an exclusive lepidic growth pattern, whereas minimally invasive adenocarcinoma (MIA) shows less than 5 mm of stromal invasion or complex papillary structures (**Fig. 5**). Surgical resection for AIS and MIA is considered curative, with a 98% to 100% 5-year survival.[27,29] In a Japanese cohort including more than 100 cases, AIS was twice as common among nonsmokers than smokers, with a predilection for women, and approximately 60% harbored an

Fig. 2. Solid-predominant *KRAS* G12V-mutated lung adenocarcinoma. Focal cytoplasmic clearing may be seen (H&E, 200×).

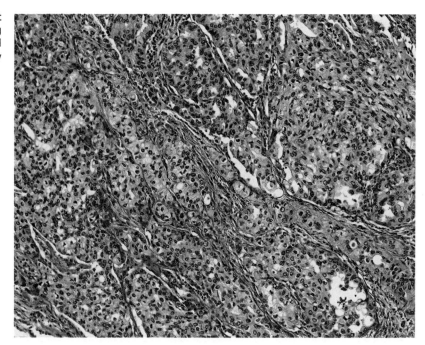

activating *EGFR* mutation. Even among smokers, *EGFR* mutations were common in this tumor subtype. In a smaller Japanese cohort, MIA was also associated with *EGFR* mutations.[29] Notably, *KRAS* mutations were detected in fewer than 3% of AIS, one of which was mucinous (see also genetics of invasive mucinous adenocarcinoma, in the next section).[43]

INVASIVE MUCINOUS ADENOCARCINOMAS

Finberg and colleagues[44] first described a strong correlation between *KRAS* activating mutations and tumors formerly referred to as "mucinous bronchioloalveolar carcinomas," now termed invasive mucinous adenocarcinomas (IMA) in the 2015 World Health Organization (WHO) classification.[2] These tumors are composed of tufts of cells with low-grade nuclei and abundant apical mucin lining lung alveoli in lepidic or papillary patterns, showing aerogenous spread with abundant extracellular mucin production. Immunohistochemical (IHC) stains often show gastrointestinal differentiation, with increased CK20, absent TTF-1, and aberrant CDX2 expression. *KRAS* mutations are detected in approximately two-thirds of IMA; transition mutations G12D and G12V predominate,[45,46] as opposed to the smoking-related transversion mutation (G12C) most commonly seen in other forms of lung adenocarcinoma.[47] Other genomic alterations in IMAs involve members of the ERBB

(HER) signaling pathway and include *NRG1* fusions and *ERBB2* exon 20 insertion mutations (**Fig. 6**).[45] Genetically engineered mouse models with conditionally activated *KRAS* and deleted *NKX2-1* (TTF-1) in lung epithelium develop tumors that show gastric differentiation based on transcriptional signatures and mucin expression patterns.[48,49] In human tumors, *NKX2.1* loss-of-function mutations are rare (<1% of lung adenocarcinomas) but are strikingly correlated with *KRAS*-mutated IMAs that show a gastrointestinal transcription factor and mucin (MUC5A and MUC6) expression pattern.[50]

THE GENOMIC BASIS OF MORPHOLOGIC HETEROGENEITY

The morphologic heterogeneity of lung adenocarcinomas, even within individual genomically defined categories, can be explained in part by other genomic alterations contributing to tumorigenesis. Even individual adenocarcinomas may display a remarkable degree of intratumoral morphologic heterogeneity, ostensibly a reflection of genomic evolution within tumoral subclones, such as acquisition of additional mutational hits in tumor suppressor genes, or changes in gene copy number status as a means of enhancing oncogenic signaling. This principle has been illustrated in *EGFR*-mutated lung adenocarcinoma[51] (**Fig. 7**).

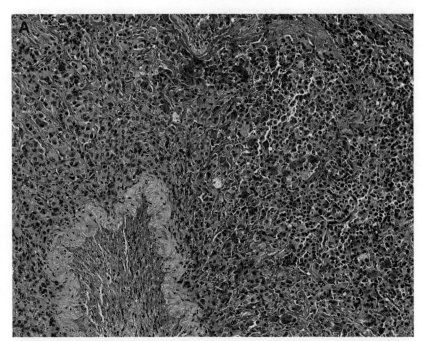

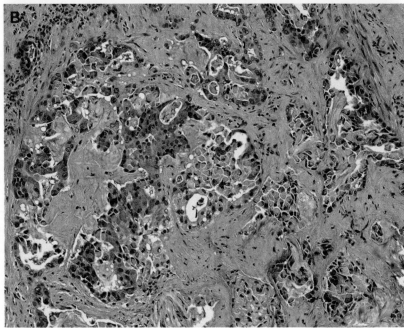

Fig. 3. Morphologic patterns of fusion-positive lung adenocarcinoma. (*A*) Solid growth pattern in an *EML4-ALK* rearranged tumor (H&E, 100×). (*B*) Cribriform pattern with extensive extracellular mucin deposition and frequent signet ring cells in an *EML4-ALK* rearranged tumor (H&E, 200×).

MET mutations in the form of intron 13/exon 14 splice site mutations and deletions lead to exon 14 skipping and deletion of a regulatory subunit of the protein.[52] These events trigger constitutive MET activation leading to malignant transformation and are associated with increased MET protein expression in tumor cells and aggressive clinicopathologic features in advanced stage tumors.[52,53] Patients with tumors containing these alterations have responded to targeted MET inhibitors, including crizotinib.[53–55] The morphologic spectrum of tumors with *MET* exon 14 skipping mutations may range from low-grade, MIA to pleomorphic carcinoma[53,56] (**Fig. 8**). Sequencing studies have demonstrated that low-grade, early-stage *MET*-mutated tumors typically have few other single nucleotide variants or copy number alterations and display low levels

Fig. 3. (*continued*). (*C*) Cribriform pattern in a *CCDC6-RET* rearranged tumor (H&E, 100×). (*D*) Solid and papillary patterns in a *CD74-ROS1* rearranged tumor (H&E 200×).

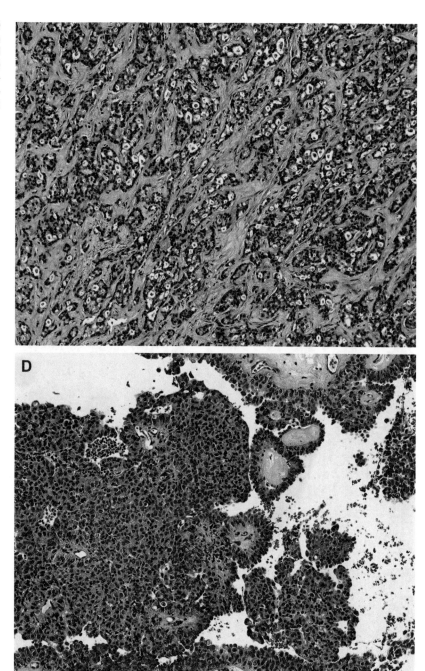

of MET protein expression. In contrast, *MET*-mutated advanced stage adenocarcinomas or pleomorphic carcinomas exhibit a large number of additional genomic alterations, including amplification and overexpression of *MET* and numerous hits in cell cycle regulatory pathways (including *MDM2* amplification, *TP53* mutation, *CDKN2A* loss).[53,55] Additional discussion of the pleomorphic carcinomas can be found in the following section: "Molecular pathology of rare tumors."

MOLECULAR PATHOLOGY OF RARE TUMORS

LARGE CELL CARCINOMA

Large cell carcinoma is defined by the 2015 WHO classification as a tumor lacking morphologic

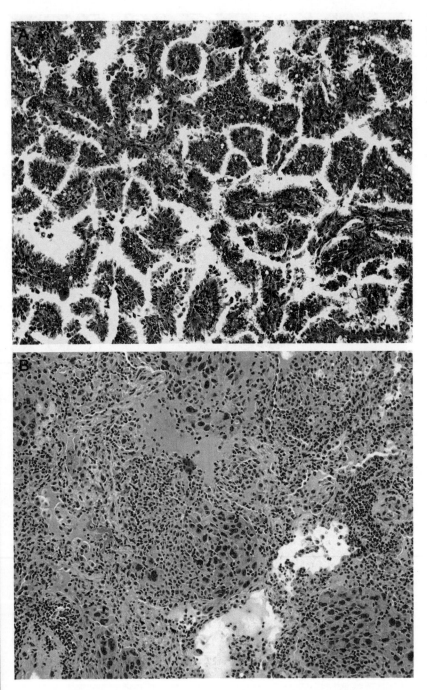

Fig. *4.* Examples of unusual morphology-genomics combinations. (*A*) Papillary predominant adenocarcinoma with *KRAS* G12C mutation (H&E, 200×). (*B*) Solid predominant adenocarcinoma with *EGFR* exon 19 ELREA746_750del mutation (H&E, 200×).

and immunophenotypic differentiation characteristic of adenocarcinoma, SQC, or neuroendocrine carcinoma. Historically, this was a morphologic category representing a heterogeneous group of tumors including poorly differentiated adenocarcinomas and SQC[57,58]; the advent of immunohistochemistry and molecular genetic pathology, however, has led to the stricter definition used today. Clinically, large cell carcinoma is significantly associated with smoking, and because the diagnosis requires exclusion of more differentiated components, it can be made only on a surgical resection. *TP53* mutations are highly enriched in this tumor type.[24] Approximately 40% of large cell carcinomas contain an activating *KRAS* mutation[57];

Fig. 5. Lowpowered histologic appearance of MIA. The blue circle designates the area of the slide selectively dissected for molecular analysis. This biopsy was taken from a smoker. The tumor is *EGFR* and *KRAS* wild type, but has 2 *NF1* mutations and a high frequency of somatic transversion substitutions consistent with smoking-induced mutagenesis (H&E, 20×).

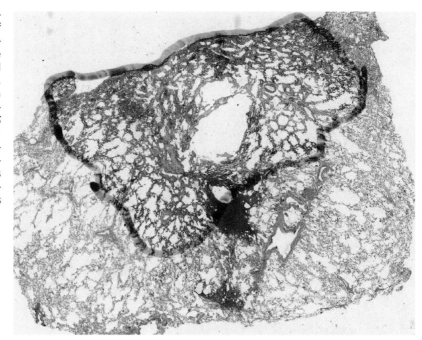

other reported alterations include mutations in *BRAF* and *NRAS*.[59,60]

ADENOSQUAMOUS CARCINOMA

Adenosquamous carcinoma contains at least 10% of both adenocarcinoma and squamous components. Genomic profiling studies indicate that adenosquamous carcinomas more closely resemble adenocarcinomas and therefore may contain targetable alterations (see **Fig. 7**).[61] In small biopsy specimens, preferential sampling of a squamous component may lead to misdiagnosis as SQC. Current guidelines generally discourage *EGFR* and *ALK* testing for SQC, however, in acknowledging this diagnostic pitfall, suggest that testing may be appropriate in never or light smokers.[8]

PLEOMORPHIC CARCINOMA

Pleomorphic carcinoma includes giant-cell and spindle-cell carcinoma, or any non–small cell carcinoma showing 10% giant-cell and/or spindle-cell elements (see **Fig.** 8B). The category of sarcomatoid carcinoma encompasses these tumors, as well as carcinosarcoma and pulmonary blastoma. All represent poorly differentiated and aggressive forms of lung carcinoma.[2] Using a whole exome sequencing approach, Liu and colleagues[56] identified *MET* exon 14 skipping mutations (see previously in this article) in 22% of

pulmonary sarcomatoid carcinomas, followed by *KRAS* in 20%. *EGFR* mutations have been reported in series of pleomorphic carcinomas from Asian populations but are rare to absent in North American and European series.[62,63] EGFR TKIs appear to have limited efficacy in this tumor type, even when an *EGFR* sensitizing mutation is present.[62] In contrast, MET-targeted inhibitors have shown promise in the context of *MET* exon 14 skipping mutations.[53,56]

LYMPHOEPITHELIAL-LIKE CARCINOMA

Lymphoepithelial-like carcinoma predominantly affects Asian populations and is invariably associated with Epstein-Barr Virus (EBV) infection. Smoking does not appear to play a role in this tumor type.[64] Programmed death-ligand-1 (PD-L1) is overexpressed in approximately 75% of lymphoepithelial-like carcinomas,[65,66] consistent with the established role of EBV in driving PD-L1 overexpression in other malignancies.[67,68] In contrast to other forms of non–small cell carcinoma in Asian populations, *EGFR* mutations are rare and *KRAS* and *ALK* do not appear to be altered in lymphoepithelial-like carcinoma.[65,66,69]

NUT CARCINOMA

Nut carcinoma (Nut midline carcinoma) is a rare, aggressive, and uniformly lethal tumor that has a predilection for head and neck, mediastinal, and

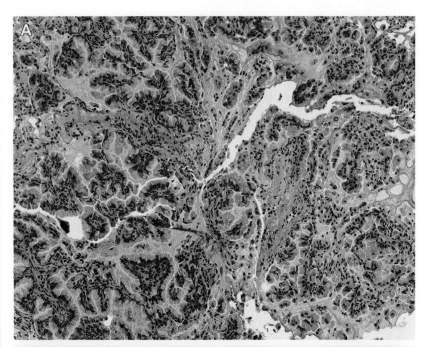

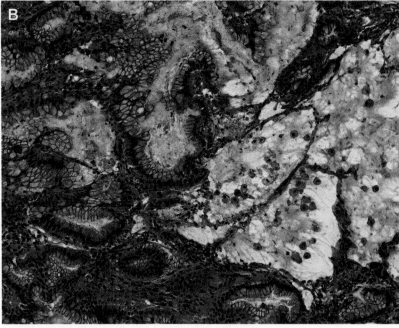

Fig. 6. (*A*) Invasive mucinous adenocarcinoma with an *ERBB2* exon 20 AYVM insertion mutation (H&E, 200×). (*B*) Invasive mucinous adenocarcinoma harboring a *KRAS* G12D and *NKX2.1* (TTF-1) loss-of-function mutation. This tumor shows characteristic low-grade nuclei with abundant apical mucin, extracellular mucin, and numerous alveolar muciphages (H&E, 200×).

pulmonary sites of origin.[70] Although this tumor was originally described in children, it presents across a wide age range. Although morphologically characterized by sheets of undifferentiated round to epithelioid cells, it is thought to represent a form of SQC, as the tumor cells typically express the squamous transcription factors p63 or p40, and in some cases focal areas of keratinizing squamous differentiation can be seen (**Fig. 9**). This tumor is frequently misdiagnosed, including as SCLC, lymphoma, and poorly differentiated non–small cell lung carcinoma.[71] Rendering the correct diagnosis is critical, as these tumors do not respond to conventional chemotherapy and have a median overall survival of less than 3 months in the lung.[70,71] The diagnosis is

Fig. 7. Copy number heterogeneity in an *EGFR*-mutated tumor. (*A*) *EGFR* ELREA746_750del-mutated lung adenosquamous carcinoma (H&E, 40×). (*B*) EGFR exon 19 deletion-specific immunohistochemistry is diffusely positive in both adenocarcinoma and squamous components (40×).

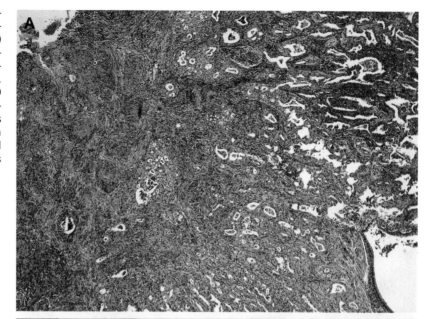

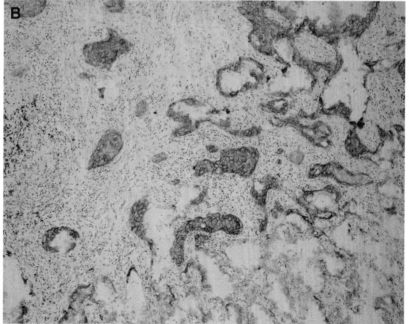

predicated on identification of Nut protein overexpression by immunohistochemistry[72] and, when possible, detection of an underlying *NUTM1* fusion. *NUTM1* commonly fuses with bromodomain family members *BRD3* or *BRD4*, leading to epigenetic dysregulation.[73] Bromodomain (BET) and histone deacetylase (HDAC) inhibitors are active against Nut carcinoma cells and are currently being investigated in clinical trials.[70]

PRIMARY SALIVARY GLAND TUMORS OF THE LUNG

Malignant salivary gland tumors have been reported to arise in the lung, including mucoepidermoid, adenoid cystic, hyalinizing clear cell, epithelial-myoepithelial, and salivary duct carcinomas. Benign salivary glands are exceptionally rare at this location. Overall, these tumors

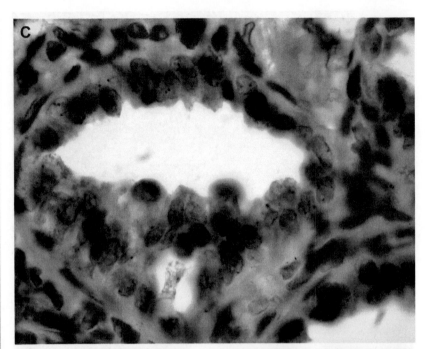

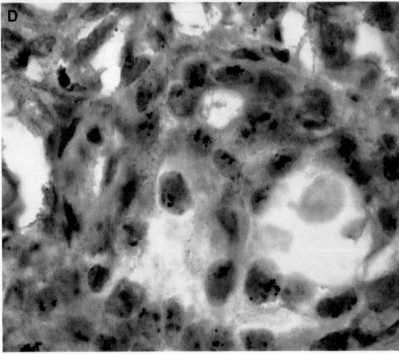

Fig. 7. (continued). (*C*) An area of low-grade acinar growth and no evidence of *EGFR* copy gain by chromogenic in situ hybridization (CISH) (1000×). (*D*) Focused amplification of *EGFR* by CISH, consistent with genomic "progression," corresponds with areas of squamous differentiation (1000×).

comprise fewer than 1% of lung tumors and are presumed to arise from the submucous glands of the central airways.[74] The molecular alterations seen in these pulmonary tumors are similar to their counterparts arising in the major salivary glands of the head and neck: *MAML2* rearrangement in mucoepidermoid carcinoma,[75] *MYB* rearrangement in adenoid cystic carcinoma,[76] and *EWSR1* rearrangement in hyalinizing clear cell carcinoma[77] and myoepithelial carcinoma.[78] Other reported

Fig. 8. Lung adenocarcinomas, both with *MET* D963_splice mutations leading to exon 14-skipping (H&E, 200×). (*A*) MIA with additional *MDM2* amplification, and few other genomic aberrations. (*B*) Pleomorphic carcinoma with additional *MET* amplification, *CDK5* amplification, *TP53* deletion, and *RB1* deletion (H&E, 200×).

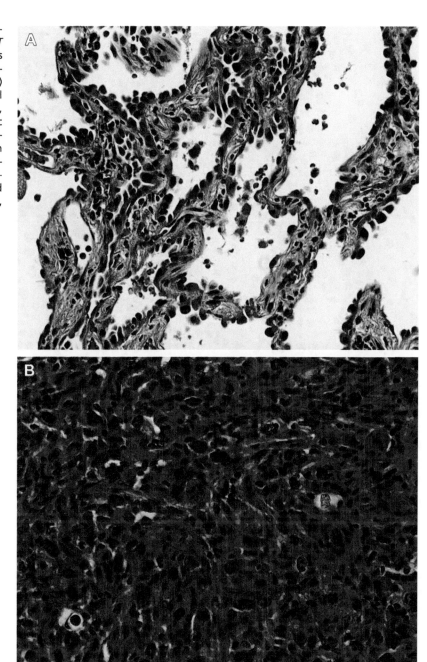

alterations include *BRAF* G464V mutation in a mucin-rich variant of pulmonary salivary duct carcinoma.[79] The reader is directed to the "Salivary Gland" section (see Seethala RR, Griffith CC: Molecular Pathology: Predictive, Prognostic, and Diagnostic Markers in Tumors – Salivary Gland, in this issue) for detailed discussion of the molecular genetics of these tumors.

PAPILLARY TUMORS OF THE LUNG/CILIATED MICRONODULAR PAPILLARY TUMOR

Lung papillomas are classified based on cell type as squamous cell, glandular, or mixed.[2] These are indolent lesions that typically present in a central airway, but have been rarely reported to arise in the periphery.[80] Lung papillomas are distinguished from non–small cell carcinomas based on absence

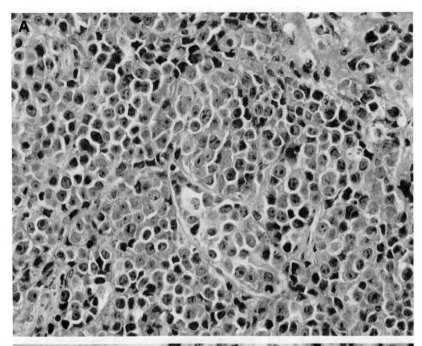

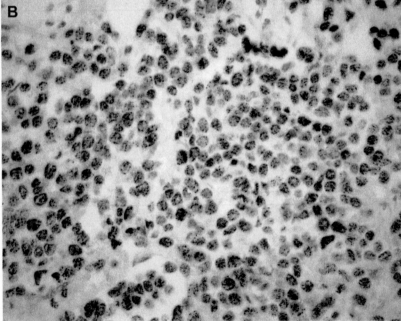

Fig. 9. (*A*) Nut carcinoma is composed of monomorphic sheets of undifferentiated round to epithelioid cells with scant, amphophilic cytoplasm, vesicular chromatin, and prominent nucleoli. Areas of squamous differentiation are often indistinct on hematoxylin-eosin–stained slides (H&E, 200×). (*B*) Nut protein overexpression in a Nut carcinoma with a *BRD4-NUT* fusion. Nut protein is located in the nucleus and has a characteristic speckled staining pattern (H&E, 200×).

of nuclear atypia and invasive architecture; some cases display a component of ciliated epithelium. These tumors should be distinguished from respiratory papillomatosis as a result of laryngeal and upper airway human papilloma virus infection.[81] Ciliated muconodular papillary tumors of the lung represent a rare, peripherally located tumor that shares morphologic features with mixed squamous cell and glandular papillomas, and indeed may exist on the spectrum with the WHO-defined papillomas.[82] Kamata and colleagues[83] reported that 8 of 10 ciliated muconodular papillary tumors harbored *BRAF* or *EGFR* activating mutations, thus confirming that these are neoplastic lesions likely representing a benign counterpart to oncogene-driven lung adenocarcinomas.

TECHNIQUES

MOLECULAR METHODS

A multiplicity of molecular methods may be used for *EGFR* testing. Traditional Sanger sequencing may be technically insensitive, particularly in samples that are contaminated by benign stromal cells. Targeted PCR-based assays are technically sensitive, but most cannot detect all of the sensitizing mutations that occur in *EGFR*, therefore multiple assays may be required[8] (**Table 3**). The most

Table 3
Techniques for detection of clinically relevant genomic alterations in lung cancer

Gene Target	Molecular Techniques	Cytogenetics	Immunohistochemistry
EGFR	• PCR-based techniques for sensitizing mutations[a] • Sizing assays for deletions (PCR + capillary electrophoresis) • Sequencing	FISH for EGFR amplification is not recommended for use in selecting patients for EGFR TKIs	• Total EGFR IHC does not predict response to TKIs • Mutation-specific IHC is highly specific but insufficiently sensitive to serve as a standalone assay; may be useful in resource-limited settings or on scant specimens[116]
KRAS	• PCR-based techniques[a] • Sequencing	n/a	n/a
ALK	• RT-PCR (from RNA) • Hybrid capture massively parallel sequencing • Anchored multiplex PCR	Breakapart FISH[b]	• Well-validated laboratory-developed ALK IHC is an acceptable screening technique, with FISH confirmation • ALK IHC kits available as standalone assays for patient selection for crizotinib therapy[b]
ROS1		Breakapart FISH	ROS1 IHC (clone D4D6) may be used as a screening tool; FISH, RT-PCR or sequencing confirmation is advised
RET		Breakapart FISH	Sensitivity and specificity of RET IHC is inadequate for clinical practice[36]
BRAF	• PCR-based techniques[a] • Sequencing	n/a	BRAF V600E IHC may be a useful screening tool but has shown mixed performance in practice[117]
ERBB2	• PCR-based techniques for exon 20 insertion mutations[a] • Sequencing	ERBB2 amplification is rare and distinct from mutations	Does not correlate with the presence of ERBB2 activating mutations[118]
MET	• Sequencing to span MET intron 13 through 14 • RT-PCR for exon14 dropout (from RNA)	MET FISH for amplification	High levels of MET protein overexpression correlate with MET amplification

Abbreviations: ALK, anaplastic lymphoma kinase; EGFR, epidermal growth factor receptor; FISH, fluorescence in situ hybridization; IHC, immunohistochemistry; KRAS, Kirsten rat sarcoma viral oncogene homolog; PCR, polymerase chain reaction; RT, reverse transcriptase; TKI, tyrosine kinase inhibitor; n/a, not applicable.
[a] Includes allele-specific/amplification refractory mutation system PCR, real-time PCR, Cycleave (Takara Bio, Otsu, Shiga, Japan), Invader (Hologic, Madison, WI), pyrosequencing, restrict fragment length polymorphism, denaturing high-performance liquid chromatography, single-stranded conformational polymorphism.
[b] Companion diagnostic assays approved by the Food and Drug Administration available.

common mutation occurs in exon 19 as a variably sized deletion involving all or part of the ELREA (Glutamic Acid Leucine Arginine Glutamic Acid Alanine) motif at amino acids 746 to 750. The variable nature of this deletion complicates detection using polymerase chain reaction (PCR)-based techniques; it is most reliably detected by sequencing or PCR followed by product sizing by capillary electrophoresis (CE).[8,84]

The limited range of mutational hotspots in *KRAS*, *BRAF*, and *ERBB2* simplifies molecular analysis of these genes (see Table 3). Although *KRAS*-driven tumors are not amenable to available targeted therapies, *KRAS* is the most commonly mutated oncogene in lung adenocarcinoma and occurs in a mutually exclusive fashion with targetable alterations. Therefore, detection of *KRAS*

mutations, nearly all of which occur at codons 12, 13, or 61, can be used to identify patients who will not respond to available targeted therapies.[8] Relatively simple and sensitive approaches to hotspot detection include pyrosequencing, multiplex PCR with primer extension and CE, and allele-specific PCR.[84,85]

ALK, *ROS1*, and *RET* fusions can be detected by reverse transcriptase PCR; historically, this approach required a priori knowledge of both fusion partners and multiplex design. Even so, this approach is prone to false negatives due to its inability to detect variant breakpoints or fusion partners. Advances in genomic target enrichment, including anchored multiplex PCR, have enabled detection of clinically relevant fusion events without knowledge of specific breakpoints or fusion partners.[86] Current guidelines, however,

A   Diagnostic supraclavicular lymph node biopsy

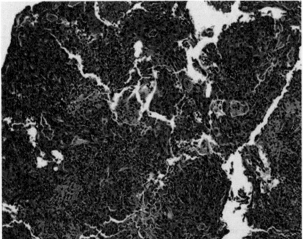

Genetic changes:

KRAS G12D
PTEN R234fs
TP53 S261_splice
BAP1 E653*

B   Intrapulmonary recurrence following chemotherapy

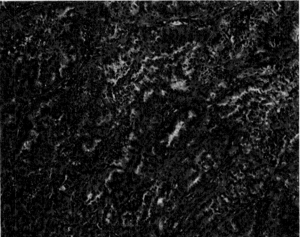

Genetic changes:

KRAS G12D
PTEN R234fs
TP53 S261_splice
BAP1 E653*
RB1 Q846*

Fig. 10. Lung adenocarcinoma diagnosed on a metastasis to a supraclavicular lymph node (H&E, 200×) (A); the patient was treated with first-line chemotherapy and relapsed after 7 months with intrapulmonary metastases (B) (H&E, 200×). The mutational profile of the tumors taken before and after treatment is similar; however, the relapse specimen likely has additional cell-cycle dysregulation secondary to an RB1 loss of function mutation. fs, frameshift; *, nonsense mutation.

recommend fluorescence in situ hybridization (FISH) as the gold standard for *ALK* rearrangement detection[8]; methodologic guidelines for *ROS1* and *RET* rearrangement testing have not yet been established.

The dropping cost of massively parallel sequencing (MPS) (so-called next-generation sequencing) has superseded much single-gene, PCR-based analysis, as this approach permits simultaneous detection of single nucleotide variants (point mutations), small insertion and/or deletions for tens or even thousands of genes. Hybrid-capture–based sequencing technology also permits analysis of copy number and structural alterations (rearrangements).[87,88] This more comprehensive approach to sequencing now

permits routine detection of both driver oncogene and tumor suppressor gene mutations. This latter category of genes may be inactivated by a variety of mechanisms, including missense, nonsense, frameshift, or splice site mutations or by gene deletion or rearrangement events. Although relative hotspots exist in common tumor suppressors, such as *TP53*, most loss-of-function events occur scattered through the gene, and thus are difficult to target using PCR-based assays. MPS, therefore, enables a much broader analysis of oncogenic pathways; incorporation of the additional information provided by this type of testing approach may begin to clarify the basis for variable responses to targeted therapies in lung cancer (**Fig. 10**).[89]

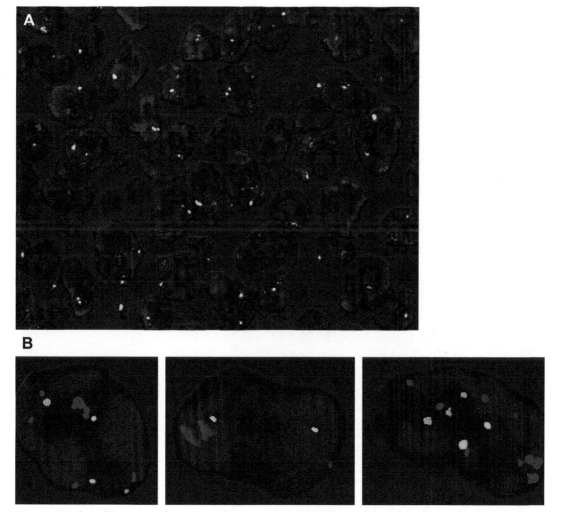

Fig. 11. FISH-based detection of gene amplification events. (*A*) *FGFR1* amplified SQC using 3 color combined breakapart and amplification probes (Cytocell, Windsor, CT). (*B*) *MET* amplified lung adenocarcinoma; *MET* (7q31.2 probe; red signal) is focally amplified on background of chromosome (CEP7 probe; green signal) copy gain (up to 6 copies) (1000×). (*Courtesy of* [A] Sanja Dacic, MD, PhD, University of Pittsburgh Medical Center, Pittsburgh, PA.)

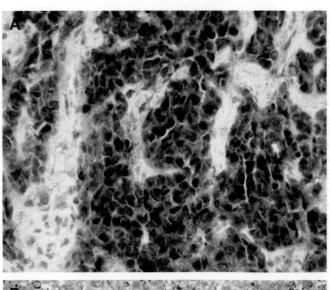

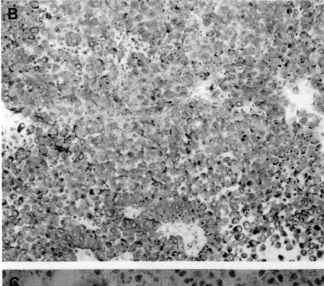

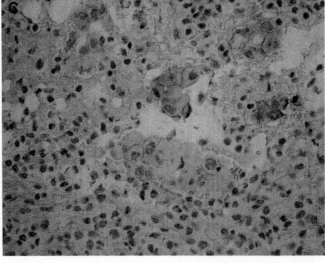

Fig. 12. Mutation-specific immunohisto-chemistry for ALK and ROS1. (*A*) ALK protein overexpression (clone D5F3) in a tumor with an *ALK* rearrangement by FISH (400×). (*B*) ROS1 protein overexpression (clone D4D6) present as diffuse cytoplasmic and focal intense dotlike staining. The tumor contains a CD74-ROS1 rearrangement (200×). (*C*) ROS1 protein overexpression (clone D4D6) presents as weak to moderate membranous and cytoplasmic staining. The tumor contains an EZR-ROS1 rearrangement (400×).

FLUORESCENCE IN SITU HYBRIDIZATION

FISH is considered the gold standard for *ALK* rearrangement detection using commercially available breakapart probes that have been approved by the Food and Drug Administration (FDA) as a companion diagnostic for crizotinib therapy.[90] Clinical trials showing efficacy of crizotinib in *ROS1* rearranged lung tumors used a variety of methods, including FISH, reverse transcriptase (RT)-PCR, and MPS to identify treatment candidates; as a result, a "gold standard" for ROS1 testing has not been identified. The clinical significance of *RET* rearrangement is less well established; however, for both *ROS1* and *RET* rearrangement detection, breakapart FISH is commonly used.

FISH is a robust approach for detection of copy number alterations; focal amplification events that have likely oncogenic activity may be therapeutic targets (**Fig. 11**). In contrast to molecular methods, such as MPS, quantitative PCR, or array CGH, all of which deliver a copy signal averaged across a mixed population of benign stroma and malignant tumor cells, FISH allows for analysis at the single-cell level and thus permits a precise quantification of copy number changes, may be more sensitive in highly contaminated tumor specimens, and additionally can detect intratumoral copy number heterogeneity.

IMMUNOHISTOCHEMISTRY

Immunohistochemistry (IHC) provides a rapid and inexpensive approach to detection of protein overexpression and may serve as a robust surrogate or screen for other more expensive and technically demanding molecular or cytogenetic assays. Clinically valuable IHC assays are discussed here. Other IHC targets with unproven clinical utility are included in **Table 3**. Mutation-specific EGFR IHC clones are available for L858R and exon 19 deletion mutations; the L858R assay is an excellent surrogate for molecular testing; the EGFR ex 19 deletion antibody is highly specific for the ELREA746_750 deletion (see **Fig. 7B**), but is relatively insensitive to other deletions at this site. Uncommon EGFR mutations cannot be queried using these antibodies.

ALK IHC using clone 5A4 or D5F3 (**Fig. 12A**) has also proven to be highly sensitive and specific as compared with *ALK* FISH; combined use of IHC and FISH and, increasingly, MPS, can increase the diagnostic yield of individual biopsies that may fail FISH testing and may help ferret out false-positive and false-negative FISH results.[91,92] A standalone ALK IHC assay has been FDA approved as a companion diagnostic for crizotinib therapy in patients with lung cancer.[93]

ROS1 IHC using clone D4D6 (**Fig. 12B, C**) is highly sensitive but less specific for *ROS1* rearrangements in lung cancer.[94,95] ROS1 is expressed in some benign cells, including macrophages and reactive alveolar epithelium, and is rarely found to be expressed at a very low level in some tumors lacking *ROS1* rearrangements.[94] As a result, ROS1 IHC is a robust screening tool; however, tumors with positive ROS1 expression should undergo confirmation using an orthogonal method, such as FISH, RT-PCR, or MPS.[96]

REFERENCES

1. Edwards BK, Noone AM, Mariotto AB, et al. Annual report to the nation on the status of cancer, 1975-2010, featuring prevalence of comorbidity and impact on survival among persons with lung, colorectal, breast, or prostate cancer. Cancer 2014; 120(9):1290–314.

2. Travis WD, Brambilla E, Burke AP, et al, editors. WHO classification of tumours of the lung, pleura, thymus and heart. 4th edition. Lyon (France): World Health Organization; 2015.

3. Meza R, Meernik C, Jeon J, et al. Lung cancer incidence trends by gender, race and histology in the United States, 1973-2010. PLoS One 2015;10(3): e0121323.

4. Wynder EL, Muscat JE. The changing epidemiology of smoking and lung cancer histology. Environ Health Perspect 1995;103(Suppl 8):143–8.

5. National Lung Screening Trial Research Team, Aberle DR, Adams AM, et al. Reduced lung-cancer mortality with low-dose computed tomographic screening. N Engl J Med 2011;365(5):395–409.

6. Aberle DR, DeMello S, Berg CD, et al. Results of the two incidence screenings in the national lung screening trial. N Engl J Med 2013;369(10):920–31.

7. Kris MG, Johnson BE, Berry LD, et al. Using multiplexed assays of oncogenic drivers in lung cancers to select targeted drugs. JAMA 2014;311(19): 1998–2006.

8. Lindeman NI, Cagle PT, Beasley MB, et al. Molecular testing guideline for selection of lung cancer patients for EGFR and ALK tyrosine kinase inhibitors: guideline from the College of American Pathologists, International Association for the Study of Lung Cancer, and Association for Molecular Pathology. J Mol Diagn 2013;15(4):415–53.

9. NCCN clinical practice guidelines: Non-small cell lung cancer version 3.2016. 2015. Available at: nccn.org. Accessed January 3, 2016.

10. Cancer Genome Atlas Research Network. Comprehensive genomic characterization of squamous cell lung cancers. Nature 2012;489(7417):519–25.

11. Peifer M, Fernandez-Cuesta L, Sos ML, et al. Integrative genome analyses identify key somatic driver mutations of small-cell lung cancer. Nat Genet 2012;44(10):1104–10.

12. Imielinski M, Berger AH, Hammerman PS, et al. Mapping the hallmarks of lung adenocarcinoma with massively parallel sequencing. Cell 2012; 150(6):1107–20.

13. Calles A, Sholl LM, Rodig SJ, et al. Immunohistochemical loss of LKB1 is a biomarker for more aggressive biology in KRAS-mutant lung adenocarcinoma. Clin Cancer Res 2015;21(12):2851–60.

14. Chen Z, Cheng K, Walton Z, et al. A murine lung cancer co-clinical trial identifies genetic modifiers of therapeutic response. Nature 2012;483(7391): 613–7.

15. Kim HR, Cho BC, Shim HS, et al. Prediction for response duration to epidermal growth factor receptor-tyrosine kinase inhibitors in EGFR mutated never smoker lung adenocarcinoma. Lung Cancer 2014;83(3):374–82.

16. Eng J, Woo KM, Sima CS, et al. Impact of concurrent PIK3CA mutations on response to EGFR tyrosine kinase inhibition in EGFR-mutant lung cancers and on prognosis in oncogene-driven lung adenocarcinomas. J Thorac Oncol 2015; 10(12):1713–9.

17. Schildhaus HU, Nogova L, Wolf J, et al. FGFR1 amplifications in squamous cell carcinomas of the lung: diagnostic and therapeutic implications. Transl Lung Cancer Res 2013;2(2):92–100.

18. Dutt A, Ramos AH, Hammerman PS, et al. Inhibitor-sensitive FGFR1 amplification in human non-small cell lung cancer. PLoS One 2011;6(6):e20351.

19. Wynes MW, Hinz TK, Gao D, et al. FGFR1 mRNA and protein expression, not gene copy number, predict FGFR TKI sensitivity across all lung cancer histologies. Clin Cancer Res 2014;20(12): 3299–309.

20. Brahmer J, Reckamp KL, Baas P, et al. Nivolumab versus docetaxel in advanced squamous-cell non-small-cell lung cancer. N Engl J Med 2015;373(2): 123–35.

21. Garon EB, Rizvi NA, Hui R, et al. Pembrolizumab for the treatment of non-small-cell lung cancer. N Engl J Med 2015;372(21):2018–28.

22. Varghese AM, Zakowski MF, Yu HA, et al. Small-cell lung cancers in patients who never smoked cigarettes. J Thorac Oncol 2014;9(6):892–6.

23. Sos ML, Dietlein F, Peifer M, et al. A framework for identification of actionable cancer genome dependencies in small cell lung cancer. Proc Natl Acad Sci U S A 2012;109(42):17034–9.

24. Clinical Lung Cancer Genome Project (CLCGP), Network Genomic Medicine (NGM). A genomics-based classification of human lung tumors. Sci Transl Med 2013;5(209):209ra153.

25. Masters GA, Temin S, Azzoli CG, et al. Systemic therapy for stage IV non-small-cell lung cancer: American Society of Clinical Oncology clinical practice guideline update. J Clin Oncol 2015; 33(30):3488–515.

26. Sholl LM, Aisner DL, Varella-Garcia M, et al. Multi-institutional oncogenic driver mutation analysis in lung adenocarcinoma: the lung cancer mutation consortium experience. J Thorac Oncol 2015; 10(5):768–77.

27. Tsuta K, Kawago M, Inoue E, et al. The utility of the proposed IASLC/ATS/ERS lung adenocarcinoma subtypes for disease prognosis and correlation of driver gene alterations. Lung Cancer 2013;81(3): 371–6.

28. Nakamura H, Saji H, Shinmyo T, et al. Association of IASLC/ATS/ERS histologic subtypes of lung adenocarcinoma with epidermal growth factor receptor mutations in 320 resected cases. Clin Lung Cancer 2015;16(3):209–15.

29. Yoshizawa A, Sumiyoshi S, Sonobe M, et al. Validation of the IASLC/ATS/ERS lung adenocarcinoma classification for prognosis and association with EGFR and KRAS gene mutations: analysis of 440 Japanese patients. J Thorac Oncol 2013;8(1): 52–61.

30. Rekhtman N, Ang DC, Riely GJ, et al. KRAS mutations are associated with solid growth pattern and tumor-infiltrating leukocytes in lung adenocarcinoma. Mod Pathol 2013;26(10):1307–19.

31. Ujiie H, Kadota K, Chaft JE, et al. Solid predominant histologic subtype in resected stage I lung adenocarcinoma is an independent predictor of early, extrathoracic, multisite recurrence and of poor postrecurrence survival. J Clin Oncol 2015; 33(26):2877–84.

32. Yeh YC, Kadota K, Nitadori J, et al. International Association for the Study of Lung Cancer/American Thoracic Society/European Respiratory Society classification predicts occult lymph node metastasis in clinically mediastinal node-negative lung adenocarcinoma. Eur J Cardiothorac Surg 2016; 49(1):e9–15.

33. Tsao MS, Marguet S, Le Teuff G, et al. Subtype classification of lung adenocarcinoma predicts benefit from adjuvant chemotherapy in patients undergoing complete resection. J Clin Oncol 2015; 33(30):3439–46.

34. Kadota K, Yeh YC, Sima CS, et al. The cribriform pattern identifies a subset of acinar predominant tumors with poor prognosis in patients with stage I lung adenocarcinoma: a conceptual proposal to classify cribriform predominant tumors as a distinct histologic subtype. Mod Pathol 2014;27(5): 690–700.

35. Nishino M, Klepeis VE, Yeap BY, et al. Histologic and cytomorphologic features of ALK-rearranged

lung adenocarcinomas. Mod Pathol 2012;25(11): 1462–72.

36. Lee SE, Lee B, Hong M, et al. Comprehensive analysis of RET and ROS1 rearrangement in lung adenocarcinoma. Mod Pathol 2015;28(4):468–79.

37. Kamata T, Yoshida A, Shiraishi K, et al. Mucinous micropapillary pattern in lung adenocarcinomas: a unique histology with genetic correlates. Histopathology 2016;68(3):356–66.

38. Planchard D, Groen H, Kim T, et al. Interim results of a phase II study of the BRAF inhibitor (BRAFi) dabrafenib (D) in combination with the MEK inhibitor trametinib (T) in patients (pts) with *BRAF* V600E mutated (mut) metastatic non-small cell lung cancer (NSCLC). J Clin Oncol 2015;33(Supp):8006.

39. Morita S, Yoshida A, Goto A, et al. High-grade lung adenocarcinoma with fetal lung-like morphology: clinicopathologic, immunohistochemical, and molecular analyses of 17 cases. Am J Surg Pathol 2013;37(6):924–32.

40. Girard N, Deshpande C, Lau C, et al. Comprehensive histologic assessment helps to differentiate multiple lung primary nonsmall cell carcinomas from metastases. Am J Surg Pathol 2009;33(12): 1752–64.

41. Girard N, Ostrovnaya I, Lau C, et al. Genomic and mutational profiling to assess clonal relationships between multiple non-small cell lung cancers. Clin Cancer Res 2009;15(16):5184–90.

42. Zhang Y, Hu H, Wang R, et al. Synchronous non-small cell lung cancers: diagnostic yield can be improved by histologic and genetic methods. Ann Surg Oncol 2014;21(13):4369–74.

43. Sato S, Motoi N, Hiramatsu M, et al. Pulmonary adenocarcinoma in situ: analyses of a large series with reference to smoking, driver mutations, and receptor tyrosine kinase pathway activation. Am J Surg Pathol 2015;39(7):912–21.

44. Finberg KE, Sequist LV, Joshi VA, et al. Mucinous differentiation correlates with absence of EGFR mutation and presence of KRAS mutation in lung adenocarcinomas with bronchioloalveolar features. J Mol Diagn 2007;9(3):320–6.

45. Shim HS, Kenudson M, Zheng Z, et al. Unique genetic and survival characteristics of invasive mucinous adenocarcinoma of the lung. J Thorac Oncol 2015;10(8):1156–62.

46. Kadota K, Yeh YC, D'Angelo SP, et al. Associations between mutations and histologic patterns of mucin in lung adenocarcinoma: invasive mucinous pattern and extracellular mucin are associated with KRAS mutation. Am J Surg Pathol 2014;38(8): 1118–27.

47. Riely GJ, Kris MG, Rosenbaum D, et al. Frequency and distinctive spectrum of KRAS mutations in never smokers with lung adenocarcinoma. Clin Cancer Res 2008;14(18):5731–4.

48. Snyder EL, Watanabe H, Magendantz M, et al. Nkx2-1 represses a latent gastric differentiation program in lung adenocarcinoma. Mol Cell 2013; 50(2):185–99.

49. Maeda Y, Tsuchiya T, Hao H, et al. Kras(G12D) and Nkx2-1 haploinsufficiency induce mucinous adenocarcinoma of the lung. J Clin Invest 2012;122(12): 4388–400.

50. Hwang DA, Sholl LM, Rojas-Rudilla V, et al. KRAS and NKX2-1 mutations in invasive mucinous adenocarcinoma of the lung. J Thorac Oncol 2016;11(4):496–503.

51. Sholl LM, Yeap BY, Iafrate AJ, et al. Lung adenocarcinoma with EGFR amplification has distinct clinicopathologic and molecular features in never-smokers. Cancer Res 2009;69(21):8341–8.

52. Kong-Beltran M, Seshagiri S, Zha J, et al. Somatic mutations lead to an oncogenic deletion of MET in lung cancer. Cancer Res 2006;66(1):283–9.

53. Awad MM, Oxnard GR, Jackman DM, et al. Met exon 14 mutations in non-small lung cancer are associated with advanced age and stage-dependent MET genomic amplification and c-met overexpression. J Clin Oncol 2016;34(7):721–30.

54. Paik PK, Drilon A, Fan PD, et al. Response to MET inhibitors in patients with stage IV lung adenocarcinomas harboring MET mutations causing exon 14 skipping. Cancer Discov 2015;5(8):842–9.

55. Frampton GM, Ali SM, Rosenzweig M, et al. Activation of MET via diverse exon 14 splicing alterations occurs in multiple tumor types and confers clinical sensitivity to MET inhibitors. Cancer Discov 2015; 5(8):850–9.

56. Liu X, Jia Y, Stoopler MB, et al. Next-generation sequencing of pulmonary sarcomatoid carcinoma reveals high frequency of actionable MET gene mutations. J Clin Oncol 2016;34(8):794–802.

57. Hwang DH, Szeto DP, Perry AS, et al. Pulmonary large cell carcinoma lacking squamous differentiation is clinicopathologically indistinguishable from solid-subtype adenocarcinoma. Arch Pathol Lab Med 2014;138(5):626–35.

58. Rekhtman N, Tafe LJ, Chaft JE, et al. Distinct profile of driver mutations and clinical features in immunomarker-defined subsets of pulmonary large-cell carcinoma. Mod Pathol 2013;26(4): 511–22.

59. Driver BR, Portier BP, Mody DR, et al. Next-generation sequencing of a cohort of pulmonary large cell carcinomas reclassified by world health organization 2015 criteria. Arch Pathol Lab Med 2015; 140(4):312–7.

60. Pelosi G, Fabbri A, Papotti M, et al. Dissecting pulmonary large-cell carcinoma by targeted next generation sequencing of several cancer genes pushes genotypic-phenotypic correlations to emerge. J Thorac Oncol 2015;10(11):1560–9.

61. Rekhtman N, Paik PK, Arcila ME, et al. Clarifying the spectrum of driver oncogene mutations in biomarker-verified squamous carcinoma of lung: lack of EGFR/KRAS and presence of PIK3CA/AKT1 mutations. Clin Cancer Res 2012;18(4):1167–76.

62. Kaira K, Horie Y, Ayabe E, et al. Pulmonary pleomorphic carcinoma: a clinicopathological study including EGFR mutation analysis. J Thorac Oncol 2010;5(4):460–5.

63. Chang YL, Wu CT, Shih JY, et al. EGFR and p53 status of pulmonary pleomorphic carcinoma: implications for EGFR tyrosine kinase inhibitors therapy of an aggressive lung malignancy. Ann Surg Oncol 2011;18(10):2952–60.

64. Chen FF, Yan JJ, Lai WW, et al. Epstein-Barr virus-associated nonsmall cell lung carcinoma: undifferentiated "lymphoepithelioma-like" carcinoma as a distinct entity with better prognosis. Cancer 1998;82(12):2334–42.

65. Chang YL, Yang CY, Lin MW, et al. PD-L1 is highly expressed in lung lymphoepithelioma-like carcinoma: a potential rationale for immunotherapy. Lung Cancer 2015;88(3):254–9.

66. Fang W, Hong S, Chen N, et al. PD-L1 is remarkably over-expressed in EBV-associated pulmonary lymphoepithelioma-like carcinoma and related to poor disease-free survival. Oncotarget 2015;6(32):33019–32.

67. Green MR, Rodig S, Juszczynski P, et al. Constitutive AP-1 activity and EBV infection induce PD-L1 in Hodgkin lymphomas and posttransplant lymphoproliferative disorders: implications for targeted therapy. Clin Cancer Res 2012;18(6):1611–8.

68. Gru AA, Haverkos BH, Freud AG, et al. The Epstein-Barr virus (EBV) in T cell and NK cell lymphomas: time for a reassessment. Curr Hematol Malig Rep 2015;10(4):456–67.

69. Wang L, Lin Y, Cai Q, et al. Detection of rearrangement of anaplastic lymphoma kinase (ALK) and mutation of epidermal growth factor receptor (EGFR) in primary pulmonary lymphoepithelioma-like carcinoma. J Thorac Dis 2015;7(9):1556–62.

70. Bauer DE, Mitchell CM, Strait KM, et al. Clinicopathologic features and long-term outcomes of NUT midline carcinoma. Clin Cancer Res 2012;18(20):5773–9.

71. Sholl LM, Nishino M, Pokharel S, et al. Primary pulmonary NUT midline carcinoma: clinical, radiographic, and pathologic characterizations. J Thorac Oncol 2015;10(6):951–9.

72. Haack H, Johnson LA, Fry CJ, et al. Diagnosis of NUT midline carcinoma using a NUT-specific monoclonal antibody. Am J Surg Pathol 2009;33(7):984–91.

73. French CA. The importance of diagnosing NUT midline carcinoma. Head Neck Pathol 2013;7(1):11–6.

74. Falk N, Weissferdt A, Kalhor N, et al. Primary pulmonary salivary gland-type tumors: a review and update. Adv Anat Pathol 2016;23(1):13–23.

75. Achcar Rde O, Nikiforova MN, Dacic S, et al. Mammalian mastermind like 2 11q21 gene rearrangement in bronchopulmonary mucoepidermoid carcinoma. Hum Pathol 2009;40(6):854–60.

76. Brill LB 2nd, Kanner WA, Fehr A, et al. Analysis of MYB expression and MYB-NFIB gene fusions in adenoid cystic carcinoma and other salivary neoplasms. Mod Pathol 2011;24(9):1169–76.

77. Garcia JJ, Jin L, Jackson SB, et al. Primary pulmonary hyalinizing clear cell carcinoma of bronchial submucosal gland origin. Hum Pathol 2015;46(3):471–5.

78. Antonescu CR, Zhang L, Chang NE, et al. EWSR1-POU5F1 fusion in soft tissue myoepithelial tumors. A molecular analysis of sixty-six cases, including soft tissue, bone, and visceral lesions, showing common involvement of the EWSR1 gene. Genes Chromosomes Cancer 2010;49(12):1114–24.

79. Fishbein GA, Grimes BS, Xian RR, et al. Primary salivary duct carcinoma of the lung, mucin-rich variant. Hum Pathol 2016;47(1):150–6.

80. Aida S, Ohara I, Shimazaki H, et al. Solitary peripheral ciliated glandular papillomas of the lung: a report of 3 cases. Am J Surg Pathol 2008;32(10):1489–94.

81. Omland T, Lie KA, Akre H, et al. Recurrent respiratory papillomatosis: HPV genotypes and risk of high-grade laryngeal neoplasia. PLoS One 2014;9(6):e99114.

82. Kamata T, Yoshida A, Kosuge T, et al. Ciliated muconodular papillary tumors of the lung: a clinicopathologic analysis of 10 cases. Am J Surg Pathol 2015;39(6):753–60.

83. Kamata T, Sunami K, Yoshida A, et al. Frequent BRAF or EGFR mutations in ciliated muconodular papillary tumors of the lung. J Thorac Oncol 2016;11(2):261–5.

84. Su Z, Dias-Santagata D, Duke M, et al. A platform for rapid detection of multiple oncogenic mutations with relevance to targeted therapy in non-small-cell lung cancer. J Mol Diagn 2011;13(1):74–84.

85. Anderson SM. Laboratory methods for KRAS mutation analysis. Expert Rev Mol Diagn 2011;11(6):635–42.

86. Zheng Z, Liebers M, Zhelyazkova B, et al. Anchored multiplex PCR for targeted next-generation sequencing. Nat Med 2014;20(12):1479–84.

87. Wagle N, Berger MF, Davis MJ, et al. High-throughput detection of actionable genomic alterations in clinical tumor samples by targeted, massively parallel sequencing. Cancer Discov 2012;2(1):82–93.

88. Cheng DT, Mitchell TN, Zehir A, et al. Memorial Sloan Kettering-integrated mutation profiling of actionable cancer targets (MSK-IMPACT): a

hybridization capture-based next-generation sequencing clinical assay for solid tumor molecular oncology. J Mol Diagn 2015;17(3):251–64.

89. Govindan R, Mandrekar SJ, Gerber DE, et al. ALCHEMIST trials: a golden opportunity to transform outcomes in early-stage non-small cell lung cancer. Clin Cancer Res 2015;21(24):5439–44.

90. FDA approval for crizotinib. Available at: http://www.cancer.gov/cancertopics/druginfo/fda-crizotinib. Accessed August 30, 2013.

91. Sholl LM, Weremowicz S, Gray SW, et al. Combined use of ALK immunohistochemistry and FISH for optimal detection of ALK-rearranged lung adenocarcinomas. J Thorac Oncol 2013;8(3):322–8.

92. Gao X, Sholl LM, Nishino M, et al. Clinical implications of variant ALK FISH rearrangement patterns. J Thorac Oncol 2015;10(11):1648–52.

93. Recently approved devices: VENTANA ALK (D5F3) CDx assay - P140025. 2015. Available at: http://www.fda.gov/MedicalDevices/ProductsandMedicalProcedures/DeviceApprovalsandClearances/RecentlyApprovedDevices/ucm454476.htm. Accessed January 3, 2016.

94. Sholl LM, Sun H, Butaney M, et al. ROS1 immunohistochemistry for detection of ROS1-rearranged lung adenocarcinomas. Am J Surg Pathol 2013;37(9):1441–9.

95. Yoshida A, Kohno T, Tsuta K, et al. ROS1-rearranged lung cancer: a clinicopathologic and molecular study of 15 surgical cases. Am J Surg Pathol 2013;37(4):554–62.

96. Mescam-Mancini L, Lantuejoul S, Moro-Sibilot D, et al. On the relevance of a testing algorithm for the detection of ROS1-rearranged lung adenocarcinomas. Lung Cancer 2014;83(2):168–73.

97. Janne PA, Shaw AT, Pereira JR, et al. Selumetinib plus docetaxel for KRAS-mutant advanced non-small-cell lung cancer: a randomised, multicentre, placebo-controlled, phase 2 study. Lancet Oncol 2013;14(1):38–47.

98. Kwak EL, Bang YJ, Camidge DR, et al. Anaplastic lymphoma kinase inhibition in non-small-cell lung cancer. N Engl J Med 2010;363(18):1693–703.

99. Shaw AT, Kim DW, Mehra R, et al. Ceritinib in ALK-rearranged non-small-cell lung cancer. N Engl J Med 2014;370(13):1189–97.

100. Shaw AT, Kim DW, Nakagawa K, et al. Crizotinib versus chemotherapy in advanced ALK-positive lung cancer. N Engl J Med 2013;368(25):2385–94.

101. Shaw AT, Gandhi L, Gadgeel S, et al. Alectinib in ALK-positive, crizotinib-resistant, non-small-cell lung cancer: a single-group, multicentre, phase 2 trial. Lancet Oncol 2016;17(2):234–42.

102. Ou SI, Ahn JS, De Petris L, et al. Alectinib in crizotinib-refractory ALK-rearranged non-small-cell lung cancer: a phase II global study. J Clin Oncol 2016;34(7):661–8.

103. Bergethon K, Shaw AT, Ou SH, et al. ROS1 rearrangements define a unique molecular class of lung cancers. J Clin Oncol 2012;30(8):863–70.

104. Lopez-Chavez A, Thomas A, Rajan A, et al. Molecular profiling and targeted therapy for advanced thoracic malignancies: a biomarker-derived, multiarm, multihistology phase II basket trial. J Clin Oncol 2015;33(9):1000–7.

105. Drilon A, Wang L, Hasanovic A, et al. Response to cabozantinib in patients with RET fusion-positive lung adenocarcinomas. Cancer Discov 2013;3(6):630–5.

106. Kohno T, Ichikawa H, Totoki Y, et al. KIF5B-RET fusions in lung adenocarcinoma. Nat Med 2012;18(3):375–7.

107. Zheng D, Wang R, Pan Y, et al. Prevalence and clinicopathological characteristics of BRAF mutations in Chinese patients with lung adenocarcinoma. Ann Surg Oncol 2015;22(Suppl 3):1284–91.

108. Arcila ME, Chaft JE, Nafa K, et al. Prevalence, clinicopathologic associations, and molecular spectrum of ERBB2 (HER2) tyrosine kinase mutations in lung adenocarcinomas. Clin Cancer Res 2012;18(18):4910–8.

109. De Greve J, Teugels E, Geers C, et al. Clinical activity of afatinib (BIBW 2992) in patients with lung adenocarcinoma with mutations in the kinase domain of HER2/neu. Lung Cancer 2012;76(1):123–7.

110. Gandhi L, Bahleda R, Tolaney SM, et al. Phase I study of neratinib in combination with temsirolimus in patients with human epidermal growth factor receptor 2-dependent and other solid tumors. J Clin Oncol 2014;32(2):68–75.

111. Wang R, Wang L, Li Y, et al. FGFR1/3 tyrosine kinase fusions define a unique molecular subtype of non-small cell lung cancer. Clin Cancer Res 2014;20(15):4107–14.

112. Liao RG, Jung J, Tchaicha J, et al. Inhibitor-sensitive FGFR2 and FGFR3 mutations in lung squamous cell carcinoma. Cancer Res 2013;73(16):5195–205.

113. Hammerman PS, Sos ML, Ramos AH, et al. Mutations in the DDR2 kinase gene identify a novel therapeutic target in squamous cell lung cancer. Cancer Discov 2011;1(1):78–89.

114. Brunner AM, Costa DB, Heist RS, et al. Treatment-related toxicities in a phase II trial of dasatinib in patients with squamous cell carcinoma of the lung. J Thorac Oncol 2013;8(11):1434–7.

115. Gupta S, Argiles G, Munster PN, et al. A phase I trial of combined ridaforolimus and MK-2206 in patients with advanced malignancies. Clin Cancer Res 2015;21(23):5235–44.

116. Brevet M, Arcila M, Ladanyi M. Assessment of EGFR mutation status in lung adenocarcinoma by immunohistochemistry using antibodies specific

to the two major forms of mutant EGFR. J Mol Diagn 2010;12(2):169–76.

117. Routhier CA, Mochel MC, Lynch K, et al. Comparison of 2 monoclonal antibodies for immunohistochemical detection of BRAF V600E mutation in malignant melanoma, pulmonary carcinoma, gastrointestinal carcinoma, thyroid carcinoma, and gliomas. Hum Pathol 2013;44(11):2563–70.

118. Li BT, Ross DS, Aisner DL, et al. HER2 amplification and HER2 mutation are distinct molecular targets in lung cancers. J Thorac Oncol 2016;11(3):414–9.

Establishing a Robust Molecular Taxonomy for Diffuse Gliomas of Adulthood

Jason T. Huse, MD, PhD

KEYWORDS

• Glioma • Glioblastoma • IDH1 • 1p/19q codeletion • Genomics • *TERT* • *ATRX*

ABSTRACT

The comprehensive molecular profiling of cancer has dramatically altered conceptions of numerous tumor types, particularly with regard to their fundamental classification. In the case of primary brain tumors, the widespread use of disease-defining biomarker sets is profoundly reshaping existing diagnostic entities that had been designated solely by histopathological criteria for decades. This review describes recent progress for diffusely infiltrating gliomas of adulthood, the most common primary brain tumor variants. More specifically, it details how routine incorporation of a handful of highly prevalent molecular alterations robustly designates refined subclasses of glioma that transcend conventional histopathological designations.

OVERVIEW

In their various forms, glial cells represent the most populous constituents of the central nervous system and exhibit characteristic histopathological features that have been extensively incorporated into the classification of brain tumors for decades. Not surprisingly, a broad spectrum of primary brain tumors exhibit morphologic features suggestive of glial histogenesis (Box 1). However, the precise term "glioma" has, over time, become inextricably associated with a defined group of diffusely infiltrating variants, most commonly seen in adulthood. Although these tumors feature considerable clinical and biological heterogeneity, their shared propensity for widespread invasion into surrounding normal brain, coupled with their notable refractoriness to conventional therapeutic approaches, effectively renders them uniformly incurable.

Gliomas are themselves segregated into distinct diagnostic entities on the basis of histopathological features thought to reflect the biological behavior of the tumors in question. Specifically, the current World Health Organization (WHO) classification system stratifies gliomas into 3 basic histigenic lineages, astrocytic, oligodendroglial, and mixed (oligoastrocytic), while also arraying them across a spectrum of malignancy encompassing grades II, III, and IV (Fig. 1).[1] Within this schema, glioblastoma (GBM), WHO grade IV, represents that most aggressive variant, with a median survival of only 15 months in affected patients. By contrast, lower-grade gliomas (LGGs: astrocytomas, oligodendrogliomas, and oligoastrocytomas), ranging in WHO grade from II to III, are typically associated with more extended overall survival, even years to decades. Nevertheless, virtually all gliomas, regardless of grade at initial diagnosis, inexorably recur and progress, ultimately acquiring the histopathological features of glioblastoma, namely microvascular proliferation and necrosis (see Fig. 1). Accordingly, the term "secondary glioblastoma" has historically been applied in instances in which hallmark GBM histopathology emerges in the context of a lower-grade, predominantly astrocytic neoplasm.

In recent years, a series of large-scale molecular-profiling studies has greatly clarified the spectrum of genomic, and in some cases epigenomic, abnormalities occurring in the various glioma subtypes, and in doing so, dramatically altered conceptions of glioma pathogenesis. These developments have also prompted a fundamental rethinking of glioma classification. Although the current WHO system, outlined previously, performs quite well from the standpoint of prognostic stratification, it suffers from inherent interobserver variability. Moreover, as shown later in this article, highly recurrent molecular aberrations are now known to designate biologically distinct

Department of Pathology, Memorial Sloan-Kettering Cancer Center, 408 East 69th Street (Z564), New York, NY 10065, USA
E-mail address: husej@mskcc.org

Surgical Pathology 9 (2016) 379–390
http://dx.doi.org/10.1016/j.path.2016.04.005
1875-9181/16/$ – see front matter © 2016 Published by Elsevier Inc.

surgpath.theclinics.com

Box 1
Primary central nervous system (CNS) neoplasms with histopathological features of at least partial glial histogenesis

Primary glial neoplasms of the CNS
- Glioblastoma
- Astrocytoma
- Anaplastic astrocytoma
- Oligodendroglioma
- Anaplastic oligodendroglioma
- Oligoastrocytoma
- Anaplastic oligoastrocytoma
- Gliomatosis cerebri
- Pilocytic astrocytoma
- Pilomyxoid astrocytoma
- Subependymal giant cell astrocytoma
- Pleomorphic xanthoastrocytoma
- Ependymoma
- Anaplastic ependymoma
- Myxopapillary ependymoma
- Subependymoma
- Astroblastoma
- Chordoid glioma of the third ventricle
- Angiocentric glioma
- Desmoplastic infantile astrocytoma/ganglioglioma
- Dysembryoplastic neuroepithelial tumor
- Ganglioglioma
- Anaplastic ganglioglioma
- Papillary glioneuronal tumor
- Rosette-forming glioneuronal tumor of the fourth ventricle
- Pituicytoma

---

disease subclasses that transcend conventional histopathological designations. This review describes how the comprehensive molecular characterization of glioma subtypes has reformed their taxonomy, impacting current clinical practice as well as the formulation of novel treatment strategies.

MUTATIONS IN ISOCITRATE DEHYDROGENASE GENES DISTINGUISH LOWER-GRADE GLIOMAS FROM GLIOBLASTOMA

Perhaps the most important insight gleaned from recent molecular profiling in gliomas is that primary GBMs fundamentally differ in their underlying biology and driving pathogenic mechanisms from LGGs and the secondary GBMs into which they evolve. This dichotomy is reflected most prominently by mutations in isocitrate dehydrogenase enzymes (IDH1 and IDH2), which have come to essentially define LGGs of adulthood. Ironically, point mutations in the *IDH1* gene were originally discovered in a study focused on primary GBM, among the first to use the unbiased approach of whole-exome sequencing.[2] The conjecture that *IDH1*-mutant tumors, approximately 11% of the sample cohort, might actually represent secondary GBMs led to subsequent investigations documenting strikingly high rates (70%–100%) of mutation in

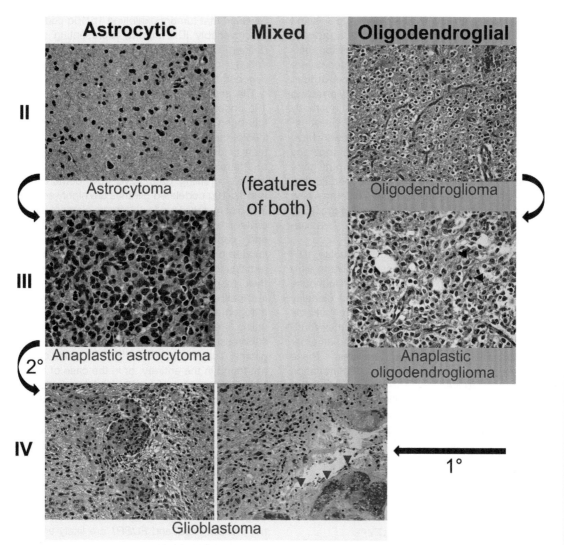

Fig. 1. Schematic of the WHO classification system for gliomas with micrographs illustrating typical morphologic features for selected entities. Histiogenic lineages (astrocytic, mixed, and oligodendroglial) are indicated along with WHO grade levels (II, III, and IV). Primary (1°) and secondary (2°) evolutionary pathways to GBM are also shown. Arrowheads indicate mitotic figures (*black*), microvascular proliferation (*green*), and necrosis (*red*). (H&E, 400× [10× eye piece and 40× objective]).

either *IDH1* or its homologue *IDH2* across all LGG subtypes.[3] Multiple studies in the time since have shown that IDH-mutant gliomas exhibit favorable prognosis and extended overall survival relative to IDH–wild-type counterparts in a manner that transcends histopathological designations and WHO grading.[4–7] Such findings suggest that IDH-mutant and IDH–wild-type tumors are entirely distinct from a pathogenic standpoint, a notion that is further supported by the largely nonoverlapping sets of molecular alterations associated with each (see later in this article). Moreover, IDH-mutant and IDH–wild-type gliomas exhibit contrasting gene expression signatures,[5,8,9] consistent with their emergence from separate pools of precursor cells.

The mechanisms by which IDH mutations actually promote gliomagenesis remain an area of active research. Glioma-associated IDH mutations invariably occur in active site arginine residues, either R132 for IDH1 or R172 for IDH2, and confer a neomorphic enzymatic activity by which α-ketoglutarate, the normal product of IDH-mediated catalysis, is converted to R(−)-2-hydroxyglutarate (2HG), an established oncometabolite that accumulates at millimolar levels in affected cells.[10] The 2HG has profound epigenomic effects, including widespread induction of both DNA and histone methylation: the glioma CpG island hypermethylator phenotype (G-CIMP), a process that fundamentally alters gene expression patterns.[11–13] Because many affected genes are involved in tissue differentiation, it has been argued that G-CIMP exerts oncogenic affects by arresting cells in stem cell and progenitor states inherently more amenable to self-renewal and propagation.[11,13] Interestingly, very recent data have shown that G-CIMP disrupts the normal topology of the genome, thereby enabling aberrant gene regulatory networks.[14] Above and beyond its dramatic effects on the epigenome, 2HG may also induce transformative behavior by antagonizing the tumor suppressive effects of hypoxia-inducible factor 1-alpha (HIF1α).[15]

Despite their broad physiologic sequelae, IDH mutations are only weakly transformative in vitro and have yet to induce glioma formation in mouse disease models.[16] Nevertheless, recent clonality studies have clearly demonstrated the remarkable degree to which recurrent LGGs retain their foundational IDH mutation, easily surpassing all other LGG-associated molecular abnormalities.[17] These data provide a strong argument that IDH mutation is absolutely required for lower-grade gliomagenesis. However, it is likely insufficient to induce full transformation in isolation, as several highly recurrent molecular abnormalities are now known to invariably co-occur with IDH mutation in LGGs (see the following section).

A DEFINED SET OF GENOMIC ABNORMALITIES ROBUSTLY DESIGNATES LOWER-GRADE GLIOMA SUBTYPE

As alluded to previously, the molecular abnormalities exhibited by IDH-mutant LGGs are highly characteristic, and contrast sharply with those seen in primary GBM. Moreover, mutually exclusive patterns of genomic alterations robustly designate distinct LGG subtypes, speaking to pathogenic mechanism as well as histogenesis. A substantial minority of LGGs, for instance, harbor concurrent loss of the short arm of chromosome 1 and the long arm of chromosome 19. So-called 1p/19q codeletion was initially discovered more than 20 years ago as a defining molecular feature of both low-grade (WHO grade II) and anaplastic (WHO grade III) oligodendroglioma.[18,19] Importantly, 1p/19q codeleted oligodendroglial neoplasms were found to exhibit favorable prognosis relative to non-codeleted counterparts, as well as robust response to cytotoxic chemotherapy.[20,21] More recent molecular profiling, subsequent to the discovery of IDH mutations,

revealed that tumors exhibiting 1p/19q codeletion are invariably IDH-mutant, representing a well-defined subclass enriched with, although not entirely composed of, gliomas demonstrating oligodendroglial histopathology.[4,8,22]

The pathogenic mechanism(s) induced by 1p/19q codeletion are not completely established. It has been thought for some time that coordinate loss of these chromosomal arms unmasks the effects of loss-of-function mutations involving genes on the remaining copies of 1p and 19q. Comprehensive profiling data have now demonstrated that 1p/19q codeleted gliomas are highly enriched for point mutations in far upstream element binding protein 1 (FUBP1), located on chromosome 1p, and capicua transcriptional repressor (CIC), located on chromosome 19q.[22,23] Although much remains to be understood about FUBP1, CIC, and their respective gene products, their previous implication as negative regulators of myc and mitogen-activated protein kinase (MAPK) signaling, respectively, is consistent with tumor suppressor function in the context of oligodendroglioma. That said, CIC and FUBP1 mutations are not found in the entirety, or in the case of FUBP1 even the majority, of 1p/19q codeleted gliomas,[8] and recent clonality studies indicate that they are likely acquired subsequent to both IDH mutation and 1p/19q codeletion in the oncogenic sequence.[24] These data would seem to suggest that oligodendrogliomas use mutations in CIC and FUBP1 to fully silence their gene products following 1p/19q codeletion-induced haploinsufficiency, and not the other way around. Therefore, additional as yet undiscovered transformative mechanisms not involving CIC and FUBP1 are likely engaged by 1p/19q codeletion.

IDH-mutant tumors with 1p/19q codeletion (IDHmut-codel) feature additional, recurrent molecular abnormalities involving genes implicated in cancer. For instance, high rates of mutations in the phosphoinositide 3-kinase (PI3K) subunits PIK3CA and PIK3R1, and in the notch 1 receptor gene (NOTCH1) have been reported.[8] Most interestingly, virtually all IDHmut-codel gliomas harbor point mutations in the promoter region of the telomerase reverse transcriptase (TERT) gene, which encodes the catalytic component of the telomerase enzyme.[25] Dividing neoplastic cells requires active maintenance of the telomere regions at the tips of their chromosomes to prevent senescence and apoptosis. This task is normally performed by telomerase, an enzymatic activity that adds repeat DNA sequences to telomeres shortened after multiple cycles of mitosis.[26] Promoter mutations in TERT, originally discovered in melanoma, bolster the gene's expression by generating novel binding sites for the activating transcription factor GA-binding protein (GABP), and in this way, increase cellular telomerase activity, presumably enabling repeated cell division.[27] The striking prevalence of *TERT* mutations in 1p/19q-codeleted gliomas argues for their absolute requirement in the pathogenesis of these tumors, regardless of the precise molecular mechanism(s) at work.

IDH-mutant gliomas that do not harbor 1p/19q codeletion and *TERT* mutation (IDHmut-noncodel) tend to exhibit either astrocytic or oligoastrocytic morphology and are characterized by high rates of inactivating mutations in both *TP53* and *ATRX* (α-thalassemia/mental retardation X-linked).[28,29] *TP53* encodes the p53 tumor suppressor, and is the most commonly mutated gene in cancer.[30] As the so-called "guardian of the genome," p53 plays a central role in the channeling of signals regarding DNA damage, cellular stress, and hypoxia toward physiologic responses that preserve genomic integrity, such as cell cycle arrest and/or apoptosis.[31] Its frequent loss in the context of astrocytic gliomas has been known for some time.[32] By contrast, *ATRX* encodes a chromatin regulatory protein only recently implicated in neoplasia.[33] Although the full spectrum of its physiologic sequelae remains unclear, ATRX deficiency appears to impact both gene expression and genomic stability.[33] That ATRX loss promotes p53-dependent cell death in neuroepithelial precursor cells may at least partially explain its invariable co-occurrence with *TP53* mutation.[34,35] Most interestingly, however, ATRX deficiency is associated with the alternative lengthening of telomeres (ALT) phenotype, a telomerase-independent telomere maintenance mechanism mediated by the cellular homologous recombination machinery.[36] Moreover, *ATRX* mutations occur with complete mutual exclusivity to the *TERT* mutations described previously.[4,8,25] In this way, virtually all IDH-mutant gliomas, whether IDHmut-codel or IDHmut-noncodel, activate pathologic telomere maintenance by way of distinct molecular events. Finally, IDHmut-noncodel gliomas exhibit increased incidence of copy number gains in the *MYC* and cyclin D2 (*CCND2*) loci, implicating direct cell cycle dysregulation in their pathogenesis.[8]

ISOCITRATE DEHYDROGENASE–WILD-TYPE GLIOMAS: GLIOBLASTOMA AND BEYOND

IDH-wild-type glioma consists primarily of GBM, with its defining histopathological characteristics. However, multiple studies have now shown that diffusely infiltrating gliomas of WHO grades II and III that do not harbor IDH mutations tend to behave aggressively, on the order of WHO grade IV neoplasms.[4–6,8,28,37,38] Although such studies may be somewhat confounded by suboptimal surgical sampling and, as a consequence, nonrepresentative histopathology, their findings support the notion that IDH–wild-type status itself designates the rapidly malignant pathogenic sequence associated with primary GBM. Moreover, comprehensive molecular analysis of IDH–wild-type tumors classified as LGGs has demonstrated genomic profiles typical of GBM (see later in this article) in many, if not most, cases.[4,5,8] Accordingly, some have argued that all IDH–wild-type gliomas should be considered the biological equivalent of primary GBM (the term "pre-GBM" has even been suggested[5]) or, at minimum, managed with the degree of aggressiveness typically associated with a WHO grade IV diagnosis. However, the significant extent to which diffuse gliomas, GBM or otherwise, exhibit histopathological similarities with comparatively indolent and biologically distinct entities, such as pilocytic astrocytoma, ganglioglioma, and pleomorphic xanthoastrocytoma (PXA), all of which are IDH–wild-type, renders problematic the implementation of such diagnostic practice at the present time. Indeed, a recent large-scale profiling study in LGG identified samples within its cohort likely representing "contaminants" from these morphologically overlapping diagnostic entities.[8]

The molecular distinctions between IDH-mutant LGGs and primary GBMs are vast. Whereas LGGs are essentially defined by a limited set of highly recurrent molecular alterations, including IDH mutation, *TERT* mutation, *ATRX* mutation, and 1p/19q codeletion, GBMs are notably heterogeneous in their underlying biology, frequently featuring numerous genomic aberrations that differ significantly from case to case.[39–41] Despite this variability, however, characteristic patterns of dysregulated molecular networks emerge when examining large GBM cohorts (**Table 1**).[39] For instance, GBMs often harbor genomic abnormalities involving receptor tyrosine kinases (RTKs), such as epidermal growth factor receptor (EGFR) and platelet-derived growth factor receptor α (PDGFRA), and/or core components of the downstream PI3K and MAPK signaling pathways. Enhanced activity in these growth factor networks has been widely implicated across several cancer types. Similarly, GBMs exhibit high rates of genomic alterations involving constituents of the retinoblastoma (RB) and p53 tumor suppressor pathways, both of which play crucial roles in cell cycle control and are disrupted in numerous

Table 1
Common molecular alterations in glioblastoma grouped by signaling network

Amplification/Gain	Deletion/Loss	Mutation
Receptor tyrosine kinases		
EGFR	—	EGFR
PDGFRA		PDGFRA
MET		MET
		FGFR3
PI3K and MAPK signaling		
—	PTEN	PIK3CA
	NF1	PIK3R1
		PTEN
		NF1
		BRAF
p53 signaling		
MDM2	TP53	TP53
MDM4		
RB signaling		
CDK4	CDKN2A/B	RB1
CDK6	CDKN2C	
	RB1	

Abbreviations: MAPK, mitogen-activated protein kinase; PI3K, phosphoinositide 3-kinase; RB, retinoblastoma.

malignancies. Recurrent dysregulation of these specific molecular networks further distinguishes GBM from LGG, while also providing compelling pathogenic links with a variety of more common cancer variants, including carcinomas of the lung, breast, and colon. However, much like IDHmut-codel gliomas, GBMs characteristically harbor *TERT* promoter mutations,[25] which are unusual in most malignant epithelial tumors.

Cytotoxic chemotherapy, specifically with the alkylating agent temozolomide (TMZ), has become a mainstay for GBM management, despite limited efficacy. It has been appreciated for some time that transcriptional repression of the O(6)-methylguanine-DNA methyltransferase gene (*MGMT*) by promoter methylation designates improved response to TMZ.[42] MGMT is a DNA repair enzyme that presumably mitigates the therapeutic impact of alkylating agents by resolving deleterious nucleotide adducts. Its widespread adoption as a predictive biomarker was initially hampered by the near universal use of TMZ as first-line therapy for GBM, regardless of *MGMT* methylation status. However, more recent data indicating that patients with MGMT-methylated tumors are more likely to exhibit the pseudoprogression,[43] a constellation of neurologic deterioration and radiographic findings mimicking tumor progression but actually representing vigorous

cytotoxic therapeutic response,[44] has provided additional rationale for routine employment in GBM management. In fact, *MGMT* promoter methylation remains to this day the only validated predictive biomarker for GBM.

Gene expression analysis has been repeatedly leveraged as a means to conceptualize the molecular heterogeneity inherent to primary GBM. As a general premise, such investigations have sought to define biological subclasses within the totality of GBM by way of robust transcriptional signatures.[45,46] In some cases, distinctions have been established that reflect histiogenic pathways, with specific gene expression profiles correlating with those of oligodendrocytes, astrocytes, neurons, or their respective progenitor cell populations. Associations have also been made with defined sets of genomic alterations, arguing that the various members of a given transcriptional subclass are fundamentally driven by shared pathogenic mechanisms. In perhaps the most well-known study focused on the gene-expression profiles within primary GBM, data from the Cancer Genome Atlas (TCGA) were used to designate 4 disease subclasses named proneural, classical, mesenchymal, and neural.[41] Moreover, proneural, classical, and mesenchymal tumors were found to be highly enriched for genomic abnormalities in *PDGFRA*, *EGFR*, and *NF1* (neurofibromin), respectively. More recent investigations probing endogenous GBM subclass, using microRNA, proteomic, or DNA methylation profiling, along with one using a systematic integrated analysis across multiple data platforms, have reaffirmed the robustness of proneural, classical, and mesenchymal categorizations, while neural tumors tend to be more widely distributed among other disease subgroups.[47–50] These findings are supportive of 3 basic subclasses within primary GBM, roughly approximating the proneural, classical, and mesenchymal designations. It is intriguing to speculate that the different GBM subclasses will selectively respond to targeted therapeutics directed against their cardinal molecular aberrations (eg, EGFR for classical tumors). However, this remains a largely untested hypothesis largely because transcriptional signatures are not routinely assessed in the clinical environment for GBM, or for that matter, any other brain tumor.

The molecular complexity of GBM is not only evident across the entire diagnostic category, but also within individual tumors themselves. Such intratumoral heterogeneity takes many forms. For instance, multiple groups have shown that distinct cell populations within the same GBM can harbor different RTK amplification events, in *EGFR*, *PDGFRA*, or *MET*.[51,52]

Intriguingly, these discrepancies appear to arise naturally, as amplified genetic loci on supernumerary double minute chromosomes repeatedly undergo independent segregation over several mitotic cycles.[52] More recently, single-cell gene-expression profiling has revealed that transcriptional subclass (eg, proneural, classical, and mesenchymal) may vary considerably from tumor cell to tumor cell within the same GBM.[53] Such heterogeneity at both genomic and transcriptional levels has obvious therapeutic implications, particularly with regard to the anticipated efficacy of agents targeting specific molecular events or dysregulated signaling networks.

MOVING TOWARD MORE BIOLOGICALLY RELEVANT GLIOMA SUBCLASSIFICATION

Comprehensive molecular characterization of the various glioma subtypes has prompted a reexamination of current classification schemes in an attempt to better define more biologically meaningful disease entities. In particular, segregating IDH-mutant from IDH–wild-type disease, regardless of morphologic characteristics, is widely favored. As discussed previously, IDH–wild-type gliomas themselves exhibit considerable genomic and transcriptional diversity; however, the lack of prognostic distinction between any molecularly defined subsets of primary GBM argues against any further stratification of this group, at least

from the standpoint of immediate clinical relevance. By contrast, within IDH-mutant tumors, those with 1p/19q codeletion (IDHmut-codel) are associated with significantly longer overall survival than those without (IDHmut-noncodel), in a manner that transcends WHO grade.[8,28,37] Therefore, a 3-pronged classification system has emerged that includes IDHmut-codel gliomas, which perform the best, IDH–wild-type gliomas, which behave the most aggressively, and IDHmut-noncodel gliomas, which exhibit an intermediate phenotype (**Fig. 2**). Importantly, this biomarker-driven scheme outperforms standard histopathology in terms of prognostic stratification.[8,37]

As indicated previously, the incidence of 1p/19q codeletion in IDH-mutant gliomas is strongly correlated (or anticorrelated) with a several other highly recurrent molecular alterations, most notably mutations in TP53, ATRX, and TERT, inviting further refinement of this basic classification system. In particular, incorporating TERT mutation along with IDH mutation and 1p/19q codeletion yields 2 additional tumor subclasses, 1 harboring none of the 3 stratifying alterations and 1 defined by the presence of both IDH and TERT mutations, but not 1p/19q codeletion (see **Fig. 2**).[4] Additional studies should further clarify whether these small subclasses truly reflect distinct biological entities or are simply unusual variants within the larger IDH–wild-type and IDHmut-noncodel categories, respectively.

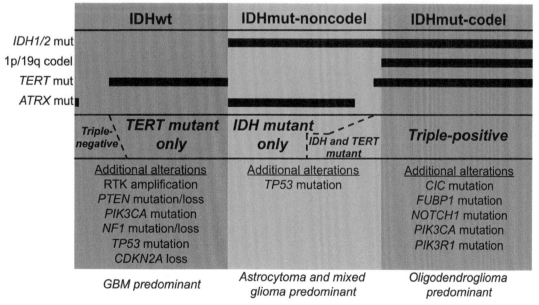

Fig. 2. Schematic showing molecular classification schemes for diffuse glioma. Subclasses designated by IDH mutation and 1p/19q codeletion are shown along with the estimated prevalence of relevant molecular alterations (*black bars*). Subclassification incorporating TERT mutation is also shown. IDHwt, IDH–wild-type.

Among the more compelling features of the recently established glioma subclasses is the relative ease with which they can be routinely designated in the clinical environment. In fact, a 2-component immunohistochemical panel consisting of IDH1 R132H and ATRX provides sufficient information to enable subclassification in most cases (Fig. 3). IDH1 R132H is the most common glioma-associated IDH mutation, present in more than 90% of all IDH-mutant tumors. As such, positivity effectively rules in IDH-mutant status,[54,55] although there is significant potential for false negatives. ATRX immunostaining normally exhibits a strong nuclear labeling pattern, which is absent in the setting of inactivating mutation.[56] Accordingly, loss of ATRX staining in tumor cells, with retention in appropriate internal controls (eg, endothelial cells and neurons), is generally consistent with IDHmut-noncodel classification, and alerts the pathologist to the likelihood of an underlying IDH mutation even when IDH1 R132H staining is negative. Including formal testing for 1p/19q codeletion with IDH1 R132H and ATRX immunohistochemistry

allows subclass assignment in almost every case.[57] In most institutions, such assessments are made by custom singleton assays, most commonly based on fluorescence in situ hybridization or polymerase chain reaction. Although these techniques perform well and typically require minimal starting material, their focused monitoring of selected loci within 1p and 19q as surrogates for the full chromosomal arms may lead to false positives, particularly in tumors harboring generalized genomic instability. For this reason, global copy number assessment by array-comparative genomic hybridization (aCGH) or molecular inversion probe array is preferable, allowing for visualization of complete 1p and 19q arm loss, albeit with somewhat larger tissue requirements on the front end. Global copy number analysis also enables the detection of other characteristic genomic abnormalities, notably concurrent gain of chromosome 7 and loss of chromosome 10, which is highly correlated with primary GBM.[57]

The availability of next-generation sequencing platforms in the clinical laboratories of most large

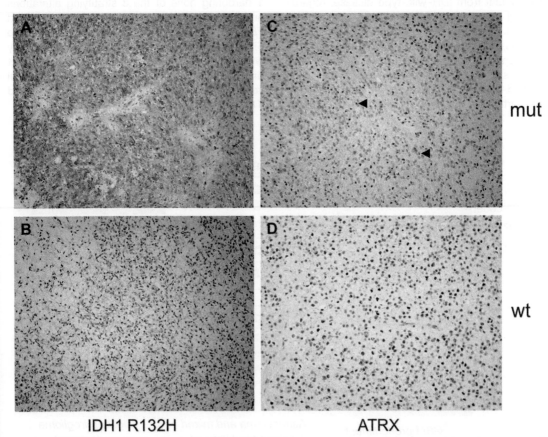

Fig. 3. Representative micrographs showing IDH1 R132H (*A, B*) and ATRX (*C, D*) immunostaining (×200 magnification). Wild-type (wt) and mutant (mut) cases are shown for each. For ATRX, retained nuclear expression in non-neoplastic cellular constituents (eg, endothelial cells, neurons, and microglia; *arrowheads* indicate examples) is required to accurately interpret a negative immunostaining pattern.

medical centers now allows multiplexed tumor genotyping for routine management of patients with cancer. In this setting,[58] including *IDH1*, *IDH2*, *TERT*, *TP53*, and *ATRX* in a larger sequencing panel of up to several hundred target genes fully enables glioma subclassification. Input tissue requirements can be substantial and turnaround time is significantly longer than for immunohistochemistry or singleton molecular assays. Moreover, parallel sequencing of healthy patient DNA to subtract out germline polymorphisms, typically from case-matched blood, introduces additional regulatory issues regarding patient privacy and protected health information. Despite these concerns, however, the ability to simultaneously assess numerous genomic biomarkers designating disease subclass, prognosis, and even likely response to targeted agents represents a paradigm-shifting advance, particularly in light of the personalized care model toward which management of patients with cancer is moving.

In its next update on brain tumor classification, WHO will include molecular criteria in the establishment of specific diagnoses, including those of the diffuse glioma subtypes.[59] For instance, IDH mutation and 1p/19q codeletion will prominently feature in the designation of LGG subtype, with specific terminology being "diffuse astrocytoma, IDH-mutant" and "oligodendroglioma, IDH-mutant and 1p/19q codeleted." Moreover, IDH–wild-type status in LGGs, particularly those of WHO grade III, will be formally correlated with GBM-like biological behavior. Nevertheless, the option for rendering diagnoses based on morphologic criteria alone (eg, astrocytoma, not otherwise specified) will be retained to enable preliminary tumor classification in medical communities in which access to molecular testing and immunohistochemistry is limited.

Of particular note, mixed glioma (oligoastrocytoma) will be largely eliminated as a diagnostic entity, given that tumors described as such do not exhibit distinct molecular features and instead neatly segregate into either the oligodendroglial (IDHmut-codel) or astrocytic (IDHmut-noncodel) glioma subclasses. That said, 2 recent reports have described cases of IDH-mutant gliomas with morphologically distinct regions featuring either 1p/19q codeletion or combined *ATRX* deficiency and *TP53* mutation.[60,61] These "true oligoastrocytomas" are entirely consistent with the dualist framework of IDH-mutant gliomagenesis presented previously and actually represent 2 independent tumors, with the signature molecular features of oligodendroglioma and astrocytoma, respectively, arising from a single IDH-mutant parental clone. The true prevalence of such composite tumors is difficult to estimate, given that gliomas are rarely subjected to complete histopathological sampling and the panel of tests required to delineate stratifying biomarkers in these cases is not widely implemented as of yet. Regardless, the mere existence of these tumors raises important questions regarding how best to treat them and how their evolution may or may not reflect the normal pathogenic sequence of IDH-mutant gliomagenesis.

MOLECULAR EVOLUTION OF GLIOMA IN THE CONTEXT OF THERAPY

Work over the past decade has emphatically confirmed that first-line therapy, consisting of ionizing radiation and/or TMZ, directly impacts the molecular landscape of diffuse glioma. Malignant glioma variants, GBM in particular, appear to shift their gene-expression patterns toward mesenchymal signatures following treatment, indicating that transcriptional subclass, to some extent, is malleable in the face of cytotoxic regimens.[62] The same cannot be said of genomically specified categories (ie, IDH–wild-type, IDHmut-codel, and IDHmut-noncodel), which appear to remain constant from primary tumor to recurrence.[17] This is not to imply, however, that important genomic alterations are not acquired during treatment. For instance, loss of DNA mismatch repair genes, such as mutS homologues 2 and 6 (*MSH2* and *MSH6*), leads to the acquisition of numerous mutations during exposure to alkylating agents. This so-called hypermutator phenotype was originally described in recurrent GBMs.[63,64] However, more recent work has uncovered analogous phenomena in lower-grade, IDH-mutant gliomas treated with TMZ.[17] Interestingly, many of the altered genes in hypermutated tumors encode core constituents of established oncogenic signaling networks, such as the PI3K and MAPK pathways. These findings suggest not only that cancer-relevant signaling is engaged by genomic evolution during treatment, but that some altered molecules are potentially targetable by agents already in clinical trials for glioma. Recent studies have also shown that, particularly for GBM, recurrence in the context of treatment can result in multiple distinct subclones featuring nonoverlapping mutational profiles.[65] Dysfunction in the p53 signaling network is predictive of such behavior. Moreover, recurrent GBMs not infrequently share few genetic alterations with case-matched primaries, indicating that they arise from cellular subclones branching off relatively early in the process of tumorigenesis.

Taken together, these data serve as a sobering reminder that current first-line treatment for glioma is far from perfect, and its application may be particularly problematic for lower-grade tumors whose natural history, even in the absence of nonsurgical therapy, can be relatively indolent. Moreover, they underscore the importance of iterative molecular profiling performed at each recurrence to assess new molecular alterations and their potential to guide clinical management.

SUMMARY AND IMPLICATIONS/IMPACT

In summary, rapid advances in the molecular characterization of glioma have fundamentally altered conceptions of glioma classification, stratifying more molecularly uniform disease entities on the basis of robust biomarker sets readily assessable in the clinical environment. These advances are already impacting the practice of neuro-oncology, providing more accurate survival prognostication. However, now that more biologically relevant disease subclasses have been designated, there is considerable hope within the field that more effective therapies targeting their respective oncogenic mechanisms will be developed. In this respect, conceiving and designing optimized clinical trials with sufficient baseline molecular profiling will be absolutely essential moving forward, not only to ensure biologically uniform study cohorts, but also to track the emergence of minor subgroups with differing responses to the therapy in question. This is truly an exciting time to be engaged in the treatment of glioma. The challenge now becomes how to optimally and expeditiously leverage the wealth of knowledge emerging from the application of modern molecular pathology.

REFERENCES

1. Louis DN, Ohgaki H, Wiestler OD, et al. WHO classification of tumours of the central nervous system. In: Bosman FT, Jaffe ES, Lakhani SR, et al, editors. Lyon (France): International Agency for Research on Cancer; 2007. p. 25–49.
2. Parsons DW, Jones S, Zhang X, et al. An integrated genomic analysis of human glioblastoma multiforme. Science 2008;321:1807–12.
3. Yan H, Parsons DW, Jin G, et al. IDH1 and IDH2 mutations in gliomas. N Engl J Med 2009;360:765–73.
4. Eckel-Passow JE, Lachance DH, Molinaro AM, et al. Glioma groups based on 1p/19q, IDH, and TERT promoter mutations in tumors. N Engl J Med 2015; 372:2499–508.
5. Gorovets D, Kannan K, Shen R, et al. IDH mutation and neuroglial developmental features define clinically distinct subclasses of lower grade diffuse astrocytic glioma. Clin Cancer Res 2012;18: 2490–501.
6. Hartmann C, Hentschel B, Wick W, et al. Patients with IDH1 wild type anaplastic astrocytomas exhibit worse prognosis than IDH1-mutated glioblastomas, and IDH1 mutation status accounts for the unfavorable prognostic effect of higher age: implications for classification of gliomas. Acta Neuropathol 2010;120:707–18.
7. Sanson M, Marie Y, Paris S, et al. Isocitrate dehydrogenase 1 codon 132 mutation is an important prognostic biomarker in gliomas. J Clin Oncol 2009;27: 4150–4.
8. Brat DJ, Verhaak RG, Aldape KD, et al. Comprehensive, integrative genomic analysis of diffuse lower-grade gliomas. N Engl J Med 2015;372:2481–98.
9. Ceccarelli M, Barthel FP, Malta TM, et al. Molecular profiling reveals biologically discrete subsets and pathways of progression in diffuse glioma. Cell 2016;164:550–63.
10. Dang L, White DW, Gross S, et al. Cancer-associated IDH1 mutations produce 2-hydroxyglutarate. Nature 2009;462:739–44.
11. Lu C, Ward PS, Kapoor GS, et al. IDH mutation impairs histone demethylation and results in a block to cell differentiation. Nature 2012;483:474–8.
12. Noushmehr H, Weisenberger DJ, Diefes K, et al. Identification of a CpG island methylator phenotype that defines a distinct subgroup of glioma. Cancer Cell 2010;17:510–22.
13. Turcan S, Rohle D, Goenka A, et al. IDH1 mutation is sufficient to establish the glioma hypermethylator phenotype. Nature 2012;483:479–83.
14. Flavahan WA, Drier Y, Liau BB, et al. Insulator dysfunction and oncogene activation in IDH mutant gliomas. Nature 2016;529:110–4.
15. Koivunen P, Lee S, Duncan CG, et al. Transformation by the (R)-enantiomer of 2-hydroxyglutarate linked to EGLN activation. Nature 2012;483:484–8.
16. Sasaki M, Knobbe CB, Itsumi M, et al. D-2-hydroxyglutarate produced by mutant IDH1 perturbs collagen maturation and basement membrane function. Genes Dev 2012;26:2038–49.
17. Johnson BE, Mazor T, Hong C, et al. Mutational analysis reveals the origin and therapy-driven evolution of recurrent glioma. Science 2014;343:189–93.
18. Kraus JA, Koopmann J, Kaskel P, et al. Shared allelic losses on chromosomes 1p and 19q suggest a common origin of oligodendroglioma and oligoastrocytoma. J Neuropathol Exp Neurol 1995;54:91–5.
19. Reifenberger G, Reifenberger J, Liu L, et al. Molecular genetic analysis of oligodendroglial tumors shows preferential allelic deletions on 19q and 1p. Am J Pathol 1994;145:1175–90.
20. Cairncross JG, Ueki K, Zlatescu MC, et al. Specific genetic predictors of chemotherapeutic response

and survival in patients with anaplastic oligodendrogliomas. J Natl Cancer Inst 1998;90:1473–9.

21. Cairncross JG, Wang M, Jenkins RB, et al. Benefit from procarbazine, lomustine, and vincristine in oligodendroglial tumors is associated with mutation of IDH. J Clin Oncol 2014;32:783–90.

22. Yip S, Butterfield YS, Morozova O, et al. Concurrent CIC mutations, IDH mutations, and 1p/19q loss distinguish oligodendrogliomas from other cancers. J Pathol 2012;226:7–16.

23. Bettegowda C, Agrawal N, Jiao Y, et al. Mutations in CIC and FUBP1 contribute to human oligodendroglioma. Science 2011;333:1453–5.

24. Suzuki H, Aoki K, Chiba K, et al. Mutational landscape and clonal architecture in grade II and III gliomas. Nat Genet 2015;47:458–68.

25. Killela PJ, Reitman ZJ, Jiao Y, et al. TERT promoter mutations occur frequently in gliomas and a subset of tumors derived from cells with low rates of self-renewal. Proc Natl Acad Sci U S A 2013;110:6021–6.

26. Artandi SE, DePinho RA. Telomeres and telomerase in cancer. Carcinogenesis 2010;31:9–18.

27. Bell RJ, Rube HT, Kreig A, et al. The transcription factor GABP selectively binds and activates the mutant TERT promoter in cancer. Science 2015; 348:1036–9.

28. Jiao Y, Killela PJ, Reitman ZJ, et al. Frequent ATRX, CIC, FUBP1 and IDH1 mutations refine the classification of malignant gliomas. Oncotarget 2012;3: 709–22.

29. Kannan K, Inagaki A, Silber J, et al. Whole exome sequencing identified ATRX mutation as a key molecular determinant in lower-grade glioma. Oncotarget 2012;3:1194–203.

30. Olivier M, Hollstein M, Hainaut P. TP53 mutations in human cancers: origins, consequences, and clinical use. Cold Spring Harb Perspect Biol 2010;2: a001008.

31. Levine AJ, Oren M. The first 30 years of p53: growing ever more complex. Nat Rev Cancer 2009;9:749–58.

32. Louis DN. The p53 gene and protein in human brain tumors. J Neuropathol Exp Neurol 1994;53:11–21.

33. Clynes D, Higgs DR, Gibbons RJ. The chromatin remodeller ATRX: a repeat offender in human disease. Trends Biochem Sci 2013;38:461–6.

34. Berube NG, Mangelsdorf M, Jagla M, et al. The chromatin-remodeling protein ATRX is critical for neuronal survival during corticogenesis. J Clin Invest 2005;115:258–67.

35. Conte D, Huh M, Goodall E, et al. Loss of Atrx sensitizes cells to DNA damaging agents through p53-mediated death pathways. PLoS One 2012;7: e52167.

36. Heaphy CM, de Wilde RF, Jiao Y, et al. Altered telomeres in tumors with ATRX and DAXX mutations. Science 2011;333:425.

37. Wiestler B, Capper D, Holland-Letz T, et al. ATRX loss refines the classification of anaplastic gliomas and identifies a subgroup of IDH mutant astrocytic tumors with better prognosis. Acta Neuropathol 2013;126:443–51.

38. Olar A, Wani KM, Alfaro-Munoz KD, et al. IDH mutation status and role of WHO grade and mitotic index in overall survival in grade II-III diffuse gliomas. Acta Neuropathol 2015;129:585–96.

39. Brennan CW, Verhaak RG, McKenna A, et al. The somatic genomic landscape of glioblastoma. Cell 2013;155:462–77.

40. TCGA. Comprehensive genomic characterization defines human glioblastoma genes and core pathways. Nature 2008;455:1061–8.

41. Verhaak RGW, Hoadley KA, Purdom E, et al. An integrated genomic analysis identifies clinically relevant subtypes of glioblastoma characterized by abnormalities in PDGFRA, IDH1, EGFR, and NF1. Cancer Cell 2009;17:98–110.

42. Hegi ME, Diserens AC, Gorlia T, et al. MGMT gene silencing and benefit from temozolomide in glioblastoma. N Engl J Med 2005;352:997–1003.

43. Brandes AA, Franceschi E, Tosoni A, et al. MGMT promoter methylation status can predict the incidence and outcome of pseudoprogression after concomitant radiochemotherapy in newly diagnosed glioblastoma patients. J Clin Oncol 2008;26: 2192–7.

44. de Wit MC, de Bruin HG, Eijkenboom W, et al. Immediate post-radiotherapy changes in malignant glioma can mimic tumor progression. Neurology 2004;63:535–7.

45. Huse JT, Aldape KD. The molecular landscape of diffuse glioma and prospects for biomarker development. Expert Opin Med Diagn 2013;7:573–87.

46. Huse JT, Phillips HS, Brennan CW. Molecular subclassification of diffuse gliomas: seeing order in the chaos. Glia 2011;59:1190–9.

47. Brennan C, Momota H, Hambardzumyan D, et al. Glioblastoma subclasses can be defined by activity among signal transduction pathways and associated genomic alterations. PLoS One 2009;4: e7752.

48. Kim TM, Huang W, Park R, et al. A developmental taxonomy of glioblastoma defined and maintained by MicroRNAs. Cancer Res 2011;71:3387–99.

49. Shen R, Mo Q, Schultz N, et al. Integrative subtype discovery in glioblastoma using iCluster. PLoS One 2012;7(4):e35236.

50. Sturm D, Witt H, Hovestadt V, et al. Hotspot mutations in H3F3A and IDH1 define distinct epigenetic and biological subgroups of glioblastoma. Cancer Cell 2012;22:425–37.

51. Snuderl M, Fazlollahi L, Le LP, et al. Mosaic amplification of multiple receptor tyrosine kinase genes in glioblastoma. Cancer Cell 2011;20:810–7.

52. Szerlip NJ, Pedraza A, Chakravarty D, et al. Intratumoral heterogeneity of receptor tyrosine kinases EGFR and PDGFRA amplification in glioblastoma defines subpopulations with distinct growth factor response. Proc Natl Acad Sci U S A 2012;109:3041–6.

53. Patel AP, Tirosh I, Trombetta JJ, et al. Single-cell RNA-seq highlights intratumoral heterogeneity in primary glioblastoma. Science 2014;344:1396–401.

54. Capper D, Weissert S, Balss J, et al. Characterization of R132H mutation-specific IDH1 antibody binding in brain tumors. Brain Pathol 2010;20:245–54.

55. Capper D, Zentgraf H, Balss J, et al. Monoclonal antibody specific for IDH1 R132H mutation. Acta Neuropathol 2009;118:599–601.

56. Liu XY, Gerges N, Korshunov A, et al. Frequent ATRX mutations and loss of expression in adult diffuse astrocytic tumors carrying IDH1/IDH2 and TP53 mutations. Acta Neuropathol 2012;124(5):615–25.

57. Reuss DE, Sahm F, Schrimpf D, et al. ATRX and IDH1-R132H immunohistochemistry with subsequent copy number analysis and IDH sequencing as a basis for an "integrated" diagnostic approach for adult astrocytoma, oligodendroglioma and glioblastoma. Acta Neuropathol 2015;129:133–46.

58. Wagle N, Berger MF, Davis MJ, et al. High-throughput detection of actionable genomic alterations in clinical tumor samples by targeted, massively parallel sequencing. Cancer Discov 2012;2:82–93.

59. Louis DN, Perry A, Burger P, et al. International Society of Neuropathology-Haarlem consensus guidelines for nervous system tumor classification and grading. Brain Pathol 2014;24(6):671–2.

60. Huse JT, Diamond EL, Wang L, et al. Mixed glioma with molecular features of composite oligodendroglioma and astrocytoma: a true "oligoastrocytoma"? Acta Neuropathol 2015;129:151–3.

61. Wilcox P, Li CC, Lee M, et al. Oligoastrocytomas: throwing the baby out with the bathwater? Acta Neuropathol 2015;129:147–9.

62. Phillips HS, Kharbanda S, Chen R, et al. Molecular subclasses of high-grade glioma predict prognosis, delineate a pattern of disease progression, and resemble stages in neurogenesis. Cancer Cell 2006;9:157–73.

63. Cahill DP, Levine KK, Betensky RA, et al. Loss of the mismatch repair protein MSH6 in human glioblastomas is associated with tumor progression during temozolomide treatment. Clin Cancer Res 2007;13: 2038–45.

64. Yip S, Miao J, Cahill DP, et al. MSH6 mutations arise in glioblastomas during temozolomide therapy and mediate temozolomide resistance. Clin Cancer Res 2009;15:4622–9.

65. Kim H, Zheng S, Amini SS, et al. Whole-genome and multisector exome sequencing of primary and post-treatment glioblastoma reveals patterns of tumor evolution. Genome Res 2015;25:316–27.

The Emerging Molecular Landscape of Urothelial Carcinoma

James P. Solomon, MD, PhD[a], Donna E. Hansel, MD, PhD[b],*

KEYWORDS

- Ancillary testing • Urothelial carcinoma • Bladder cancer • Molecular pathology • Subclassification
- Targeted therapy • Immunotherapy

Key points

- Urothelial carcinoma is one of the most complex and heterogeneous neoplasms at the molecular level, with alterations seen in a wide variety of molecular pathways and cancer-related genes.

- Historically, a two-pathway model has been used in which low-grade papillary urothelial carcinoma is associated with alterations in cell growth and proliferation pathways, whereas high-grade and invasive urothelial carcinoma is associated with alterations in cell-cycle regulation, although there now seems to be overlap in a subset of pathways.

- Currently, research groups are further subdividing urothelial carcinoma based on molecular signatures, but these are still in the research stage and application to clinical and treatment decisions remains elusive.

ABSTRACT

Although there have been many recent discoveries in the molecular alterations associated with urothelial carcinoma, current understanding of this disease lags behind many other malignancies. Historically, a two-pathway model had been applied to distinguish low- and high-grade urothelial carcinoma, although significant overlap and increasing complexity of molecular alterations has been recently described. In many cases, mutations in *HRAS* and *FGFR3* that affect the MAPK and PI3K pathways seem to be associated with noninvasive low-grade papillary tumors, whereas mutations in *TP53* and *RB* that affect the G1-S transition of the cell cycle are associated with high-grade in situ and invasive carcinoma. However, recent large-scale analyses have identified overlap in these pathways relative to morphology, and in addition, many other variants in a wide variety of oncogenes and tumor-suppressor genes have been identified. New technologies including next-generation sequencing have enabled more detailed analysis of urothelial carcinoma, and several groups have proposed molecular classification systems based on these data, although consensus is elusive. This article reviews the current understanding of alterations affecting oncogenes and tumor-suppressor genes associated with urothelial carcinoma, and their application in the context of morphology and classification schema.

OVERVIEW

With more than 73,000 projected new cases and 16,000 deaths in 2015, bladder cancer is one of the most common cancers in the United States. It is the fourth most common new cancer diagnosis in men and the twelfth most common new cancer diagnosis in women.[1] Presenting signs often include

Conflicts of Interest: None.
Financial Conflicts of Interest: None.
[a] Department of Pathology, University of California, San Diego, 200 West Arbor Drive, La Jolla, CA 92103, USA;
[b] Division of Anatomic Pathology, Department of Pathology, University of California, San Diego, 9500 Gilman Drive, MC 0612, La Jolla, CA 92093, USA
* Corresponding author.
E-mail address: dhansel@ucsd.edu

Surgical Pathology 9 (2016) 391–404
http://dx.doi.org/10.1016/j.path.2016.04.004
1875-9181/16/$ – see front matter © 2016 Elsevier Inc. All rights reserved.

surgpath.theclinics.com

either macroscopic or microscopic hematuria in otherwise asymptomatic patients, whereas a subset of patients describe urinary urgency or dysuria. In the United States, the most common risk factor is smoking, although other risk factors include exposure to arsenic or nitrosylating agents.[2] There is no current screening test for bladder cancer. However, patients who present with hematuria and are suspected to harbor bladder cancer undergo an evaluation that includes cystoscopy, computed tomography imaging and urogram, and urine cytology analysis.[2]

Definitive diagnosis of urothelial carcinoma, the most common form of bladder cancer, requires pathologic evaluation of biopsy or transurethral resection material obtained at the time of cystoscopy. Once diagnosed, clinical management is chiefly determined by the histologic and morphologic characteristics of the tumor, including the tumor stage and grade.[2,3] Briefly, tumor stage is determined by the presence and depth of invasion, with invasion into the muscularis propria (detrusor muscle) representing a critical determinant of surgical versus nonsurgical management (Fig. 1).

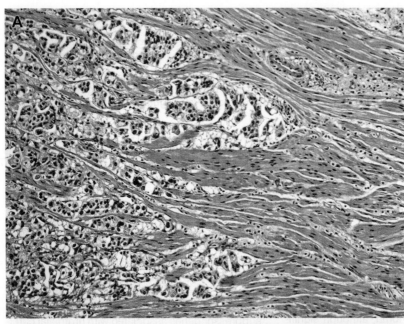
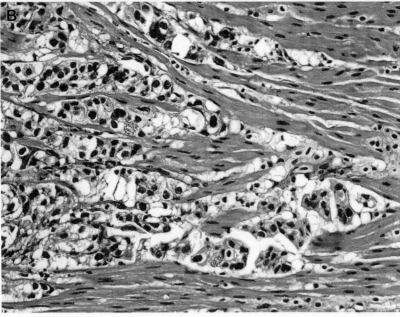

Fig. 1. Muscle-invasive urothelial carcinoma invading through and dissecting muscle bundles of the muscularis propria (detrusor muscle) (*A,* hematoxylin-eosin, original magnification ×100; *B,* hematoxylin-eosin, original magnification ×200).

Tumors are graded using architectural features and extent of nuclear atypia using one of two analogous grading systems. Flat lesions are graded, in increasing grade, as urothelial proliferation of uncertain malignant potential, urothelial dysplasia, and urothelial carcinoma in situ, through assessment of polarity, regularity of nuclear outlines, nuclear to cytoplasmic ratio, presence of nucleoli, and height of mitotic figures within the urothelium. Papillary lesions, characterized by papillary architecture and branching fibrovascular cores, are graded as urothelial papilloma, papillary urothelial neoplasm of low malignant potential, low-grade papillary carcinoma, and high-grade papillary carcinoma.[3] Distinction between high-grade lesions (urothelial carcinoma in situ and high-grade papillary urothelial carcinoma) and lower grade categories is critical in determining clinical management, and examples of the histologic differences are demonstrated in **Figs. 2** and **3**. Whereas low-grade lesions typically undergo removal or ablation and follow-up, high-grade or in situ lesions are managed through bacille Calmette-Guérin instillation or intravesical chemotherapy and more frequent follow-up.

Fig. 2. Low-grade papillary urothelial carcinoma is characterized by fibrovascular cores with overlying urothelium with cells that show loss of polarity, nuclear hyperchromasia, and minimal variability in nuclear size and shape (*A,* hematoxylin-eosin, original magnification ×100; *B,* hematoxylin-eosin, original magnification ×400).

A

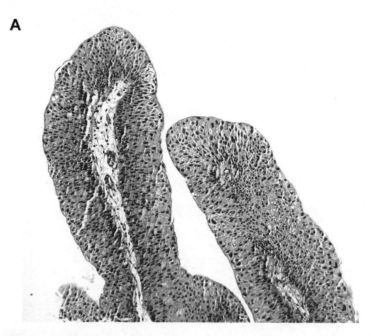

B

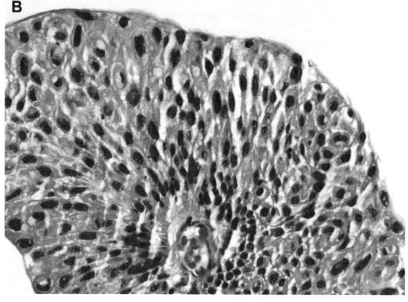

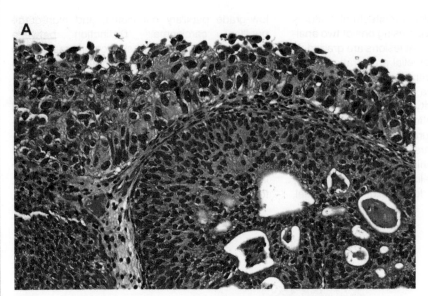

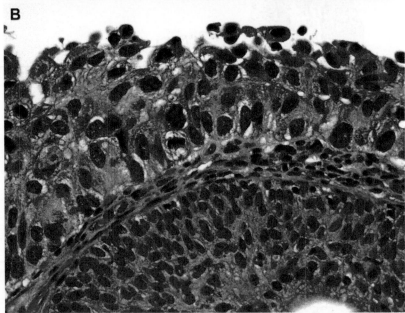

Fig. 3. High-grade lesions (high-grade papillary urothelial carcinoma or flat carcinoma in situ) are characterized by urothelium with disorganization, marked nuclear pleomorphism, nuclear hyperchromasia, and mitoses. Here, flat carcinoma in situ overlies uninvolved Brunn nests (*A,* hematoxylin-eosin, original magnification ×200; *B,* hematoxylin-eosin, original magnification ×400).

Unlike many other solid tumors, which routinely use ancillary molecular testing to provide additional prognostic information, the prognostic outcomes of urothelial carcinoma are currently based primarily on clinical and pathologic characteristics. The most important of these includes stage and grade, multifocality, tumor size, and presence of concurrent carcinoma in situ.[3] Moreover, the use of molecular diagnostics, such as targeted gene analysis or whole genome sequencing, to guide treatment decisions has not been routinely applied in bladder cancer. In recent years, neoadjuvant chemotherapy has emerged as a treatment modality in advanced bladder cancer before cystectomy, and some studies have demonstrated significant survival benefits in this setting. At the same time, neoadjuvant and adjuvant chemotherapy may be completely ineffective for a subset of populations,[4] and there is currently limited information to determine which patients would be cured by cystectomy alone and which would benefit from chemotherapy. Application of molecular diagnostics in this setting may be of value.[5]

Overall, the molecular understanding of urothelial carcinoma pathogenesis has lagged behind that of many other solid tumors, and application of ancillary molecular testing would be potentially useful in the setting of diagnosis, prognosis, and therapy. This article discusses the current understanding of genomic changes in urothelial carcinoma, focusing on grade and stage subdivisions.

MOLECULAR PATHOLOGY FEATURES OF UROTHELIAL CARCINOMA

TWO-PATHWAY MODEL

Traditionally, the biologic development of urothelial carcinoma was thought to progress along one of two pathways. One pathway leads to a low-grade, papillary, noninvasive carcinoma, whereas the other leads to high-grade, flat, and ultimately invasive carcinoma.[6] The histologic appearance, biology, and clinical outcomes of these tumors often seem to be distinct, although intermixed patterns of low- and high-grade tumors has now been recognized, and is termed grade heterogeneity. Whereas many studies have attempted to molecularly classify tumors along grade-specific pathways, the likelihood is that the molecular landscape of noninvasive disease is more complex than originally assumed. In muscle-invasive urothelial carcinoma, for example, the frequency of somatic mutation is 7.7 per megabase, a frequency that is only higher in lung carcinoma and

melanoma,[7] and mutations often include chromatin remodeling genes, further increasing the biologic heterogeneity and complexity of urothelial carcinoma.[7] Finally, there seems to be a high degree of epigenetic modifications that further complicate the understanding of urothelial tumorigenesis and classification. In this background context, we discuss low-grade noninvasive disease and high-grade in situ and invasive carcinoma in the context of existing and emerging molecular knowledge.

NONINVASIVE LOW-GRADE PAPILLARY CARCINOMA

Low-grade noninvasive papillary carcinomas are thought to initially derive from alterations in pathways that affect cell growth and proliferation, causing urothelial hyperplasia and vascular ingrowth. Low-grade papillary carcinoma is often multifocal and tends to recur following resection, but rarely progresses to invasive disease. Contributing molecular pathways include the mitogen-activated protein kinase (MAPK) pathway that controls cell growth, proliferation, differentiation, and apoptosis,[8] and the phosphoinositide 3-kinase (PI3K) pathway, which affects downstream protein kinase B (AKT) and mammalian target of rapamycin (mTOR) and is important in protein synthesis and cell-cycle regulation (**Fig. 4**).[9]

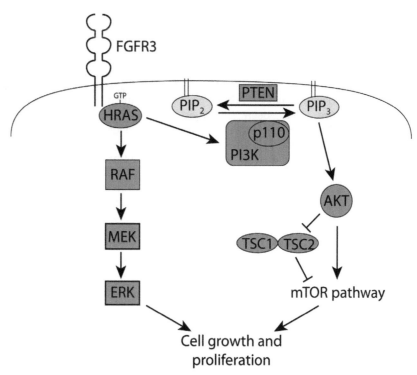

Fig. 4. Pathways associated with low-grade papillary urothelial carcinoma. FGFR3 activation and phosphorylation activates HRas, which activates the MAPK pathway (*left*) through a kinase cascade, and the PI3K/AKT/mTOR pathway (*right*), leading to cell growth and proliferation. FGFR3, fibroblast growth factor receptor 3; mTOR, mammalian target of rapamycin.

Alterations in *HRAS*, a proto-oncogene in the Ras GTPase family, were discovered early in the study of urothelial carcinoma, with point mutations in codons 12 and 13 being the most prevalent alterations.[10] These mutations constitutively activate the Ras protein by eliminating its GTPase activity and rendering it unable to hydrolyze GTP to GDP.[11,12] This constitutive activation is sufficient to activate the MAPK and PI3K pathways. In a transgenic mouse model where mutated *HRAS* is expressed in urothelium under the uroplakin II promotor, it was shown that these mutations in *HRAS* alone are sufficient to cause urothelial proliferation and papillary tumors.[13]

Upstream of the Ras protein is fibroblast growth factor receptor 3 (FGFR3), a tyrosine kinase receptor. Normally, the ligands that bind to this receptor are fibroblast growth factors, which are secreted glycoproteins in the extracellular matrix that effect dimerization, transphosphorylation, and activation of the receptor and the downstream MAPK pathway.[14] Multiple splice variants of FGFR3 have been identified that modulate activity, including a secreted form lacking the transmembrane domain that has been shown to sequester fibroblast growth factors.[15] In urothelial carcinoma, several pathologic mutations exist that enable constitutive activation of FGFR3 independent of ligand binding. The downstream pathways affected are identical to those activated by *HRAS* mutations; in fact, *HRAS* and *FGFR3* mutations seem to be mutually exclusive.[16] Either *HRAS* or *FGFR3* mutations are identified in up to 82% of nonmuscle invasive bladder cancer. However, there seems to be upregulation of FGFR3 signaling even in tumors without genetic somatic mutations.[17] Both decreased levels of the secreted splice variant that normally inhibits FGFR3 signaling and downregulation of FGFR3 inhibitory microRNAs have also been observed.[15,18] Chromosomal alterations affecting FGFR3 activity can also occur. Specifically, fusion of the N-terminus of FGFR3, which includes the kinase domain, to transforming acid coiled coil 3 or to BAI1-associated protein 2-like 1 constitutively activates the MAPK pathway. Cell lines containing these fusions were demonstrated to be sensitive to FGFR inhibitors, suggesting that patients with these alterations could be candidates for molecular targeted therapy.[19]

The PI3K pathway can also be activated by mutations in *PIK3CA*, which encodes p110α, the catalytic subunit of class I PI3K. The prevalence of alterations in *PIK3CA* in urothelial carcinoma ranges from 25% to 48%, and these alterations are often associated with noninvasive low-grade papillary carcinoma.[9,20–23] Mutations in *PIK3CA* are also associated with concurrent *FGFR3* mutations, and they likely have distinct downstream functional consequences. In addition, mutations in other PI3K pathway members frequently occur, including alterations in *PTEN* and *TSC1*.[22] Because alterations in these genes may interact to synergistically increase PI3K pathway activity and affect the clinical outcome, pathway components should be assessed in parallel. Given the broad variety of mutations seen in noninvasive low-grade papillary urothelial carcinoma, molecular targeted therapies against the PI3K pathway may be challenging in application. It should also be noted that molecular variants affecting the PI3K pathway may not be entirely specific for low-grade tumors and are often seen in muscle-invasive tumors.

HIGH-GRADE IN SITU AND INVASIVE DISEASE

High-grade noninvasive and invasive carcinomas have traditionally been considered to arise from molecular pathways distinct from low-grade urothelial carcinoma. These carcinomas generally have a worse prognosis and are associated with a high risk of progressive invasion and metastasis. They are also much more diverse at the molecular level.

High-grade noninvasive and invasive carcinomas frequently contain mutations in proteins that affect cell-cycle regulation, especially those involved in the G1/S checkpoint (Fig. 5). *TP53*, one of the most well-studied genes in human cancer, is located on chromosome 17p13.1 and encodes the p53 protein. Normally, this protein regulates the G1/S checkpoint by activating the transcription of the CDKN1A gene to generate the p21 protein, a cyclin-dependent kinase inhibitor.[24] Many urothelial carcinomas show loss of a single 17p allele coupled with mutations in the second *TP53*-containing allele, leading to loss of p53 activity and rendering tumor cells unable to regulate the G1/S cell-cycle transition.[25] Mutated p53 proteins have an aberrantly long half-life and accumulate in the nucleus, leading to a useful ancillary immunohistochemical tool to assess p53 status. Nuclear overexpression of p53 has been demonstrated to be an adverse prognostic factor in urothelial carcinoma.[26–28] Contribution of both gene and protein level alterations seem to also add combined value in predicting urothelial carcinoma prognosis, because a recent study demonstrated that *TP53* mutations and altered p53 expression in combination carries a worse prognosis than either genetic mutations or altered protein expression

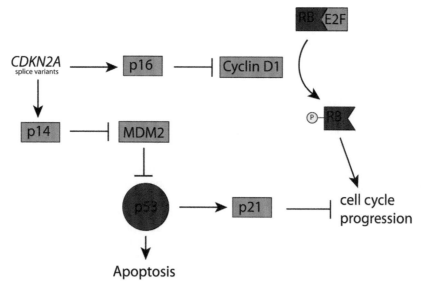

Fig. 5. Pathways associated with high-grade and invasive urothelial carcinoma are involved with cell-cycle regulation. Normally, p53 is inactivated by being bound and sequestered by Mdm2. The activation of p53 through DNA damage or cell stress activates expression of p21, halting cell-cycle progression. Activation of p53 can also lead to an apoptotic response. The Rb protein is normally bound to E2F, regulating the G1/S phase of the cell cycle. On phosphorylation by cyclin-dependent kinases, Rb releases E2F, leading to cell-cycle progression. Upstream of both the Rb and p53 pathways is the *CDKN2A* gene, which codes for the p16 and p14 tumor-suppressor proteins. Mutations in any of these genes result in cell-cycle dysregulation and are associated with high-grade and invasive urothelial carcinoma.

individually.[29] In addition, it is hypothesized that the cisplatin-based chemotherapeutic regimens are more effective in patients with p53 alterations, possibly because the S-phase and mitosis becomes uncoupled in cells lacking p21.[30] However, a recent prospective phase III study failed to demonstrate any significantly increased benefit from chemotherapy in patients with p53-positive tumors.[31]

The MDM2 protein serves to regulate p53 through an autoregulatory feedback loop. Increased p53 levels upregulate transcription of *MDM2*, leading to increased levels of MDM2 that mediate proteosomal degradation of p53. Amplification of *MDM2* has been identified in urothelial carcinoma, and is associated with increasing tumor grade and stage.[32]

Mutations in *CDKN2A* are also commonly seen in high-grade urothelial carcinoma. Two splice variants of *CDKN2A* exist that encode the p14 and p16 proteins. The p14 protein inhibits transcription of *MDM2*, and mutations that affect the activity or expression level of p14 leads to an increase in MDM2 protein expression.[24] Inactivating mutations in *CDKN2A* are an independent predictor of progression to muscle-invasive carcinoma and

may be associated with concurrent mutations in *FGFR3*.[33]

Genetic alterations that affect the retinoblastoma (RB) pathway also commonly occur in high-grade urothelial carcinoma. The RB protein, translated from the *RB* gene located on 13q14, regulates the G1-S phase of the cell cycle and also plays a role in apoptosis. Under nonneoplastic conditions, dephosphorylated RB is bound to the E2F transcription factor, sequestering it. The RB protein can be phosphorylated by cyclin-dependent kinase complexes, including cyclin D1/CDK4/6 and cyclin E/CDK2. Once phosphorylated, RB releases E2F, allowing it to promote the expression of genes necessary for transition to the DNA synthesis phase of the cell cycle. Some of the genetic alterations seen in urothelial carcinoma affect the RB protein itself, including deletion of chromosome 13q. Amplification of the genes involved in the cyclin-dependent kinase complexes also frequently occurs in urothelial carcinoma. For example, one study showed that low cyclin D1 and E1 expression were present in 44% and 55% of urothelial carcinoma cystectomy specimens, respectively, and resulted in altered expression of the RB protein. Low cyclin E1 expression was also associated with increased bladder cancer–specific mortality.[34]

Whereas alterations in genes that regulate the G1/S transition are frequently found in noninvasive and invasive high-grade lesions, several of these have also been reported in low-grade papillary urothelial carcinoma. However, the relationship and derivation of low-grade papillary urothelial carcinoma and high-grade in situ lesions remains undefined and is the focus of ongoing investigation. Many mutations described here have also been generally associated with a worse prognosis, although their utility seems to be enhanced when used in combination. One study showed the utility of a "molecular signature" that combined mutational status of multiple genes including FGFR3, PIK3CA, KRAS, HRAS, NNRAS, TP53, CDKN2A, and TSC1. This molecular signature correlated with histologic categories, and the investigators could accurately predict whether a tumor fit into the noninvasive low-grade papillary or high-grade in situ and invasive groups. The molecular signature also seemed to provide independent prognostic information.[35]

In addition to these specific genetic alterations affecting cell-cycle regulation, complex chromosomal abnormalities have also been demonstrated. Karyotyping of muscle-invasive bladder cancer shows aneuploidy with many chromosomal alterations including chromothripsis. One hypothesis for the relatively frequent chromosomal changes in urothelial carcinoma implicates nonhomologous end-joining as a mechanism for error-prone double-strand break repair. A defective replication-licensing complex may also be involved.[4] Such complex chromosomal abnormalities highlight the complexity and heterogeneity of invasive urothelial carcinoma.

Recently, the Cancer Genome Atlas Project completed one of the most comprehensive molecular analyses in bladder cancer, examining 131 cases of muscle-invasive urothelial carcinoma. Tumors were histologically categorized and evaluated via whole genome sequencing, whole exome sequencing, DNA copy number, complete mRNA and microRNA expression, DNA methylation, and protein expression and phosphorylation. Many of the previously described genes were consistently mutated, including TP53, PIK3CA, RB1, FGFR3, and TSC1. In addition, a few pathways were identified as consistently dysregulated. Mutations in the p53/RB tumor-suppressor pathway were seen in 93% of tumors, and alterations in the PI3K/AKT/mTOR and RTK/RAS signaling were seen in 72%. Finally, alterations that impact epigenetic changes, including alterations in genes affecting chromatin remodeling and histone modification, were seen up to 89% of tumors, more than in any other cancer studied,[7] which serves to further emphasize the complex biology of urothelial carcinoma.

EARLY SHARED ALTERATIONS

Common to both pathways of development of urothelial carcinoma are alterations in chromosome 9, and these alterations may represent a unifying early event in urothelial neoplastic transformation. Loss of heterozygosity in chromosome 9 is commonly seen in many bladder tumors, and it is even often seen in hyperplastic and histologically unremarkable urothelium in patients with urothelial carcinoma.[36] Mutations in chromosome 9 may be associated with increasing genetic instability that ultimately leads to other tumorigenic mutations.[37] One gene present on chromosome 9 that is often altered is the tuberous sclerosis gene, TSC1, located at 9q34. Normally, its physiologic function is to create a protein complex, the TSC1-TSC2 complex, which negatively regulates the mTOR branch of the PI3K pathway. Not only is TSC1 altered by deletions, but mutations in the gene have also been seen in many instances of urothelial carcinoma and bladder cancer cell lines.[38,39] Also located on chromosome 9 is CDKN2A, described previously.

FURTHER SUBDIVIDING INVASIVE DISEASE BASED ON MOLECULAR PATHWAYS

Overall, the two-pathway model today seems to be an oversimplification of the complexity of urothelial carcinoma, because it primarily stratifies tumors based on grade and focuses especially on noninvasive carcinoma. Bladder carcinoma has been demonstrated to have more genetic alterations than many other solid tumors, and epigenetic changes also seem to play a vital role in its biology.[7] Given the molecular heterogeneity in muscle-invasive urothelial carcinoma, several groups have attempted to further subclassify these lesions.

The first group to use molecular classifiers in urothelial carcinoma used gene expression profiles to create a "molecular taxonomy" and defined five major subtypes: (1) urobasal A, (2) urobasal B, (3) genomically unstable, (4) squamous-cell carcinoma–like, and (5) infiltrated. These subtypes showed differential expression of cytokeratins, FGFR3 mutation status, cell adhesion genes, and genes that regulate the cell-cycle. It was also demonstrated that the different subtypes have significantly different clinical outcomes and provide independent prognostic information.[40] The group also described the use of defined histologic and immunohistochemical methods to help subdivide lesions into urobasal, genomically unstable, or squamous-cell carcinoma–like categories.[41]

Another study undertook a meta-analysis of high-grade muscle invasive urothelial carcinomas and described molecular subsets of "luminal" and "basal-like" urothelial carcinoma, using similar terminology to breast cancer.[42] The luminal subtype showed high levels of uroplakin and CK20, markers that are usually expressed in umbrella cells. GATA3 and estrogen receptor signaling were also seen preferentially in the luminal subtype. Overall, the histology of these luminal tumors is most similar to the superficial papillary subtype of urothelial carcinoma, and indeed, FGFR3 upregulation is also often seen. Basal-like tumors, however, expressed genes that are more characteristic of urothelial basal cells, including CD44 and basal-cell-associated cytokeratins. The basal-like tumors were also shown to have a significantly worse prognosis.[42]

A third study used whole genome mRNA expression profiling to similarly describe three subtypes of high-grade muscle-invasive urothelial carcinoma: (1) basal, (2) luminal, and (3) p53-like. Luminal urothelial carcinoma was characterized by FGFR3 mutations, sensitivity to FGFR inhibitors, and also showed activation of the peroxisome proliferator-activated receptor-gamma pathway. Basal-type tumors showed expression of basal cell markers including p63, often had squamous differentiation, and also had a more aggressive disease at presentation. In addition to the luminal and basal subtypes that are similar to those described in the previous study, the p53-like class of urothelial carcinomas was described as being resistant to neoadjuvent chemotherapy.[43]

In addition to other methodologies used to subclassify urothelial carcinoma, epigenetic alterations have been the subject of a recent study that assessed genome-wide analysis of DNA methylation in 98 cases of urothelial carcinoma. In this study, the investigators identified differentially methylated DNA regions and could subclassify tumors into groups that were associated with clinical and pathologic features, and gene expression profiles. They were also able to identify a putative epigenetic switch in HOX genes that may have additional prognostic information or be a molecularly actionable target.[44]

THERAPEUTIC AND PROGNOSTIC TARGETS

The standard of care in advanced disease involves cisplatin-based chemotherapeutic regimens. Recent studies have demonstrated that use of a methotrexate/vinblastine/adriamycin/cisplatin regimen or a cisplatin/gemcitabine regimen is better than cisplatin alone, whereas adding paclitaxel does not seem to improve overall survival.[45] However, resistance to platinum-based chemotherapy often occurs, and second-line chemotherapies are currently virtually nonexistent in bladder cancer.[46]

Advances in the understanding of the molecular mechanisms of urothelial cancer have led to many studies to evaluate targeted therapies. Some studies are examining agents that target pathways described previously, and include FGFR3 inhibitors[47] and PI3K/mTOR inhibitors.[48,49] Other studies currently underway are evaluating inhibitors that have been effective in other types of cancer, including epidermal growth factor receptor, vascular endothelial growth factor, and human epidermal growth factor receptor 2 (HER2)/neu inhibitors.[46,50–53] A subset of molecular targets currently under evaluation are highlighted in Table 1.[7,46–60] HER2/neu deserves special mention because although overexpression is seen in more than half of urothelial carcinomas by immunohistochemistry, genomic alterations including amplifications and translocations are less common, seen in less than 20% of cases. The micropapillary variant of urothelial carcinoma, a variant growth pattern associated with adverse outcomes (Fig. 6), is associated with significantly higher rates of HER2/neu protein amplification by immunohistochemistry and increased prevalence of ERBB2 genomic alterations, making it a potential therapeutic target in this variant.[59–64]

Finally, multiple studies are examining the role of immunotherapy in urothelial carcinoma treatment. Whereas bacille Calmette-Guérin has been used for several decades as an immunotherapeutic agent in non–muscle invasive high-grade bladder cancer, newer and more specific immune-modulating agents shown to be effective in other types of cancers are currently being examined. The programmed cell death (PD)-1 receptor on activated T cells transmits an inhibitory signal to the T cell when it binds to its ligand, PD-L1. The ligand PD-L1, upregulated in many types of cancers, can be masked by administration of an anti-PD-L1 monoclonal antibody, thus

Table 1
Subset of genes with prognostic information and/or targeted treatment possibilities

Gene	General Function	Prevalence[a]	Prognostic Effects and Treatment Possibilities	References
FGFR3	Tyrosine kinase receptor upstream of Ras that activates MAP kinase pathway	• FGFR3 mutations or amplifications seen in 15%–20% of high-grade urothelial carcinomas • FGFR3-TACC fusions seen in up to 5%	• Usually associated with low-grade and papillary tumors • One trial of inhibitors has shown to have poor efficacy	[47]
mTOR	Part of the PI3K/Akt/mTOR pathway, kinase that regulates many processes including cell-cycle progression and cell proliferation	• Genomic alterations in genes of the mTOR signaling pathway are seen in up to 65% of muscle-invasive urothelial carcinomas	• Multiple trials of mTOR inhibition by everolimus have shown moderate efficacy and may be related to TSC1 mutational status	[48,49]
EGFR	Member of ErbB family of receptor tyrosine kinases (ErbB1), mediates multiple signal transduction cascades resulting in cell proliferation	• Overexpression is common, seen in approximately 70% of tumors, but genomic amplifications only seen in about 10%	• Overexpression is a poor prognostic factor • Trials of gefitinib, an orally active selective EGFR inhibitor, have shown modest results • An anti-EGFR monoclonal antibody, cetuximab, does not seem to be effective	[50,51,59]
HER2	Member of ErbB family of cell surface receptor tyrosine kinases (ErbB2)	• In all tumors, overexpression is observed in about half of cases, whereas genomic alterations are seen in 6%–14% • Translocations are uncommon, but have been reported • In micropapillary tumors, about 40% of tumors may have mutations or amplifications of ERBB2	• Associated with micropapillary variant (see text) • Overexpression may be a poor prognostic factor • Trastuzumab does not seem to significantly improve outcomes in patients with urothelial carcinoma overexpressing HER2	[52,59–64]
VEGF receptor	Tyrosine kinase that stimulates nitric oxide synthase and activation of PI3K/Akt/mTOR pathway, associated with angiogenesis	Members of angiogenesis pathways including VEGF receptors are altered in up to 20% of muscle-invasive urothelial carcinomas	• Poor prognostic factor • Multiple VEGF and VEGFR inhibitors are currently being studied, with variable efficacy	[53]

Abbreviations: EGFR, epidermal growth factor receptor; HER2, human epidermal growth factor receptor 2; TACC, transforming acid coiled coil; VEGF, vascular endothelial growth factor.

[a] Prevalence of mutational status and copy number variants obtained from www.cbioportal.org (see Refs.[57,58]).
Data from Refs.[7,54–56]

retaining the ability of T cells to recognize and attack tumor cells. Another molecule, cytotoxic T-lymphocyte associated antigen-4 is a T-cell surface molecule that binds to B7 on antigen-presenting cells to downregulate T-cell activation and prevent their clonal expansion. Cytotoxic T-lymphocyte associated antigen-4 blockade through administration of a specific monoclonal

Fig. 6. Micropapillary urothelial carcinoma is characterized by papillary structures that lack fibrovascular cores, and the presence of this growth pattern is an adverse prognostic factor (*A*, hematoxylin-eosin, original magnification ×200; *B*, hematoxylin-eosin, original magnification ×400).

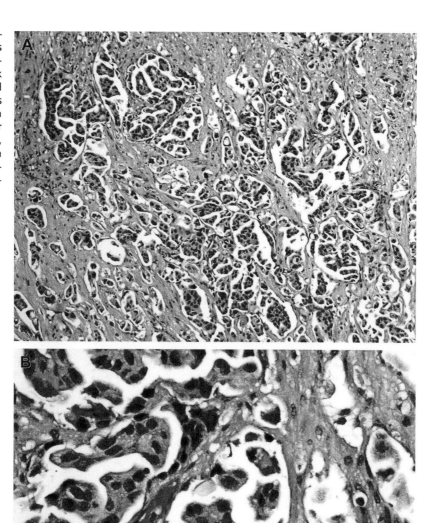

antibody can retain the antitumor activity of cytotoxic T cells. Both therapeutic strategies are being actively studied in primary and metastatic urothelial carcinoma.[46]

SUMMARY

The historical two-pathway model proposed to subdivide urothelial carcinoma may be somewhat limited in its application. New and rapidly evolving subclassifications that combine histology, gene mutation status, protein biomarker expression, and epigenetic information may be necessary. Despite numerous proposed classification systems, the heterogeneity and molecular instability inherent to urothelial carcinoma needs to be evaluated in the context of subclassification model application. Also, the intratumoral heterogeneity of these classifiers remains to be determined. Therefore, future studies that focus on linking patient outcomes to selected molecular and phenotypic alterations may provide the most rapid benefit to patient care.

REFERENCES

1. American Cancer Society. Cancer facts and figures: 2015. Atlanta (GA): American cancer society; 2015.

2. Morgan TM, Cookson MS, Netto G, et al. Bladder cancer overview and staging. In: Hansel DE, McKenney JK, Stephenson AJ, et al, editors. The urinary tract: a comprehensive guide to patient diagnosis and management. New York: Springer Science and Business Media; 2012. p. 83–112.

3. Soloway M, Khoury S, editors. Bladder cancer: second international consultation on bladder cancer – Vienna. 2nd edition. Bristol (United Kingdom): International Consultation on Urological Disease; 2012.

4. Morrison CD, Liu PY, Woloszynska-Read A, et al. Whole-genome sequencing identifies genomic heterogeneity at a nucleotide and chromosomal level in bladder cancer. Proc Natl Acad Sci U S A 2014; 111(6):E672–81.

5. Shah JB, McConkey DJ, Dinney CPN. New strategies in muscle-invasive bladder cancer: on the road to personalized medicine. Clin Cancer Res 2011;17(9):2608–12.

6. Wu XR. Urothelial tumorigenesis: a tale of divergent pathways. Nat Rev Cancer 2005;5(9):713–25.

7. Cancer Genome Atlas Research Network. Comprehensive molecular characterization of urothelial bladder carcinoma. Nature 2014;507:315–22.

8. Dhillon AS, Hagan S, Rath O, et al. MAP kinase signalling pathways in cancer. Oncogene 2007;26(22): 3279–90.

9. Dueñas M, Martínez-Fernández M, García-Escudero R, et al. PIK3CA gene alterations in bladder cancer are frequent and associate with reduced recurrence in non-muscle invasive tumors. Mol Carcinog 2015; 54(7):566–76.

10. Boulalas I, Zaravinos A, Karyotis I, et al. Activation of RAS family genes in urothelial carcinoma. J Urol 2009;181(5):2312–9.

11. Reddy EP, Reynolds RK, Santos E, et al. A point mutation is responsible for the acquisition of transforming properties by the T24 human bladder-carcinoma oncogene. Nature 1982;300(5888):149–52.

12. Tabin CJ, Bradley SM, Bargmann CI, et al. Mechanism of activation of a human oncogene. Nature 1982;300(5888):143–9.

13. Zhang ZT, Pak J, Huang HY, et al. Role of Ha-ras activation in superficial papillary pathway of urothelial tumor formation. Oncogene 2001;20(16): 1973–80.

14. Turner N, Grose R. Fibroblast growth factor signalling: from development to cancer. Nat Rev Cancer 2010;10(2):116–29.

15. Tomlinson DC, L'Hote CG, Kennedy W, et al. Alternative splicing of fibroblast growth factor receptor 3 produces a secreted isoform that inhibits fibroblast growth factor-induced proliferation and is repressed in urothelial carcinoma cell lines. Cancer Res 2005; 65(22):10441–9.

16. Jebar AH, Hurst CD, Tomlinson DC, et al. FGFR3 and Ras gene mutations are mutually exclusive genetic events in urothelial cell carcinoma. Oncogene 2005;24(33):5218–25.

17. Tomlinson DC, Baldo O, Hamden P, et al. FGFR3 protein expression and its relationship to mutation status and prognostic variables in bladder cancer. J Pathol 2007;213(1):91–8.

18. Catto JWF, Miah S, Owen HC, et al. Distinct microRNA alterations characterize high- and low-grade bladder cancer. Cancer Res 2009;69(21):8472–81.

19. Williams SV, Hurst CD, Knowles MA. Oncogenic FGFR3 gene fusions in bladder cancer. Hum Mol Genet 2013;22(4):795–803.

20. Matsushita K, Cha EK, Matsumoto K, et al. Immunohistochemical biomarkers for bladder cancer prognosis. Int J Urol 2011;18(9):616–29.

21. Ross RL, Askham JM, Knowles MA. PIK3CA mutation spectrum in urothelial carcinoma reflects cell context-dependent signaling and phenotypic outputs. Oncogene 2013;32(6):768–76.

22. Platt FM, Hurst CD, Taylor CF, et al. Spectrum of phosphatidylinositol 3-kinase pathway gene alterations in bladder cancer. Clin Cancer Res 2009; 15(19):6008–17.

23. Lopez-Knowles E, Hernandez S, Malats N, et al. PIK3CA mutations are an early genetic alteration associated with FGFR3 mutations in superficial papillary bladder tumors. Cancer Res 2006;66(15): 7401–4.

24. Mitra AP, Hansel DE, Cote RJ. Prognostic value of cell-cycle regulation biomarkers in bladder cancer. Semin Oncol 2012;39(5):524–33.

25. Dalbagni G, Presti JC, Reuter VE, et al. Molecular-genetic alterations of chromosome 17 and p53 nuclear overexpression in human bladder cancer. Diagn Mol Pathol 1993;2(1):4–13.

26. Esrig D, Elmajian D, Groshen S, et al. Accumulation of nuclear p53 and tumor progression in bladder cancer. N Engl J Med 1994;331(19):1259–64.

27. Sarkis AS, Dalbagni G, Cordoncardo C, et al. Nuclear overexpression of p53-protein in transitional cell bladder carcinoma: a marker for disease progression. J Natl Cancer Inst 1993;85(1):53–9.

28. Serth J, Kuczyk MA, Bokemeyer C, et al. p53 immunohistochemistry as an independent prognostic factor for superficial transitional-cell carcinoma of the bladder. Br J Cancer 1995;71(1):201–5.

29. George B, Datar RH, Wu L, et al. p53 gene and protein status: the role of p53 alterations in predicting outcome in patients with bladder cancer. J Clin Oncol 2007;25(34):5352–8.

30. Waldman T, Lengauer C, Kinzler KW, et al. Uncoupling of S phase and mitosis induced by anticancer agents in cells lacking p21. Nature 1996;381(6584): 713–6.

31. Stadler WM, Lerner SP, Groshen S, et al. Phase III study of molecularly targeted adjuvant therapy in locally advanced urothelial cancer of the bladder based on p53 status. J Clin Oncol 2011;29(25):3443–9.

32. Simon R, Struckmann K, Schraml P, et al. Amplification pattern of 12q13-q15 genes (MDM2, CDK4, GLI) in urinary bladder cancer. Oncogene 2002; 21(16):2476–83.

33. Rebouissou S, Herault A, Letouze E, et al. CDKN2A homozygous deletion is associated with muscle invasion in FGFR3-mutated urothelial bladder carcinoma. J Pathol 2012;227(3):315–24.

34. Shariat SF, Ashfaq R, Sagalowsky AI, et al. Correlation of cyclin D1 and E1 expression with bladder cancer presence, invasion, progression, and metastasis. Hum Pathol 2006;37(12):1568–76.

35. Lindgren D, Frigyesi A, Gudjonsson S, et al. Combined gene expression and genomic profiling define two intrinsic molecular subtypes of urothelial carcinoma and gene signatures for molecular grading and outcome. Cancer Res 2010;70(9):3463–72.

36. Knowles MA, Hurst CD. Molecular biology of bladder cancer: new insights into pathogenesis and clinical diversity. Nat Rev Cancer 2015;15(1):25–41.

37. Netto GJ, Cheng L. Emerging critical role of molecular testing in diagnostic genitourinary pathology. Arch Pathol Lab Med 2012;136(4):372–90.

38. Knowles MA, Habuchi T, Kennedy W, et al. Mutation spectrum of the 9q34 tuberous sclerosis gene TSC1 in transitional cell carcinoma of the bladder. Cancer Res 2003;63(22):7652–6.

39. Hornigold N, Devlin J, Davies AM, et al. Mutation of the 9q34 gene TSC1 in sporadic bladder cancer. Oncogene 1999;18(16):2657–61.

40. Sjodahl G, Lauss M, Lovgren K, et al. A molecular taxonomy for urothelial carcinoma. Clin Cancer Res 2012;18(12):3377–86.

41. Sjodahl G, Lovgren K, Lauss M, et al. Toward a molecular pathologic classification of urothelial carcinoma. Am J Pathol 2013;183(3):681–91.

42. Damrauer JS, Hoadley KA, Chism DD, et al. Intrinsic subtypes of high-grade bladder cancer reflect the hallmarks of breast cancer biology. Proc Natl Acad Sci U S A 2014;111(8):3110–5.

43. Choi W, Porten S, Kim S, et al. Identification of distinct basal and luminal subtypes of muscle-invasive bladder cancer with different sensitivities to frontline chemotherapy. Cancer Cell 2014;25(2): 152–65.

44. Aine M, Sjodahl G, Eriksson P, et al. Integrative epigenomic analysis of differential DNA methylation in urothelial carcinoma. Genome Med 2015;7(1):23.

45. Bellmunt J, von der Maase H, Mead GM, et al. Randomized phase III study comparing paclitaxel/cisplatin in patients with locally advanced or metastatic urothelial cancer without prior systemic therapy: EORTC Intergroup Study 30987. J Clin Oncol 2012;30(10):1107–13.

46. Seront E, Machiels JP. Molecular biology and targeted therapies for urothelial carcinoma. Cancer Treat Rev 2015;41(4):341–53.

47. Milowsky MI, Dittrich C, Martinez ID, et al. Final results of a multicenter, open-label phase II trial of dovitinib (TKI258) in patients with advanced urothelial carcinoma with either mutated or nonmutated FGFR3. J Clin Oncol 2013;31(6) supplement: 255.

48. Iyer G, Hanrahan AJ, Milowsky MI, et al. Genome sequencing identifies a basis for everolimus sensitivity. Science 2012;338(6104):221.

49. Milowsky MI, Iyer G, Regazzi AM, et al. Phase II study of everolimus in metastatic urothelial cancer. BJU Int 2013;112(4):462–70.

50. Philips GK, Halabi S, Sanford BL, et al. A phase II trial of cisplatin (C), gemcitabine (G) and gefitinib for advanced urothelial tract carcinoma: results of Cancer and Leukemia Group B (CALGB) 90102. Ann Oncol 2009;20(6):1074–9.

51. Wong YN, Litwin S, Vaughn D, et al. Phase II trial of cetuximab with or without paclitaxel in patients with advanced urothelial tract carcinoma. J Clin Oncol 2012;30(28):3545–51.

52. Hussain MHA, MacVicar GR, Petrylak DP, et al. Trastuzumab, paclitaxel, carboplatin, and gemcitabine in advanced human epidermal growth factor receptor-2/neu-positive urothelial carcinoma: results of a multicenter phase II National Cancer Institute trial. J Clin Oncol 2007;25(16):2218–24.

53. Balar AV, Apolo AB, Ostrovnaya I, et al. Phase II study of gemcitabine, carboplatin, and bevacizumab in patients with advanced unresectable or metastatic urothelial cancer. J Clin Oncol 2013;31(6): 724–30.

54. Guo GW, Sun XJ, Chen C, et al. Whole-genome and whole-exome sequencing of bladder cancer identifies frequent alterations in genes involved in sister chromatid cohesion and segregation. Nat Genet 2013;45(12):1459–63.

55. Iyer G, Al-Ahmadie H, Schultz N, et al. Prevalence and co-occurrence of actionable genomic alterations in high-grade bladder cancer. J Clin Oncol 2013;31(25):3133.

56. Kim PH, Cha EK, Sfakianos JP, et al. Genomic predictors of survival in patients with high-grade urothelial carcinoma of the bladder. Eur Urol 2015;67(2): 198–201.

57. Gao JJ, Aksoy BA, Dogrusoz U, et al. Integrative analysis of complex cancer genomics and clinical

profiles using the cBioPortal. Sci Signal 2013;6(269): pl1.

58. Cerami E, Gao JJ, Dogrusoz U, et al. The cBio cancer genomics portal: an open platform for exploring multidimensional cancer genomics data. Cancer Discov 2012;2(5):401–4.

59. Carlsson J, Wester K, De La Torre M, et al. EGFR-expression in primary urinary bladder cancer and corresponding metastases and the relation to HER2-expression. On the possibility to target these receptors with radionuclides. Radiol Oncol 2015; 49(1):50–8.

60. Li JH, Jackson CL, Yang DF, et al. Comparison of tyrosine kinase receptors HER2, EGFR, and VEGFR expression in micropapillary urothelial carcinoma with invasive urothelial carcinoma. Target Oncol 2015;10(3):355–63.

61. Chen PCH, Yu HJ, Chang YH, et al. Her2 amplification distinguishes a subset of non-muscle-invasive bladder cancers with a high risk of progression. J Clin Pathol 2013;66(2):113–9.

62. Ross JS, Wang K, Gay LM, et al. A high frequency of activating extracellular domain ERBB2 (HER2) mutation in micropapillary urothelial carcinoma. Clin Cancer Res 2014;20(1):68–75.

63. Ching CB, Amin MB, Tubbs RR, et al. HER2 gene amplification occurs frequently in the micropapillary variant of urothelial carcinoma: analysis by dual-color in situ hybridization. Mod Pathol 2011;24(8):1111–9.

64. Hansel DE, Swain E, Dreicer R, et al. HER2 overexpression and amplification in urothelial carcinoma of the bladder is associated with MYC coamplification in a subset of cases. Am J Clin Pathol 2008; 130(2):274–81.

Molecular Pathology
Predictive, Prognostic, and Diagnostic Markers in Uterine Tumors

Lauren L. Ritterhouse, MD, PhD, Brooke E. Howitt, MD*

KEYWORDS

- Endometrial • Endometrioid • Serous • Adenocarcinomas • POLE • MSI

ABSTRACT

This article focuses on the diagnostic, prognostic, and predictive molecular biomarkers in uterine malignancies, in the context of morphologic diagnoses. The histologic classification of endometrial carcinomas is reviewed first, followed by the description and molecular classification of endometrial epithelial malignancies in the context of histologic classification. Taken together, the molecular and histologic classifications help clinicians to approach troublesome areas encountered in clinical practice and evaluate the utility of molecular alterations in the diagnosis and subclassification of endometrial carcinomas. Putative prognostic markers are reviewed. The use of molecular alterations and surrogate immunohistochemistry as prognostic and predictive markers is also discussed.

question of appropriate prospective screening selection criteria in endometrial carcinomas. Various screening algorithms using mismatch repair immunohistochemistry (IHC) on tumor samples with or without *MLH1* promoter methylation testing are reviewed here.

HISTOLOGIC SUBTYPES OF ENDOMETRIAL CARCINOMA (WORLD HEALTH ORGANIZATION CLASSIFICATION)

Although the focus of this article is on the molecular features of uterine tumors, the histologic subtypes must first be considered in order to make the context useful for practicing pathologists. A comprehensive review of the morphologic characteristics of endometrial carcinomas is beyond the scope of this article; however, a brief description of the histologic types as defined by the World Health Organization (WHO)[1] is covered here as a basis for the remainder of the article.

ENDOMETRIOID ENDOMETRIAL ADENOCARCINOMA

Endometrioid is the most common histotype of endometrial carcinoma, and resembles normal proliferative-type endometrial glands with smooth luminal borders. Endometrioid carcinomas are divided into 3 grades based on architectural and nuclear features (International Federation of Gynecology and Obstetrics [FIGO] classification). Grade 1 (FIGO) endometrioid adenocarcinoma is composed of entirely, or greater than 95%, gland-forming tumor cells, with little or no (≤5%) solid growth of tumor cells, and notably lacks significant (severe) nuclear atypia (**Fig. 1A**).

OVERVIEW

Knowledge of the molecular features of endometrial carcinoma has rapidly expanded in the last 5 to 10 years. This increase in information has led to several helpful diagnostic markers as well as future potential prognostic/predictive markers, both molecular and immunohistochemical. Recent large-scale genomic studies have led to a shift in the paradigm of endometrial carcinoma classification, with the recognition of distinct molecular groups that only partially overlap with the histologic classification. Historically, endometrial carcinomas had been divided into biological type I and type II pathways, but these are quickly being replaced. Hereditary endometrial cancer, namely Lynch syndrome, has raised the controversial

Department of Pathology, Brigham and Women's Hospital, Harvard Medical School, 75 Francis Street, Boston, MA 02115, USA
* Corresponding author.
E-mail address: bhowitt@partners.org

Surgical Pathology 9 (2016) 405–426
http://dx.doi.org/10.1016/j.path.2016.04.006
1875-9181/16/$ – see front matter © 2016 Elsevier Inc. All rights reserved.

surgpath.theclinics.com

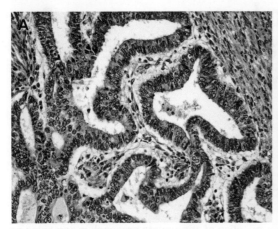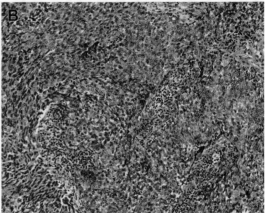

Fig. 1. Endometrial endometrioid adenocarcinoma (H&E, 400×). (*A*) Grade 1 endometrioid adenocarcinoma is composed entirely or predominantly (95% or more) of gland-forming architecture. (*B*) Grade 3 endometrioid adenocarcinoma shows solid growth in at least 50% of the tumor, as shown here.

Importantly, areas of squamous differentiation must be excluded when determining the percentage solid growth. Grade 2 (FIGO) endometrioid adenocarcinoma is composed of a mixture of solid and gland-forming tumor, with the solid component comprising less than 50% but more than 5% of the tumor. Alternatively, an architecturally grade 1 adenocarcinoma may be upgraded to FIGO grade 2 based on the presence of significant severe nuclear atypia. Grade 3 (FIGO) endometrioid adenocarcinoma has poorly formed glands and is composed of greater than 50% solid growth (Fig. 1B), or the presence of severe nuclear atypia in the setting of an architecturally grade 2 tumor. However, the precise definition of severe nuclear atypia is not well defined and is subjective, leading to poor interobserver reproducibility.[2–6] Most clinicians accept striking cytologic atypia visible at 10× objective, present in most tumor cells, as the threshold for upgrading a tumor.[7,8]

SEROUS CARCINOMA OF THE ENDOMETRIUM

Serous carcinoma is the second most common type of endometrial cancer, and is histologically similar to the high-grade serous carcinomas of the ovary and fallopian tube. Serous carcinomas have a high nuclear to cytoplasmic (N/C) ratio with conspicuous nuclear atypia and high mitotic index (Fig. 2). Architecturally, they are typically papillary and micropapillary, forming characteristic slitlike spaces because of the lack of polarity. Irregular, infiltrative myometrial invasion is common. Serous carcinomas may also show glandular and/or solid growth patterns, causing diagnostic confusion with endometrioid carcinoma (discussed in more detail later).

CLEAR CELL CARCINOMA OF THE ENDOMETRIUM

Clear cell carcinoma is a high-grade malignancy, with similar morphologic features to its ovarian counterpart. Clear cell carcinomas are characterized by a variety of architectural patterns, including papillary, tubulocystic, and solid (Fig. 3). Hyalinized stroma is frequently seen and can be a helpful clue to making the diagnosis. The cells are large with typically clear to palely eosinophilic cytoplasm and have a bumpy hobnail appearance with large hyperchromatic and irregular nuclei and prominent nucleoli.

MUCINOUS ADENOCARCINOMA

Mucinous adenocarcinomas are very similar to endometrioid carcinomas; when greater than 50% of the tumor shows conspicuous mucinous differentiation, it is termed mucinous adenocarcinoma. These tumors are typically low grade and low stage, and may have strikingly bland cytomorphology.

UNDIFFERENTIATED/DEDIFFERENTIATED CARCINOMA

Undifferentiated carcinomas lack any amount of gland formation. Architecturally they are composed of solid sheets or vague nests, and the cells appear discohesive.[1] Hematopoietic malignancies, carcinosarcoma, and/or mesenchymal tumors are often considered in the differential diagnosis of undifferentiated carcinoma.[9] Dedifferentiated carcinomas are biphasic tumors consisting of an undifferentiated component as well as a (grade 1–2) well-differentiated endometrioid carcinoma component, and more than half of them show mismatch repair (MMR) deficiency by IHC.[9]

Fig. 2. Serous carcinoma of the endometrium. Serous carcinomas are characterized by papillary and glandular architecture with uneven luminal borders, exfoliation, and fracture planes within the epithelium. There is typically a high nuclear/cytoplasmic ratio with conspicuous nuclear atypia (H&E, 400×).

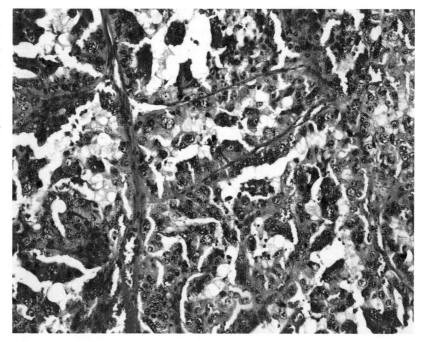

NEUROENDOCRINE CARCINOMA

Neuroendocrine carcinomas of the endometrium are rare, representing less than 1% of all endometrial malignancies and existing primarily as case reports and small case series in the literature.[10–12] Morphologically they have the same characteristics as neuroendocrine tumors outside the female genital tract, with high-grade neuroendocrine carcinomas much more common than low-grade neuroendocrine tumors/carcinomas.

Immunohistochemical expression of a neuroendocrine marker (chromogranin, synaptophysin) in at least 10% of tumor cells, although arbitrary, is generally accepted as supportive in the appropriate morphologic context.[1]

MIXED-TYPE ENDOMETRIAL CARCINOMA

Mixed endometrial adenocarcinoma is a tumor composed of 2 distinct histologic subtypes, with at least 1 of them being a high-grade histotype

Fig. 3. Clear cell carcinoma of the endometrium. Clear cell carcinomas typically show a range of architectural growth patterns (papillary, tubulocystic, solid), with hobnailing of the cells, often clear cytoplasm, and variable nuclear atypia (H&E, 400×).

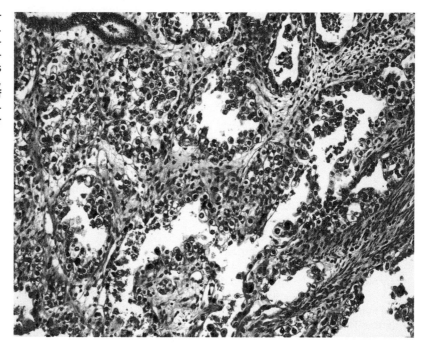

and the minor component comprising at least 5% of the tumor. The most common example of this is mixed endometrioid and serous carcinoma. Some clinicians classify tumors showing morphologic features that are intermediate between 2 histotypes as a mixed carcinoma; however, this might be more appropriately labeled ambiguous or indeterminate, as discussed later.

CARCINOSARCOMA

Carcinosarcoma is morphologically a mixed type tumor, composed of both malignant glands and malignant stroma. However, almost all of these tumors have been shown to be monoclonal, arising from a common origin,[13,14] and it is now accepted that carcinosarcoma is a type of epithelial malignancy, with overlapping features of both serous and endometrioid carcinomas.[1,15–21]

MORPHOLOGICALLY AMBIGUOUS HIGH-GRADE ENDOMETRIAL ADENOCARCINOMA

Although not a formal WHO-defined histologic category, it is well recognized that some high-grade endometrial carcinomas have ambiguous or indeterminate morphology,[22–34] with overlapping features suggestive of endometrioid, serous, and/or clear cell carcinoma, and this accounts for much of the interobserver variability in endometrial carcinoma classification.

MOLECULAR SUBTYPES OF ENDOMETRIAL CARCINOMA AND THEIR MORPHOLOGIC CORRELATES

Historically, endometrial carcinomas had been divided into 2 major biological pathways: type I, which was associated with estrogenic stimulation

and composed of primarily endometrioid endometrial carcinoma; and type II, which were the high-grade (serous and clear cell) tumors unassociated with estrogen exposure and typically occurring in a background of atrophy.[35] In this context, endometrioid tumors were thought to commonly have *PTEN* mutations and microsatellite instability (MSI),[36–40] whereas the serous carcinomas frequently harbored *TP53* mutations[36,41,42] as well as *PIK3CA*, *FBXW7*, and *PPP2R1A* mutations.[42,43] Before 2013, using expression analyses, targeted next-generation sequencing, and small series of cases with whole-exome sequencing, multiple studies reported and appreciated more than 2 molecular categories of endometrial carcinoma.[26,44]

In 2013, The Cancer Genome Atlas (TCGA) released the first large-scale in-depth molecular study of endometrial carcinomas using whole-exome and RNA sequencing.[45] This study led to a shift in the paradigm of endometrial carcinoma classification with the description of 4 broad molecular categories (independent of histology) based on both the number of single-nucleotide variations (SNVs), presence or absence of microsatellite instability (MSI), and number of somatic copy number alterations (also referred to as copy number variations [CNVs]): (1) ultramutated (*POLE* mutated), (2) hypermutated secondary to MSI, (3) low copy number, and (4) high copy number (serous-like).[45] Many of the tumors in these TCGA categories correspond with particular histologic types or grades, and can be associated with mutations in specific genes (**Table 1**). One limitation of the TCGA study is that it included only endometrioid and serous histotypes. Several reviews covering the molecular features of endometrial cancer have since been published.[46–48]

Table 1
TCGA molecular subtypes of endometrial carcinoma

	Histotypes	Number of Mutations	Copy Number Alterations	Specific Genes Recurrently Altered
POLE-mutated (ultramutated)	Endometrioid (G3 > G2-1)	Very high	Low	*POLE, PTEN, PIK3R1, PIK3CA, FBXW7, KRAS, TP53*
MSI (hypermutated)	Endometrioid (G2-3 > G1)	High	Variable (low to intermediate)	*PTEN, KRAS, ARID1A*
Copy number: low	Endometrioid (G1-2 > G3)	Low	Variable (low to intermediate)	*CTNNB1, PTEN*
Copy number: high (serous-like)	Serous	Low	High	*TP53, FBXW7, PPP2R1A*

ULTRAMUTATED (*POLE*-MUTATED) ENDOMETRIAL CARCINOMAS

POLE is the gene encoding the catalytic subunit of DNA polymerase epsilon, involved in DNA replication and proofreading/DNA repair. *POLE*-mutated (ultramutated) carcinomas were notable for very large numbers of SNVs (mutations) (often with a mutation frequency of >100 mutations/Mb) with a particular preference for C > A transversion mutations.[49,50] These tumors were associated with good outcome, a finding that has been corroborated in multiple subsequent studies[51–54]; however, in 1 large study there was no association with improved clinical outcome based on *POLE* mutation status.[55] Genes that are commonly mutated in the *POLE*-mutated tumors include *PTEN* (94%), *PIK3R1* (65%), *PIK3CA* (71%), *FBXW7* (82%), and *KRAS* (53%). *TP53* notably was mutated in 35% of *POLE*-mutated cases. The most common histologic subtype in this category is endometrioid, with nearly two-thirds being either grade 2 or grade 3 endometrioid, and the minority of cases being mixed endometrial carcinomas including an endometrioid component.[45,56] Subsequent studies of the morphologic features of *POLE*-mutated tumors have described high-grade and often ambiguous histology[53,56]; in 1 study a subset of tumors histologically classified as serous carcinoma harbored *POLE* mutations[41] and in another study serous carcinomas (likely misclassified) with *POLE*-related hypermutation were associated with a better clinical outcome.[57] However, a subset of *POLE*-mutated tumors are grade 1 endometrioid carcinomas lacking these distinct morphologic features. Other morphologic features present in *POLE*-mutated tumors include increased tumor-infiltrating lymphocytes (TILs), which are predominantly CD3+ CD8+ T lymphocytes.[53,56,58,59]

MICROSATELLITE INSTABILITY ENDOMETRIAL CANCERS (HYPERMUTATED)

Tumors with MSI have an underlying defect in one of the mismatch repair genes (*MSH2*, *MSH6*, *MLH1*, or *PMS2*) or *EPCAM* deletions, which may be either germline in nature (discussed in more detail later) or somatic/acquired. Most of the somatic alterations resulting in MSI are caused by promoter hypermethylation of *MLH1*, resulting in complete silencing of the gene. Regardless of the route to MSI, tumors with MSI are characterized by a high number of SNVs, and a characteristic molecular signature, characterized predominantly by C > T at NpCpG sites and small indels (typically 1 base pair).[60]

Unlike the *MLH1*-methylated colorectal carcinomas, which are highly associated with *BRAF* mutation, endometrial carcinomas with *MLH1* methylation do not commonly harbor *BRAF* mutations.[61] However, *KRAS* mutations are common in this subgroup and additional studies have shown that *KRAS* mutations are more common in MSI carcinomas than in MSS,[40,62–64] whereas others have found no difference.[65,66] *KRAS* mutations are mutually exclusive with both *CTNNB1* and *FGFR2* mutations, which occur in a subset of these endometrioid carcinomas. Some studies have shown that *PTEN* may be mutated (or show loss of protein expression by IHC) at slightly higher frequencies in MSI than in MSS tumors,[66–71] although other studies found no difference.[37,72,73] *ARID1A* is more frequently lost in MSI tumors[32] and has been implicated in the pathogenesis of non–Lynch syndrome–associated, *MLH1*-methylated cases of MSI endometrial tumors, suggesting that *ARID1A* mutation may precede and contribute to the development of MSI via epigenetic

POLE-mutated Carcinomas (Key Points)		
Molecular Features	**Morphologic Features**	**Clinical Features**
Ultramutated (mutation rate: median 232 mutations per megabase pair) Minimal CNVs Mutation in the exonuclease domain of *POLE* (hotspots at P286R and V411L)	Endometrioid, but may have high-grade features bordering on serous carcinoma (nuclear atypia) Eosinophilic cytoplasm	5%–10% of all endometrial carcinomas
Increased C > A transversion frequency *PTEN, PIK3R1, PIK3CA, FBXW7, KRAS* mutations common	Lymphovascular invasion common Increased tumor infiltrating lymphocytes levels	Wide age range
TP53 mutation common, rare indels Increased predicted neoantigen load Majority are microsatellite stable	Higher mitotic index	Favorable prognosis in most studies

MSI (Hypermutated) (Key Points)		
Molecular Features	**Morphologic Features**	**Clinical Features**
Hypermutated (mutation rate: median 18 mutations per megabase pair; 10-fold higher than MSS tumors)	Most are endometrioid FIGO grades 2–3 > grade 1	Subset are caused by hereditary endometrial cancer (Lynch syndrome)
C > T at NpCpG sites and numerous small indels		20%–40% of all endometrial carcinomas
Low numbers of CNVs		
KRAS, ARID1A, ARID5B, PTEN mutations		Prognosis similar to low copy number; intermediate between POLE-mutated and high-copy-number carcinomas
RPL22 deletions	Increased TIL levels	
Few TP53, FBXW7, CTNNB1, PPP2R1A mutations		
MLH1 hypermethylation		
Increased predicted neoantigen load		

silencing of MLH1 in a subset of endometrial carcinomas.[74]

LOW-COPY-NUMBER ENDOMETRIAL CARCINOMAS

This class of tumors has low numbers of copy number alterations as well as smaller numbers of mutations. Most tumors in this category are endometrioid in histotype and grade 1 to 2. Only 1 gene is more frequently mutated in this group compared with MSI tumors: CTTNB1.[45] Other commonly mutated genes in the low-copy-number group include PTEN, PIK3CA, PIK3R1, and ARID1A. As in other subgroups, PIK3CA and PIK3R1 mutations seem to be mutually exclusive. Although some clinicians have debated whether MSI carries prognostic significance, most large studies show that MSS (low-copy-number) endometrial cancers have similar clinical outcomes to those with MSI.[75,76]

HIGH-COPY-NUMBER ENDOMETRIAL CARCINOMA (SEROUS-LIKE)

This category of tumors is characterized by genomic instability as shown by high numbers of CNVs, and smaller numbers of SNVs. TP53 is the most common gene mutated among this subgroup (>90%), followed by PIK3CA (>40%), FBXW7 (22%), and PPP2R1A (22%). Most histotypes in this molecular category are serous carcinomas, but some endometrioid carcinomas (predominantly grade 2–3) are also included in this molecular category. Previous studies showed high rates of TP53, PPP2R1A, PIK3CA, and FBXW7 mutations[41,43,77] and a subset of endometrial serous carcinomas with CCNE1, PIK3CA, or ERBB2 amplification.[41,43,78] CCNE1 amplification and FBXW7 mutation seem to be mutually exclusive,[42] suggesting redundant function of these genes in serous carcinogenesis.

Low Copy Number (Key Points)		
Molecular Features	**Morphologic Features**	**Clinical Features**
Few CNVs	Endometrioid, predominantly FIGO grade 1–2	Most common subtype
Few SNVs		Prognosis similar to MSI; intermediate between POLE-mutated and high-copy-number carcinomas
Frequent CTNNB1 mutations (>50%), only gene mutated more frequently in low-copy-number group compared with MSI group		
PTEN, PIK3CA, PIK3R1, and ARID1A are common mutations		

High Copy Number (Key Points)		
Molecular Features	Morphologic Features	Clinical Features
High levels of CNVs Low numbers of SNVs	Predominantly serous histotype	Older age at presentation
Frequent TP53 mutation (>90%) PIK3CA, FBXW7, PPP2R1A mutations	Prominent nuclear atypia	Higher stage at presentation
Uncommon PTEN and ARID1A mutations Gene amplifications: CCNE1, MYC, PIK3CA, CDKN2A, ERBB2	Subset of high-grade endometrioid carcinoma (FIGO grade 3)	Poor prognosis

Molecular Features of Clear Cell Carcinomas

Clear cell carcinoma histologies were not included in the recent TCGA publication, and no large-scale in-depth genomic analyses have been performed on endometrial clear cell carcinomas of pure histotype. One study evaluated uterine clear cell carcinomas for PTEN and TP53 mutations and MSI and found a small percentage of cases harboring mutations in PTEN or TP53, as well as MSI[79]; however, this study also included some mixed carcinomas. Another study performed limited targeted exonic sequencing of a small series of endometrial pure clear cell carcinomas[80] and found that, in general, clear cell carcinomas seemed to be molecularly more similar to serous carcinoma than endometrioid carcinoma, with the most frequent mutations occurring in TP53, PPP2R1A, FBXW7, and SPOP. ARID1A mutations were identified in only 2 out of 14 cases and both PTEN and CTNNB1 alterations were absent in all cases. Another study found frequent loss of PTEN IHC expression, ERBB2 alterations, and PIK3CA mutations in pure clear cell carcinomas.[81] One group performed TERT promoter mutation analysis in ovarian and endometrial clear cell carcinomas, finding this alteration present in a subset of cases.[82] Although this finding seemed to carry prognostic significance, with shorter disease-free survival and shorter overall survival, for endometrial tumors there was no clear association with any clinicopathologic feature, but in the ovary it was associated with older age and intact ARID1A expression.[82] This study, and previous IHC-based studies, suggest that endometrial clear cell carcinoma may be a molecularly heterogeneous histotype, and biologically different from ovarian clear cell carcinoma, and this deserves further in-depth study.

Molecular Features of Undifferentiated/Dedifferentiated Carcinomas

According to one targeted study, undifferentiated/dedifferentiated carcinomas and their associated lower-grade endometrioid components (when present) have frequent mutations in PIK3CA, CTNNB1, TP53, FBXW7, and PPP2R1A, but lack mutations in KRAS.[83] Studies that selectively microdissect and analyze low-grade and undifferentiated components show clonal relatedness of the 2 components, with additional mutations present in the undifferentiated component. PTEN expression was also lost in a subset. In this small study CTNNB1 and TP53 mutations were mutually exclusive.[83]

Molecular Features of Neuroendocrine Carcinomas

Only very small studies and case reports exist on the molecular alterations present in neuroendocrine carcinomas of the endometrium. One study evaluating neuroendocrine carcinomas associated with adenocarcinomas of the ovary and uterus identified several copy number losses, many of which were shared with adenocarcinoma, but in general found a higher number of alterations in the neuroendocrine component.[84] Another case report sequenced 1 tumor for KIT and PDGFRA hotspot mutations, but found no alterations.[85]

Molecular Features of Mucinous Endometrial Carcinoma

No large-scale studies of mucinous carcinomas investigating a wide range of genes have been done to date; however, several studies have reported increased frequency of KRAS mutation in mucinous carcinomas of the endometrium.[86–88]

Molecular Features of Carcinosarcomas

Few studies have focused on a broad examination of molecular alterations in endometrial carcinosarcomas. PIK3CA/PTEN and related pathways alterations, TP53, KRAS, CTNNB1 mutations, MMR alterations, and EGFR amplification or mutation are present in a subset of uterine carcinosarcomas.[15,89–96] One study used targeted next generation sequencing (NGS) to classify carcinosarcomas molecularly into endometrioid-like and serous-like, based on the presence of TP53,

PTEN, ARID1A, KRAS, PIK3CA, and PPP2R1A mutations.[97] One large whole-exome sequencing study of carcinosarcomas confirmed most of these findings and also found recurrent mutations in SPOP, ARID1B, MLL3, BAZ1A, and FBXW7, and in particular noted the very high frequency of alterations in chromatin remodeling genes (ARID1A, ARID1B, MLL3, SPOP, and BAZ1A).[98]

PROBLEMATIC AREAS IN THE CLASSIFICATION OF ENDOMETRIAL CARCINOMAS AND THE USE OF MOLECULAR ALTERATIONS AS DIAGNOSTIC TOOLS

Most of the diagnostic difficulty in endometrial carcinomas is within the category of high-grade carcinomas (endometrioid, serous, clear cell). Grade 3 endometrioid carcinomas are unique in that they appear in significant proportions in all 4 TCGA-defined molecular classes of tumors (albeit with smaller numbers in the low-copy-number category), including a quarter of the high-copy-number (serous-like) category.[45] Furthermore, there is not always agreement, even between expert gynecologic pathologists, on the distinction between grade 3 endometrioid and serous carcinoma[99] or clear cell carcinoma,[23] suggesting that grade 3 endometrioid carcinomas are both molecularly and histologically heterogeneous. Some clinicians have proposed that endometrioid carcinomas with serous-like molecular profiles might be best treated in a similar fashion to serous carcinomas.[45]

High-grade endometrial carcinomas that are difficult to classify histologically may be aided by the use of p53 IHC[26,29] to help classify or give prognostic information. Note that p53 overexpression may be seen in endometrioid carcinomas, so the presence of p53 overexpression (or complete absence of staining) does not indicate a diagnosis of serous carcinoma. In contrast, a heterogeneous wild-type pattern of p53 staining strongly argues against a serous carcinoma. Some clinicians have suggested that p16 is a more specific marker for serous carcinoma (vs endometrioid).[100–102] Studies have suggested that canonical hotspot TP53 mutations are more common in serous carcinomas and that PTEN mutations more frequently coexist with TP53 mutations in high-grade endometrioid carcinomas.[45,103] Importantly, PTEN IHC is superior to molecular sequencing in detecting loss of function.[104]

Some groups have proposed a combination of molecular and immunohistochemical assays to aid in the classification of endometrial carcinomas.[105] One group showed good prognostic value of POLE mutational analysis, p53 IHC, and MMR IHC.[105]

Another area of difficulty in classifying endometrial carcinomas concerns those tumors that show mixed histologic types. One recent study evaluated molecular alterations in both components of various mixed endometrial carcinomas and found identical mutations present in both morphologic components in at least 89% of cases.[106] Other studies have similarly shown shared driver mutations in mixed carcinomas of the endometrium, suggesting that almost all of these mixed carcinomas represent either morphologic variation of 1 tumor histotype, or a common clonal origin for both histotypes.[79] A study examining gene expression profiles of mixed endometrial tumors found that both the endometrioid and serous components of mixed carcinomas had expression profiles similar to those from pure serous tumors. This finding suggests that regions within mixed tumors that have endometrioid histology have molecular profiles more reminiscent of serous tumors and prompts the question as to whether these tumors should be treated as serous carcinomas.[107]

Key Points				
IHC AND MOLECULAR FEATURES OF ENDOMETRIAL CARCINOMA				
Molecular Classification	IHC			
	MMR	**p53**	**p16**	**PTEN**
POLE mutated	Intact[a]	Variable; mostly wild-type pattern	Variable	Loss
MSI	Loss	Mostly wild-type pattern	Loss	Variable
High copy number	Intact	Null or diffuse/strong	Diffuse/strong	Intact
Low copy number	Intact	Wild-type pattern	Variable	Variable
[a] MMR IHC loss has been reported in some studies.[55]				

PROGNOSTIC BIOMARKERS IN UTERINE CARCINOMAS

It remains to be determined whether the molecular category of grade 3 endometrioid carcinomas will help with prognosis and treatment decisions. For instance, a grade 3 endometrioid tumor with a high-copy-number (serous-like) molecular profile may be better treated in similar fashion to serous carcinomas. Likewise, given the apparent outstanding prognosis of the *POLE*-mutated carcinomas,[45,51–54] these tumors may warrant less aggressive therapy.

Hormonal receptor expression has long been associated with prognosis in endometrial carcinoma. Several studies have shown that endometrial carcinomas lacking both estrogen receptor (ER) and progesterone receptor (PR) expression have a worse prognosis.[108–112] Similar to the hormonal classification system used in breast carcinomas, stratifying endometrial carcinomas based on ER/PER/HER2 status has been performed in an attempt to assess prognosis, with the ER+/PR+/HER2− and ER−/PR−/HER2+ groups showing more benign and aggressive behaviors, respectively.[112–114] However, many of these data are retrospective; large prospective studies examining the prognostic utility of ER/PR staining in endometrial carcinomas are lacking.

Several investigations have found that alterations in p53 are associated with a worse prognosis in endometrial carcinomas. A study of endometrioid and nonendometrioid carcinomas found that patients with p53 alterations (by expression or mutation analysis) had a shorter survival than those without.[115] In addition, the presence of p53 overexpression as detected by IHC in mixed and morphologically ambiguous endometrial carcinomas has been shown to be significantly associated with a worse progression-free and disease-specific survival and was found to be as clinically informative as expert gynecologic pathology consultation.[29] This study found p53 overexpression in approximately half of all ambiguous cases and concluded that p53 status was able to provide prognostic information as well as aid in clinical decision making.

A subset of endometrial carcinomas, all of which had at least a component of endometrioid histology, have coexisting *TP53* and *PIK3CA* alterations, and patients with concurrent p53 and PI3K-AKT alterations had shorter survival than patients with p53 alterations alone; however, this association was only seen with exon 20 *PIK3CA* hotspot mutations.[115] In another study, the presence of *PIK3CA* hotspot mutations in exons 9 and 20 was found to be associated with a shorter disease-specific survival within grade 3 endometrioid but not serous carcinomas.[116] These studies show that alterations in *PIK3CA* are likely to be associated with a worse prognosis in high-grade or mixed endometrioid endometrial carcinomas, particularly when associated with *TP53* mutations.

CCNE1 amplification has been associated with a worse prognosis in ovarian high-grade serous carcinoma.[117] One study found that a small subset of endometrioid endometrial carcinomas also show *CCNE1* amplification and in a multivariate analysis this was found to be an independent prognostic factor for overall survival, but not for progression-free survival.[118]

Key Points
PROGNOSTIC BIOMARKERS IN UTERINE CANCER

Biomarker	Histotype	Prognosis
High copy number	Serous, high-grade endometrioid	Worse PFS
ER/PR loss	All	Worse DSS
TP53 mutation	All	Worse PFS and DSS
PIK3CA mutation	High-grade endometrioid	Worse DSS
CCNE1 amplification	High-grade endometrioid	Worse PFS and OS
POLE mutation	Endometrioid	Improved PFS

Abbreviations: DSS, disease-specific survival; OS, overall survival; PFS, progression-free survival.

PREDICTIVE BIOMARKERS IN UTERINE CANCER

The TCGA project identified 2 groups of endometrioid endometrial carcinomas that showed a high mutation frequency, one being MSI tumors and the other being tumors with mutations in the exonuclease domain of *POLE*.[45] Studies have shown that both of those highly mutated groups are associated with high neoantigen loads and increased number of TILs.[53,58,59] In addition, 1 study also showed that this increase in TILs was accompanied by an overexpression of PD-1 and PD-L1 in the immune cells.[58] The data presented from these studies suggest that both MSI and *POLE*-mutated endometrial carcinomas may be excellent candidates for immunotherapies.

Hormonal therapy may be attempted in endometrial carcinomas, and pathologists are often asked to perform IHC for ER and PR on endometrial cancer samples. Studies are conflicting regarding the predictive value of the ER/PR IHC staining and clinical response to hormonal therapy.[111] There are more data supporting PR as a predictor of response,[119] but tumor heterogeneity is considered to be a challenge and in many cases performing IHC on the primary tumor is irrelevant when the patient has had a tumor recurrence. One study showed a possible role for endometrial stromal expression of PR in predicting response to hormonal therapy.[120] Prospective studies specifically evaluating the ER/PR status in recurrent tumors and the response to hormonal therapy are lacking.

A large percentage of endometrial carcinomas harbor mutations in the PI3K pathway, and it has been estimated that approximately 60% of high-risk endometrial carcinomas harbor mutations in the PI3K-AKT pathway, highlighting the value of this pathway as a therapeutic target.[121] In studies with endometrial cancer–derived xenografts, it has been shown that AKT inhibitors have strong antitumor effects, and the effect was most pronounced in cancer cell lines with mutations in *PIK3CA* or *PIK3R1*.[122] The results of current clinical trials will shed further light on the efficacy of targeting the PI3K pathway in endometrial carcinomas.

Because targeting HER2 (*ERBB2*) has had significant success in breast cancer, there has been considerable interest in targeting HER2 in other tumor types. HER2 has been shown to be amplified in up to 44% of serous carcinomas,[41,45] and several case reports have reported responses to trastuzumab in endometrial carcinomas with HER2 overexpression.[123–125] However, clinical trials of HER2-directed therapies to date have had very limited success.[126,127]

HEREDITARY ENDOMETRIAL CARCINOMA

Lynch syndrome (LS; hereditary nonpolyposis colorectal carcinoma syndrome) is an inherited syndrome characterized by a predisposition to developing colorectal and endometrial carcinomas (most commonly), as well as gastric, small intestine, ovarian, upper urinary tract, pancreas, and others.[128–132] It is caused by germline mutations in one of the mismatch repair genes (*MSH2*, *MSH6*, *MLH1*, or *PMS2*) or occasionally by deletions involving the 3′ end of *EPCAM*, which results in epigenetic silencing of *MSH2*.[133,134] Another form of hereditary endometrial carcinoma is Cowden syndrome, caused by germline mutations in *PTEN*, accounting for only rare instances of endometrial carcinoma.[135–137] In most studies, less than 5% of all endometrial carcinomas are associated with Lynch syndrome[138–140]; however, when considering endometrial carcinomas in women less than 50 years of age, it accounts for nearly 10% of all endometrial carcinomas.[141–144] Half of women with undiagnosed Lynch syndrome present with endometrial cancer as a first malignancy; overall, endometrial carcinoma may be more common than colorectal carcinoma in

---

Key Points
FOR PREDICTIVE BIOMARKERS IN ENDOMETRIAL CANCER

Biomarker	Tumor Type	Therapy	Status
PD-1 and PD-L1	POLE mutated, MSI	PD-1 targeted	Investigational
PR	PR positive	Progesterones	Routinely tested in practice
PI3K-AKT alterations	All	PI3K targeted	Investigational
HER2 amplification	Serous	HER2 targeted	Limited efficacy

women with Lynch syndrome.[145–147] As a result, there is an opportunity for the detection of women at risk for having Lynch syndrome using IHC for mismatch repair (MMR) proteins on endometrial carcinoma samples.[138]

Although universal screening of colorectal carcinoma with MMR IHC is now widely accepted as the standard of care,[148–150] there is no consensus on the guidelines for screening pathology samples of endometrial carcinoma for Lynch syndrome.[138,151–155] The Revised Bethesda Guidelines and Amsterdam II criteria use patient age and family history to determine which cases should be screened with MMR IHC and/or MSI testing.[156,157] Most Lynch syndrome–associated endometrial carcinomas are of endometrioid histotype, and some groups have described morphologic features associated with Lynch syndrome and MSI,[152,158–160] including undifferentiated histology, origin in the lower uterine segment,[161] and increased TILs (typically >40 per high-power field)[162]; however, both the clinical and pathologic

features may fail to identify a significant number of Lynch syndrome–associated cases[151,163–168] and the morphologic features are not specific for Lynch syndrome or for MSI.[139] Recently, some investigators have proposed universal screening for Lynch syndrome in all endometrial cancer,[138,163–165,169,170] whereas others have proposed continued use of morphologic and clinical history (patient age and family/personal history of cancers), or expanding the screening to all patients less than 60 years of age[139,171] or 70 years of age.[172] One group has recommended universal MSH6 IHC screening and selective screening of MSH2, MLH1, and PMS2 by patient age, pathologic features, and clinical history.[152] Regardless of how the patient population to screen is chosen, the subsequent screening algorithm is similar (Fig. 4): the first step is performing MMR IHC using either the 4-antibody panel, or a 2-antibody panel (MSH6/PMS2).[173,174] If all MMR expression is retained, then the patient is considered unlikely to have Lynch syndrome, but may still be sent to

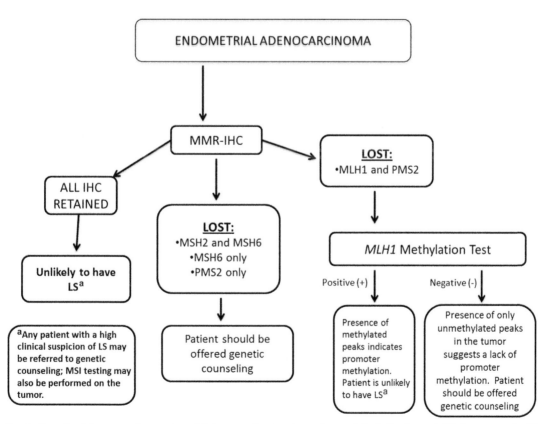

Fig. 4. Flowchart for screening endometrial cancer for Lynch syndrome (LS). Patients are initially screened with MMR IHC, and based on those results are either considered low risk for LS, high risk for LS, or requiring additional testing in the form of MLH1 promoter testing to further classify. Patients with high clinical suspicion for LS may still undergo genetic counseling and germline testing. MSI testing on the tumor may also be performed in conjunction with the MMR IHC.

a genetic counselor in the setting of significant clinical suspicion (age, family/personal cancer history) for a hereditary process. If expression of MSH6 alone, MSH2 and MSH6, or PMS2 alone is lost in tumor cells, then the patient should be referred to a genetic counselor for consideration of germline testing. In the setting of MLH1 and PMS2 loss of expression, the most common explanation is somatic inactivation of *MLH1* via promoter hypermethylation, which should be confirmed by molecular methods, including polymerase chain reaction (PCR)–based *MLH1* promoter methylation analysis (Fig. 5A). If MLH1 promoter methylation is detected, the patient is unlikely to have Lynch syndrome. In the absence of *MLH1* promoter methylation, patients should be referred to genetic counseling because there is a high likelihood of carrying a germline mutation in *MLH1*. PCR-based testing for MSI may be performed in conjunction with MMR IHC (Fig. 5B), using PCR-based technology to examine the microsatellite markers necessary to determine the

presence of MSI. This technology may increase the sensitivity of detection of rare patients with Lynch syndrome whose tumors show intact MMR protein expression despite loss of normal protein function.

Before implementation of Lynch syndrome screening, there are several things to consider. First, the interpretation of the MMR IHC has some pitfalls. With all of the MMR stains, clinicians should see clear staining in stroma, inflammatory cells, and endothelial cells, which serve as internal positive controls. In order to call an MMR stain negative, positive internal control staining should be seen in the same area where loss of MMR staining is seen in the tumor. Second, any degree of positive nuclear staining in tumor cells is considered intact (not lost). It is common to see weak staining, heterogeneous patterns, or variable intensity in the MMR IHC.[175] Third, in order to effectively screen, clinicians must have the ability to perform *MLH1* promoter methylation testing, whether this is performed on site or at a reference

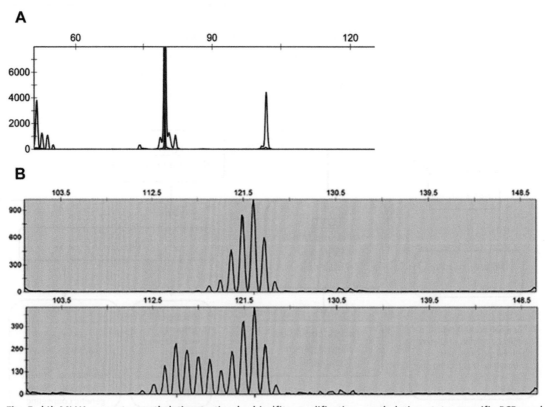

Fig. 5. (*A*) *MLH1* promoter methylation testing by bisulfite modification, methylation status–specific PCR, and capillary gel electrophoresis. Both methylated (*left*) and unmethylated (*right*) peaks are detected, indicating the presence of *MLH1* promoter methylation. (*B*) MSI testing by a PCR-based assay targeting microsatellite loci and capillary gel electrophoresis on normal control (*top*) and tumor (*bottom*) tissues. The tumor sample shows alterations in the length of the repetitive sequence, which is visualized by the presence of additional peaks compared with the normal sample, thus indicating the presence of MSI at this locus.[175]

laboratory. In addition, there has to be clinical infrastructure in place to follow up on nonmethylated, abnormal MMR IHC cases, and the capacity for genetic counseling to advise any additional patients identified by MMR IHC screening.[167] One last obstacle in performing MMR screening on endometrial cancers is the question of reimbursement. Once there is consensus on who should be tested, insurance policies are likely to reimburse for these tests, similar to the Lynch syndrome screening in colorectal carcinomas. To this effect, the recent large NRG and Gynecologic Oncology Group study of more than 1000 patients supported universal screening of all endometrial cancers regardless of patient age, tumor morphology, or clinical history, suggesting that consensus may be forthcoming.[151]

MOLECULAR FEATURES OF MALIGNANT UTERINE MESENCHYMAL NEOPLASMS

UTERINE LEIOMYOSARCOMA

Uterine leiomyosarcoma (LMS) can be divided into several overlapping molecular subcategories. Although in general LMS is considered to arise *de novo* there is some molecular evidence for LMS arising from preexisting leiomyomas in a subset of cases.[176,177] LMS has markedly complex karyotypes; recurrent karyotypic alterations include gain of 1q, 17p, and Xp. Loss of heterozygosity for 10q (containing *PTEN*) and/or 13q (containing *RB1*) is present in greater than 50% of LMS.[178–181] *TP53* mutations may also be present in greater than 50% of cases,[182,183] but p53 IHC is not a reliable marker for LMS because many other tumors harbor mutations in *TP53*, including some atypical leiomyomas. Moreover, in smooth muscle tumors, *TP53* mutations and p53 expression carry no definite prognostic significance.[184,185] Between 2% and 20% of LMS harbors *MED12* mutations,[182,186–192] a gene that is frequently altered in leiomyomas; however, many of the *MED12* mutations present in LMS are not hotspot mutations found in benign uterine leiomyomas.[182] HMGA2 overexpression is seen in ~35% of LMS and seems to be mutually exclusive with *MED12* mutation.[186] Neither *HMGA2* nor *MED12* alterations have been associated with differences in prognosis in LMS. *ATRX* and *DAXX* are mutated frequently in LMS and are associated with alterative lengthening of telomere (ALT) phenotype contributing to the pathogenesis of uterine LMS in up to 60% of cases.[188,193] ATRX has been shown to be a prognostic biomarker in uterine LMS; loss of expression of ATRX/DAXX by IHC, or mutations in these genes, is associated with ALT phenotype

and adverse clinical outcome compared with those tumors with intact ATRX/DAXX expression.[188,193,194]

ENDOMETRIAL STROMAL SARCOMA

Endometrial stromal sarcomas (ESS) are divided into 2 categories: low-grade ESS and high-grade ESS.[1] ESS is a tumor associated with highly recurrent translocations resulting in gene fusions. *JAZF1-SUZ12* is the most common gene fusion in low-grade (LG) ESS (25% to >90% of cases).[195–202] This gene fusion reflects the chromosomal translocation t(7;17) (p15;q21) frequently observed in ESS via karyotype evaluation.[203,204] Another translocation involving *JAZF1*, t(6;7) (p21;p15), resulting in a *JAZF1-PHF1* gene fusion, is present in up to 28% of ESS as well as in an ESS cell line.[201,205–207] PHF1, a polycomb repressor gene, has been shown to produce a gene fusion with *MEAF6* (on 1p34)[208,209] in a smaller subset of ESS. ESSs with *PHF1* rearrangement are enriched for sex cord–like differentiation,[207] but also may show typical morphology.[210] A small subset of ESS may have *PHF1* rearrangements resulting in fusion with genes other than *JAZF1* or *MEAF6*, most notably *EPC1* on 10p11.[205] One ESS has been shown to harbor a *BCOR-ZC3H7B* gene fusion and 2 cases harbored a *MBTD1-CXorf67* gene fusion.[211,212] The presence of recurrent gene fusions involving the polycomb genes *PHF1*, *EPC1*, *MBTD1*, or *SUZ12*, even in the absence of a *JAZF1* abnormality, suggests that polycomb genes likely play a significant role in ESS pathogenesis. Several other less common cytogenetic aberrations have been described in ESS, which are outside the scope of this article but have recently been reviewed.[213] A subset of LGESS with conventional cytogenetics has no evidence of chromosomal rearrangements as well as no evidence of *JAZF1* or *PHF1* gene fusions by reverse-transcription PCR (RT-PCR) or fluorescence in situ hybridization (FISH), suggesting that some of the molecular alterations in ESS have not yet been discovered.

Recurrent *YWHAE-FAM22A/B* gene fusions have recently been described in a subset of ESS, which are associated with a clinical outcome intermediate between that of LGESS and undifferentiated uterine sarcoma, and has led to the reintroduction of the high-grade ESS (HGESS) diagnostic category in the most recent WHO classification.[1,214,215] These tumors lack the typical morphology of LGESS in that they do not resemble nonneoplastic endometrium, lack CD10 expression, and have uniform high-grade atypia. In some cases, HGESS is associated with more typical-appearing areas of

LGESS.[216,217] *YWHAE* rearrangements have not been found in other gynecologic tumors, and FISH and/or RT-PCR studies may serve as a useful adjunct to the histologic diagnosis.[218,219] CyclinD1 IHC may be used as a marker for *YWHAE* rearrangement,[202,217] although this is not entirely sensitive or specific, particularly when considering undifferentiated endometrial carcinoma and tumors outside the gynecologic tract (clear cell sarcoma of kidney),[220–223] so confirmatory FISH studies are advised.

Other malignant mesenchymal tumors of the uterus include adenosarcoma and undifferentiated uterine sarcoma. Neither of these tumors has highly recurrent molecular aberrations and they may represent heterogeneous groups of tumors. Adenosarcomas have frequent low-level amplification of *MDM2* and *CDK4*, *MYBL1* and *TERT* amplification, infrequent *TP53* mutation, and occasional gene fusions involving NCOA2/3.[224,225] Undifferentiated uterine sarcomas (UUS) are a heterogeneous group of tumors and may represent various dedifferentiated forms of uterine sarcomas given that small subsets of UUS harbor genetic alterations characteristic of LMS or gene fusions reported in ESS.[226] Regardless, these tumors have lost any evidence of clear differentiation and tend to be histologically pleomorphic and cytogenetically complex. *TP53* mutations are common in UUS, in contrast with endometrial stromal neoplasms,[226] suggesting that those UUS with *TP53* mutation have no relationship with ESS or alternatively have acquired a secondary *TP53* mutation.

REFERENCES

1. Kurman RJ, Carcangiu ML, Herrington CS, et al. WHO classification of tumours of female reproductive organs. 4th edition. Lyon (France): IARC; 2014.
2. Scholten AN, Smit VT, Beerman H, et al. Prognostic significance and interobserver variability of histologic grading systems for endometrial carcinoma. Cancer 2004;100(4):764–72.
3. Lax SF, Kurman RJ, Pizer ES, et al. A binary architectural grading system for uterine endometrial endometrioid carcinoma has superior reproducibility compared with FIGO grading and identifies subsets of advance-stage tumors with favorable and unfavorable prognosis. Am J Surg Pathol 2000; 24(9):1201–8.
4. Sagae S, Saito T, Satoh M, et al. The reproducibility of a binary tumor grading system for uterine endometrial endometrioid carcinoma, compared with FIGO system and nuclear grading. Oncology 2004;67(5–6):344–50.
5. Guan H, Semaan A, Bandyopadhyay S, et al. Prognosis and reproducibility of new and existing binary grading systems for endometrial carcinoma compared to FIGO grading in hysterectomy specimens. Int J Gynecol Cancer 2011;21(4):654–60.
6. Gemer O, Uriev L, Voldarsky M, et al. The reproducibility of histological parameters employed in the novel binary grading systems of endometrial cancer. Eur J Surg Oncol 2009;35(3):247–51.
7. Zaino RJ, Kurman RJ, Diana KL, et al. The utility of the revised International Federation of Gynecology and Obstetrics histologic grading of endometrial adenocarcinoma using a defined nuclear grading system. A Gynecologic Oncology Group study. Cancer 1995;75(1):81–6.
8. Conlon N, Leitao MM Jr, Abu-Rustum NR, et al. Grading uterine endometrioid carcinoma: a proposal that binary is best. Am J Surg Pathol 2014; 38(12):1583–7.
9. Tafe LJ, Garg K, Chew I, et al. Endometrial and ovarian carcinomas with undifferentiated components: clinically aggressive and frequently underrecognized neoplasms. Mod Pathol 2010;23(6):781–9.
10. Albores-Saavedra J, Martinez-Benitez B, Luevano E. Small cell carcinomas and large cell neuroendocrine carcinomas of the endometrium and cervix: polypoid tumors and those arising in polyps may have a favorable prognosis. Int J Gynecol Pathol 2008;27(3):333–9.
11. van Hoeven KH, Hudock JA, Woodruff JM, et al. Small cell neuroendocrine carcinoma of the endometrium. Int J Gynecol Pathol 1995;14(1):21–9.
12. Huntsman DG, Clement PB, Gilks CB, et al. Small-cell carcinoma of the endometrium. A clinicopathological study of sixteen cases. Am J Surg Pathol 1994;18(4):364–75.
13. Wada H, Enomoto T, Fujita M, et al. Molecular evidence that most but not all carcinosarcomas of the uterus are combination tumors. Cancer Res 1997;57(23):5379–85.
14. Abeln EC, Smit VT, Wessels JW, et al. Molecular genetic evidence for the conversion hypothesis of the origin of malignant mixed mullerian tumours. J Pathol 1997;183(4):424–31.
15. de Jong RA, Nijman HW, Wijbrandi TF, et al. Molecular markers and clinical behavior of uterine carcinosarcomas: focus on the epithelial tumor component. Mod Pathol 2011;24(10):1368–79.
16. Gorai I, Yanagibashi T, Taki A, et al. Uterine carcinosarcoma is derived from a single stem cell: an in vitro study. Int J Cancer 1997;72(5):821–7.
17. McCluggage WG. Malignant biphasic uterine tumours: carcinosarcomas or metaplastic carcinomas? J Clin Pathol 2002;55(5):321–5.
18. de Brito PA, Silverberg SG, Orenstein JM. Carcinosarcoma (malignant mixed mullerian (mesodermal) tumor) of the female genital tract: immunohistochemical and ultrastructural analysis of 28 cases. Hum Pathol 1993;24(2):132–42.

19. Jin Z, Ogata S, Tamura G, et al. Carcinosarcomas (malignant mullerian mixed tumors) of the uterus and ovary: a genetic study with special reference to histogenesis. Int J Gynecol Pathol 2003;22(4): 368–73.

20. Lopez-Garcia MA, Palacios J. Pathologic and molecular features of uterine carcinosarcomas. Semin Diagn Pathol 2010;27(4):274–86.

21. Kernochan LE, Garcia RL. Carcinosarcomas (malignant mixed Mullerian tumor) of the uterus: advances in elucidation of biologic and clinical characteristics. J Natl Compr Canc Netw 2009; 7(5):550–6, [quiz: 7].

22. Soslow RA. Endometrial carcinomas with ambiguous features. Semin Diagn Pathol 2010;27(4): 261–73.

23. Fadare O, Parkash V, Dupont WD, et al. The diagnosis of endometrial carcinomas with clear cells by gynecologic pathologists: an assessment of interobserver variability and associated morphologic features. Am J Surg Pathol 2012;36(8): 1107–18.

24. Gilks CB, Oliva E, Soslow RA. Poor interobserver reproducibility in the diagnosis of high-grade endometrial carcinoma. Am J Surg Pathol 2013;37(6): 874–81.

25. Han G, Sidhu D, Duggan MA, et al. Reproducibility of histological cell type in high-grade endometrial carcinoma. Mod Pathol 2013;26(12):1594–604.

26. Lomo L, Nucci MR, Lee KR, et al. Histologic and immunohistochemical decision-making in endometrial adenocarcinoma. Mod Pathol 2008;21(8): 937–42.

27. Darvishian F, Hummer AJ, Thaler HT, et al. Serous endometrial cancers that mimic endometrioid adenocarcinomas: a clinicopathologic and immunohistochemical study of a group of problematic cases. Am J Surg Pathol 2004;28(12):1568–78.

28. Murali R, Soslow RA, Weigelt B. Classification of endometrial carcinoma: more than two types. Lancet Oncol 2014;15(7):e268–78.

29. Garg K, Leitao MM Jr, Wynveen CA, et al. p53 overexpression in morphologically ambiguous endometrial carcinomas correlates with adverse clinical outcomes. Mod Pathol 2010;23(1):80–92.

30. Han G, Soslow RA, Wethington S, et al. Endometrial carcinomas with clear cells: a study of a heterogeneous group of tumors including interobserver variability, mutation analysis, and immunohistochemistry with HNF-1beta. Int J Gynecol Pathol 2015;34(4):323–33.

31. Chiang S, Soslow RA. Updates in diagnostic immunohistochemistry in endometrial carcinoma. Semin Diagn Pathol 2014;31(3):205–15.

32. Allo G, Bernardini MQ, Wu RC, et al. ARID1A loss correlates with mismatch repair deficiency and intact p53 expression in high-grade endometrial carcinomas. Mod Pathol 2014; 27(2):255–61.

33. Clarke BA, Gilks CB. Endometrial carcinoma: controversies in histopathological assessment of grade and tumour cell type. J Clin Pathol 2010;63(5):410–5.

34. Soslow RA. High-grade endometrial carcinomas - strategies for typing. Histopathology 2013;62(1): 89–110.

35. Bokhman JV. Two pathogenetic types of endometrial carcinoma. Gynecol Oncol 1983;15(1):10–7.

36. Lax SF, Kendall B, Tashiro H, et al. The frequency of p53, K-ras mutations, and microsatellite instability differs in uterine endometrioid and serous carcinoma: evidence of distinct molecular genetic pathways. Cancer 2000;88(4):814–24.

37. Tashiro H, Blazes MS, Wu R, et al. Mutations in PTEN are frequent in endometrial carcinoma but rare in other common gynecological malignancies. Cancer Res 1997;57(18):3935–40.

38. Catasus L, Machin P, Matias-Guiu X, et al. Microsatellite instability in endometrial carcinomas: clinicopathologic correlations in a series of 42 cases. Hum Pathol 1998;29(10):1160–4.

39. Burks RT, Kessis TD, Cho KR, et al. Microsatellite instability in endometrial carcinoma. Oncogene 1994;9(4):1163–6.

40. Duggan BD, Felix JC, Muderspach LI, et al. Microsatellite instability in sporadic endometrial carcinoma. J Natl Cancer Inst 1994;86(16):1216–21.

41. Zhao S, Choi M, Overton JD, et al. Landscape of somatic single-nucleotide and copy-number mutations in uterine serous carcinoma. Proc Natl Acad Sci U S A 2013;110(8):2916–21.

42. Le Gallo M, O'Hara AJ, Rudd ML, et al. Exome sequencing of serous endometrial tumors identifies recurrent somatic mutations in chromatin-remodeling and ubiquitin ligase complex genes. Nat Genet 2012;44(12):1310–5.

43. Kuhn E, Wu RC, Guan B, et al. Identification of molecular pathway aberrations in uterine serous carcinoma by genome-wide analyses. J Natl Cancer Inst 2012;104(19):1503–13.

44. Alkushi A, Clarke BA, Akbari M, et al. Identification of prognostically relevant and reproducible subsets of endometrial adenocarcinoma based on clustering analysis of immunostaining data. Mod Pathol 2007;20(11):1156–65.

45. Cancer Genome Atlas Research Network, Kandoth C, Schultz N, et al. Integrated genomic characterization of endometrial carcinoma. Nature 2013;497(7447):67–73.

46. Hong B, Le Gallo M, Bell DW. The mutational landscape of endometrial cancer. Curr Opin Genet Dev 2015;30:25–31.

47. Le Gallo M, Bell DW. The emerging genomic landscape of endometrial cancer. Clin Chem 2014; 60(1):98–110.

48. O'Hara AJ, Bell DW. The genomics and genetics of endometrial cancer. Adv genomics Genet 2012; 2012(2):33–47.

49. Rayner E, van Gool IC, Palles C, et al. A panoply of errors: polymerase proofreading domain mutations in cancer. Nat Rev Cancer 2016;16(2):71–81.

50. Shinbrot E, Henninger EE, Weinhold N, et al. Exonuclease mutations in DNA polymerase epsilon reveal replication strand specific mutation patterns and human origins of replication. Genome Res 2014;24(11):1740–50.

51. McConechy MK, Talhouk A, Leung S, et al. Endometrial carcinomas with POLE exonuclease domain mutations have a favorable prognosis. Clin Cancer Res 2016, [Epub ahead of print].

52. Church DN, Stelloo E, Nout RA, et al. Prognostic significance of POLE proofreading mutations in endometrial cancer. J Natl Cancer Inst 2015; 107(1):402.

53. Hussein YR, Weigelt B, Levine DA, et al. Clinico-pathological analysis of endometrial carcinomas harboring somatic POLE exonuclease domain mutations. Mod Pathol 2015;28(4):505–14.

54. Meng B, Hoang LN, McIntyre JB, et al. POLE exonuclease domain mutation predicts long progression-free survival in grade 3 endometrioid carcinoma of the endometrium. Gynecol Oncol 2014;134(1):15–9.

55. Billingsley CC, Cohn DE, Mutch DG, et al. Polymerase varepsilon (POLE) mutations in endometrial cancer: clinical outcomes and implications for Lynch syndrome testing. Cancer 2015;121(3): 386–94.

56. Bakhsh S, Kinloch M, Hoang LN, et al. Histopathological features of endometrial carcinomas associated with POLE mutations: implications for decisions about adjuvant therapy. Histopathology 2015;68(6):916–24.

57. Santin AD, Bellone S, Centritto F, et al. Improved survival of patients with hypermutation in uterine serous carcinoma. Gynecol Oncol Rep 2015;12: 3–4.

58. Howitt BE, Shukla SA, Sholl LM, et al. Association of polymerase e-mutated and microsatellite-instable endometrial cancers with neoantigen load, number of tumor-infiltrating lymphocytes, and expression of PD-1 and PD-L1. JAMA Oncol 2015;1(9):1319–23.

59. van Gool IC, Eggink FA, Freeman-Mills L, et al. POLE proofreading mutations elicit an antitumor immune response in endometrial cancer. Clin Cancer Res 2015;21(14):3347–55.

60. Alexandrov LB, Nik-Zainal S, Wedge DC, et al. Signatures of mutational processes in human cancer. Nature 2013;500(7463):415–21.

61. Metcalf AM, Spurdle AB. Endometrial tumour BRAF mutations and MLH1 promoter methylation as predictors of germline mismatch repair gene mutation

status: a literature review. Fam Cancer 2014;13(1): 1–12.

62. Lagarda H, Catasus L, Arguelles R, et al. K-ras mutations in endometrial carcinomas with microsatellite instability. J Pathol 2001;193(2):193–9.

63. Thoury A, Descatoire V, Kotelevets L, et al. Evidence for different expression profiles for c-Met, EGFR, PTEN and the mTOR pathway in low and high grade endometrial carcinomas in a cohort of consecutive women. Occurrence of PIK3CA and K-Ras mutations and microsatellite instability. Histol Histopathol 2014;29(11):1455–66.

64. Byron SA, Gartside M, Powell MA, et al. FGFR2 point mutations in 466 endometrioid endometrial tumors: relationship with MSI, KRAS, PIK3CA, CTNNB1 mutations and clinicopathological features. PLoS One 2012;7(2):e30801.

65. Stewart CJ, Amanuel B, Grieu F, et al. KRAS mutation and microsatellite instability in endometrial adenocarcinomas showing MELF-type myometrial invasion. J Clin Pathol 2010;63(7):604–8.

66. Peterson LM, Kipp BR, Halling KC, et al. Molecular characterization of endometrial cancer: a correlative study assessing microsatellite instability, MLH1 hypermethylation, DNA mismatch repair protein expression, and PTEN, PIK3CA, KRAS, and BRAF mutation analysis. Int J Gynecol Pathol 2012;31(3):195–205.

67. Bussaglia E, del Rio E, Matias-Guiu X, et al. PTEN mutations in endometrial carcinomas: a molecular and clinicopathologic analysis of 38 cases. Hum Pathol 2000;31(3):312–7.

68. Bilbao C, Rodriguez G, Ramirez R, et al. The relationship between microsatellite instability and PTEN gene mutations in endometrial cancer. Int J Cancer 2006;119(3):563–70.

69. Peiro G, Lohse P, Mayr D, et al. Insulin-like growth factor-I receptor and PTEN protein expression in endometrial carcinoma. Correlation with bax and bcl-2 expression, microsatellite instability status, and outcome. Am J Clin Pathol 2003;120(1):78–85.

70. An HJ, Kim KI, Kim JY, et al. Microsatellite instability in endometrioid type endometrial adenocarcinoma is associated with poor prognostic indicators. Am J Surg Pathol 2007;31(6):846–53.

71. Zhou XP, Kuismanen S, Nystrom-Lahti M, et al. Distinct PTEN mutational spectra in hereditary non-polyposis colon cancer syndrome-related endometrial carcinomas compared to sporadic microsatellite unstable tumors. Hum Mol Genet 2002; 11(4):445–50.

72. Djordjevic B, Barkoh BA, Luthra R, et al. Relationship between PTEN, DNA mismatch repair, and tumor histotype in endometrial carcinoma: retained positive expression of PTEN preferentially identifies sporadic non-endometrioid carcinomas. Mod Pathol 2013;26(10):1401–12.

73. Cohn DE, Basil JB, Venegoni AR, et al. Absence of PTEN repeat tract mutation in endometrial cancers with microsatellite instability. Gynecol Oncol 2000; 79(1):101–6.

74. Bosse T, ter Haar NT, Seeber LM, et al. Loss of ARID1A expression and its relationship with PI3K-Akt pathway alterations, TP53 and microsatellite instability in endometrial cancer. Mod Pathol 2013;26(11):1525–35.

75. Zighelboim I, Goodfellow PJ, Gao F, et al. Microsatellite instability and epigenetic inactivation of MLH1 and outcome of patients with endometrial carcinomas of the endometrioid type. J Clin Oncol 2007;25(15):2042–8.

76. Diaz-Padilla I, Romero N, Amir E, et al. Mismatch repair status and clinical outcome in endometrial cancer: a systematic review and meta-analysis. Crit Rev Oncol Hematol 2013;88(1):154–67.

77. Nagendra DC, Burke J 3rd, Maxwell GL, et al. PPP2R1A mutations are common in the serous type of endometrial cancer. Mol Carcinog 2012; 51(10):826–31.

78. Kuhn E, Bahadirli-Talbott A, Shih Ie M. Frequent CCNE1 amplification in endometrial intraepithelial carcinoma and uterine serous carcinoma. Mod Pathol 2014;27(7):1014–9.

79. An HJ, Logani S, Isacson C, et al. Molecular characterization of uterine clear cell carcinoma. Mod Pathol 2004;17(5):530–7.

80. Hoang LN, McConechy MK, Meng B, et al. Targeted mutation analysis of endometrial clear cell carcinoma. Histopathology 2015;66(5): 664–74.

81. Bae HS, Kim H, Young Kwon S, et al. Should endometrial clear cell carcinoma be classified as type II endometrial carcinoma? Int J Gynecol Pathol 2015; 34(1):74–84.

82. Huang HN, Chiang YC, Cheng WF, et al. Molecular alterations in endometrial and ovarian clear cell carcinomas: clinical impacts of telomerase reverse transcriptase promoter mutation. Mod Pathol 2015; 28(2):303–11.

83. Kuhn E, Ayhan A, Bahadirli-Talbott A, et al. Molecular characterization of undifferentiated carcinoma associated with endometrioid carcinoma. Am J Surg Pathol 2014;38(5):660–5.

84. Mhawech-Fauceglia P, Odunsi K, Dim D, et al. Array-comparative genomic hybridization analysis of primary endometrial and ovarian high-grade neuroendocrine carcinoma associated with adenocarcinoma: mystery resolved? Int J Gynecol Pathol 2008;27(4):539–46.

85. Terada T. Large cell neuroendocrine carcinoma with sarcomatous changes of the endometrium: a case report with immunohistochemical studies and molecular genetic study of KIT and PDGFRA. Pathol Res Pract 2010;206(6):420–5.

86. He M, Jackson CL, Gubrod RB, et al. KRAS mutations in mucinous lesions of the uterus. Am J Clin Pathol 2015;143(6):778–84.

87. Alomari A, Abi-Raad R, Buza N, et al. Frequent KRAS mutation in complex mucinous epithelial lesions of the endometrium. Mod Pathol 2014;27(5): 675–80.

88. Xiong J, He M, Jackson C, et al. Endometrial carcinomas with significant mucinous differentiation associated with higher frequency of k-ras mutations: a morphologic and molecular correlation study. Int J Gynecol Cancer 2013;23(7):1231–6.

89. Bashir S, Jiang G, Joshi A, et al. Molecular alterations of PIK3CA in uterine carcinosarcoma, clear cell, and serous tumors. Int J Gynecol Cancer 2014;24(7):1262–7.

90. Biscuola M, Van de Vijver K, Castilla MA, et al. Oncogene alterations in endometrial carcinosarcomas. Hum Pathol 2013;44(5):852–9.

91. Growdon WB, Roussel BN, Scialabba VL, et al. Tissue-specific signatures of activating PIK3CA and RAS mutations in carcinosarcomas of gynecologic origin. Gynecol Oncol 2011;121(1):212–7.

92. South SA, Hutton M, Farrell C, et al. Uterine carcinosarcoma associated with hereditary nonpolyposis colorectal cancer. Obstet Gynecol 2007;110(2 Pt 2):543–5.

93. Taylor NP, Zighelboim I, Huettner PC, et al. DNA mismatch repair and TP53 defects are early events in uterine carcinosarcoma tumorigenesis. Mod Pathol 2006;19(10):1333–8.

94. Hoang LN, Ali RH, Lau S, et al. Immunohistochemical survey of mismatch repair protein expression in uterine sarcomas and carcinosarcomas. Int J Gynecol Pathol 2014;33(5):483–91.

95. Lancaster JM, Risinger JI, Carney ME, et al. Mutational analysis of the PTEN gene in human uterine sarcomas. Am J Obstet Gynecol 2001;184(6): 1051–3.

96. Amant F, de la Rey M, Dorfling CM, et al. PTEN mutations in uterine sarcomas. Gynecol Oncol 2002; 85(1):165–9.

97. McConechy MK, Ding J, Cheang MC, et al. Use of mutation profiles to refine the classification of endometrial carcinomas. J Pathol 2012;228(1): 20–30.

98. Jones S, Stransky N, McCord CL, et al. Genomic analyses of gynaecologic carcinosarcomas reveal frequent mutations in chromatin remodelling genes. Nat Commun 2014;5:5006.

99. Hussein YR, Broaddus R, Weigelt B, et al. The genomic heterogeneity of FIGO grade 3 endometrioid carcinoma impacts diagnostic accuracy and reproducibility. Int J Gynecol Pathol 2016;35(1): 16–24.

100. Yemelyanova A, Ji H, Shih Ie M, et al. Utility of p16 expression for distinction of uterine serous

carcinomas from endometrial endometrioid and endocervical adenocarcinomas: immunohisto-chemical analysis of 201 cases. Am J Surg Pathol 2009;33(10):1504–14.

101. Alkushi A, Kobel M, Kalloger SE, et al. High-grade endometrial carcinoma: serous and grade 3 endometrioid carcinomas have different immunophenotypes and outcomes. Int J Gynecol Pathol 2010; 29(4):343–50.

102. Chiesa-Vottero AG, Malpica A, Deavers MT, et al. Immunohistochemical overexpression of p16 and p53 in uterine serous carcinoma and ovarian high-grade serous carcinoma. Int J Gynecol Pathol 2007;26(3):328–33.

103. Schultheis AM, Martelotto LG, De Filippo MR, et al. TP53 Mutational spectrum in endometrioid and serous endometrial cancers. Int J Gynecol Pathol 2016;35(4):289–300.

104. Djordjevic B, Hennessy BT, Li J, et al. Clinical assessment of PTEN loss in endometrial carcinoma: immunohistochemistry outperforms gene sequencing. Mod Pathol 2012;25(5):699–708.

105. Talhouk A, McConechy MK, Leung S, et al. A clinically applicable molecular-based classification for endometrial cancers. Br J Cancer 2015; 113(2):299–310.

106. Kobel M, Meng B, Hoang LN, et al. Molecular analysis of mixed endometrial carcinomas shows clonality in most cases. Am J Surg Pathol 2016;40(2): 166–80.

107. Lawrenson K, Pakzamir E, Liu B, et al. Molecular analysis of mixed endometrioid and serous adenocarcinoma of the endometrium. PLoS One 2015; 10(7):e0130909.

108. Trovik J, Wik E, Werner HM, et al. Hormone receptor loss in endometrial carcinoma curettage predicts lymph node metastasis and poor outcome in prospective multicentre trial. Eur J Cancer 2013;49(16):3431–41.

109. Suthipintawong C, Wejaranayang C, Vipupinyo C. Prognostic significance of ER, PR, Ki67, c-erbB-2, and p53 in endometrial carcinoma. J Med Assoc Thai 2008;91(12):1779–84.

110. Huvila J, Talve L, Carpen O, et al. Progesterone receptor negativity is an independent risk factor for relapse in patients with early stage endometrioid endometrial adenocarcinoma. Gynecol Oncol 2013;130(3):463–9.

111. Singh M, Zaino RJ, Filiaci VJ, et al. Relationship of estrogen and progesterone receptors to clinical outcome in metastatic endometrial carcinoma: a Gynecologic Oncology Group Study. Gynecol Oncol 2007;106(2):325–33.

112. Zhang Y, Zhao D, Gong C, et al. Prognostic role of hormone receptors in endometrial cancer: a systematic review and meta-analysis. World J Surg Oncol 2015;13:208.

113. Lapinska-Szumczyk SM, Supernat AM, Majewska HI, et al. Immunohistochemical characterisation of molecular subtypes in endometrial cancer. Int J Clin Exp Med 2015;8(11):21981–90.

114. Lapinska-Szumczyk S, Supernat A, Majewska H, et al. HER2-positive endometrial cancer subtype carries poor prognosis. Clin Transl Sci 2014;7(6): 482–8.

115. Catasus L, Gallardo A, Cuatrecasas M, et al. Concomitant PI3K-AKT and p53 alterations in endometrial carcinomas are associated with poor prognosis. Mod Pathol 2009;22(4):522–9.

116. McIntyre JB, Nelson GS, Ghatage P, et al. PIK3CA missense mutation is associated with unfavorable outcome in grade 3 endometrioid carcinoma but not in serous endometrial carcinoma. Gynecol Oncol 2014;132(1):188–93.

117. Cancer Genome Atlas Research N. Integrated genomic analyses of ovarian carcinoma. Nature 2011;474(7353):609–15.

118. Nakayama K, Rahman MT, Rahman M, et al. CCNE1 amplification is associated with aggressive potential in endometrioid endometrial carcinomas. Int J Oncol 2016;48(2):506–16.

119. Yamazawa K, Hirai M, Fujito A, et al. Fertility-preserving treatment with progestin, and pathological criteria to predict responses, in young women with endometrial cancer. Hum Reprod 2007;22(7): 1953–8.

120. Janzen DM, Rosales MA, Paik DY, et al. Progesterone receptor signaling in the microenvironment of endometrial cancer influences its response to hormonal therapy. Cancer Res 2013;73(15):4697–710.

121. Stelloo E, Bosse T, Nout RA, et al. Refining prognosis and identifying targetable pathways for high-risk endometrial cancer; a TransPORTEC initiative. Mod Pathol 2015;28(6):836–44.

122. Yu Y, Savage RE, Eathiraj S, et al. Targeting AKT1-E17K and the PI3K/AKT pathway with an allosteric AKT inhibitor, ARQ 092. PLoS One 2015;10(10): e0140479.

123. Santin AD, Bellone S, Roman JJ, et al. Trastuzumab treatment in patients with advanced or recurrent endometrial carcinoma overexpressing HER2/neu. Int J Gynaecol Obstet 2008; 102(2):128–31.

124. Jewell E, Secord AA, Brotherton T, et al. Use of trastuzumab in the treatment of metastatic endometrial cancer. Int J Gynecol Cancer 2006;16(3): 1370–3.

125. Villella JA, Cohen S, Smith DH, et al. HER-2/neu overexpression in uterine papillary serous cancers and its possible therapeutic implications. Int J Gynecol Cancer 2006;16(5):1897–902.

126. Fleming GF, Sill MW, Darcy KM, et al. Phase II trial of trastuzumab in women with advanced or recurrent, HER2-positive endometrial carcinoma: a

Gynecologic Oncology Group study. Gynecol Oncol 2010;116(1):15–20.

127. Leslie KK, Sill MW, Lankes HA, et al. Lapatinib and potential prognostic value of EGFR mutations in a Gynecologic Oncology Group phase II trial of persistent or recurrent endometrial cancer. Gynecol Oncol 2012;127(2):345–50.

128. Bonadona V, Bonaiti B, Olschwang S, et al. Cancer risks associated with germline mutations in MLH1, MSH2, and MSH6 genes in Lynch syndrome. JAMA 2011;305(22):2304–10.

129. Lynch HT, Lynch PM, Lanspa SJ, et al. Review of the Lynch syndrome: history, molecular genetics, screening, differential diagnosis, and medicolegal ramifications. Clin Genet 2009;76(1):1–18.

130. Aarnio M, Sankila R, Pukkala E, et al. Cancer risk in mutation carriers of DNA-mismatch-repair genes. Int J Cancer 1999;81(2):214–8.

131. Vasen HF, Watson P, Mecklin JP, et al. New clinical criteria for hereditary nonpolyposis colorectal cancer (HNPCC, Lynch syndrome) proposed by the International Collaborative group on HNPCC. Gastroenterology 1999;116(6):1453–6.

132. Win AK, Young JP, Lindor NM, et al. Colorectal and other cancer risks for carriers and noncarriers from families with a DNA mismatch repair gene mutation: a prospective cohort study. J Clin Oncol 2012;30(9):958–64.

133. Kempers MJ, Kuiper RP, Ockeloen CW, et al. Risk of colorectal and endometrial cancers in EPCAM deletion-positive Lynch syndrome: a cohort study. Lancet Oncol 2011;12(1):49–55.

134. Ligtenberg MJ, Kuiper RP, Geurts van Kessel A, et al. EPCAM deletion carriers constitute a unique subgroup of Lynch syndrome patients. Fam Cancer 2013;12(2):169–74.

135. Tan MH, Mester JL, Ngeow J, et al. Lifetime cancer risks in individuals with germline PTEN mutations. Clin Cancer Res 2012;18(2):400–7.

136. Ngeow J, Stanuch K, Mester JL, et al. Second malignant neoplasms in patients with Cowden syndrome with underlying germline PTEN mutations. J Clin Oncol 2014;32(17):1818–24.

137. Daniels MS. Genetic testing by cancer site: uterus. Cancer J 2012;18(4):338–42.

138. Mills AM, Liou S, Ford JM, et al. Lynch syndrome screening should be considered for all patients with newly diagnosed endometrial cancer. Am J Surg Pathol 2014;38(11):1501–9.

139. Ferguson SE, Aronson M, Pollett A, et al. Performance characteristics of screening strategies for Lynch syndrome in unselected women with newly diagnosed endometrial cancer who have undergone universal germline mutation testing. Cancer 2014;120(24):3932–9.

140. Egoavil C, Alenda C, Castillejo A, et al. Prevalence of Lynch syndrome among patients with newly diagnosed endometrial cancers. PLoS One 2013; 8(11):e79737.

141. Berends MJ, Wu Y, Sijmons RH, et al. Toward new strategies to select young endometrial cancer patients for mismatch repair gene mutation analysis. J Clin Oncol 2003;21(23): 4364–70.

142. Lu KH, Schorge JO, Rodabaugh KJ, et al. Prospective determination of prevalence of lynch syndrome in young women with endometrial cancer. J Clin Oncol 2007;25(33):5158–64.

143. Shih KK, Garg K, Levine DA, et al. Clinicopathologic significance of DNA mismatch repair protein defects and endometrial cancer in women 40 years of age and younger. Gynecol Oncol 2011;123(1): 88–94.

144. Burleigh A, Talhouk A, Gilks CB, et al. Clinical and pathological characterization of endometrial cancer in young women: identification of a cohort without classical risk factors. Gynecol Oncol 2015;138(1):141–6.

145. Lu KH, Broaddus RR. Gynecologic cancers in lynch syndrome/HNPCC. Fam Cancer 2005;4(3): 249–54.

146. Lu KH, Dinh M, Kohlmann W, et al. Gynecologic cancer as a "sentinel cancer" for women with hereditary nonpolyposis colorectal cancer syndrome. Obstet Gynecol 2005;105(3):569–74.

147. Stoffel E, Mukherjee B, Raymond VM, et al. Calculation of risk of colorectal and endometrial cancer among patients with Lynch syndrome. Gastroenterology 2009;137(5):1621–7.

148. Stoffel EM, Mangu PB, Gruber SB, et al. Hereditary colorectal cancer syndromes: American Society of Clinical Oncology Clinical Practice Guideline endorsement of the familial risk-colorectal cancer: European Society for Medical Oncology Clinical Practice Guidelines. J Clin Oncol 2015;33(2): 209–17.

149. Stoffel EM, Mangu PB, Limburg PJ. Hereditary colorectal cancer syndromes: American Society of Clinical Oncology clinical practice guideline endorsement of the familial risk-colorectal cancer: European Society for Medical Oncology clinical practice guidelines. J Oncol Pract 2015;11(3): e437–41.

150. Evaluation of Genomic Applications in Practice and Prevention (EGAPP) Working Group. Recommendations from the EGAPP Working Group: genetic testing strategies in newly diagnosed individuals with colorectal cancer aimed at reducing morbidity and mortality from Lynch syndrome in relatives. Genet Med 2009;11(1):35–41.

151. Goodfellow PJ, Billingsley CC, Lankes HA, et al. Combined microsatellite instability, MLH1 methylation analysis, and immunohistochemistry for Lynch syndrome screening in endometrial cancers

from GOG210: an NRG Oncology and Gynecologic Oncology Group study. J Clin Oncol 2015;33(36): 4301–8.

152. Rabban JT, Calkins SM, Karnezis AN, et al. Association of tumor morphology with mismatch-repair protein status in older endometrial cancer patients: implications for universal versus selective screening strategies for Lynch syndrome. Am J Surg Pathol 2014;38(6):793–800.

153. Kalloger SE, Allo G, Mulligan AM, et al. Use of mismatch repair immunohistochemistry and microsatellite instability testing: exploring Canadian practices. Am J Surg Pathol 2012;36(4): 560–9.

154. Lu KH, Ring KL. One size may not fit all: the debate of universal tumor testing for Lynch syndrome. Gynecol Oncol 2015;137(1):2–3.

155. Mills AM, Longacre TA. Lynch syndrome screening in the gynecologic tract: current state of the art. Am J Surg Pathol 2016;40(4):e35–44.

156. Lancaster JM, Powell CB, Kauff ND, et al. Society of Gynecologic Oncologists Education Committee statement on risk assessment for inherited gynecologic cancer predispositions. Gynecol Oncol 2007; 107(2):159–62.

157. Walsh CS, Blum A, Walts A, et al. Lynch syndrome among gynecologic oncology patients meeting Bethesda guidelines for screening. Gynecol Oncol 2010;116(3):516–21.

158. Garg K, Shih K, Barakat R, et al. Endometrial carcinomas in women aged 40 years and younger: tumors associated with loss of DNA mismatch repair proteins comprise a distinct clinicopathologic subset. Am J Surg Pathol 2009;33(12): 1869–77.

159. Garg K, Leitao MM Jr, Kauff ND, et al. Selection of endometrial carcinomas for DNA mismatch repair protein immunohistochemistry using patient age and tumor morphology enhances detection of mismatch repair abnormalities. Am J Surg Pathol 2009;33(6):925–33.

160. Shia J, Holck S, Depetris G, et al. Lynch syndrome-associated neoplasms: a discussion on histopathology and immunohistochemistry. Fam Cancer 2013;12(2):241–60.

161. Westin SN, Lacour RA, Urbauer DL, et al. Carcinoma of the lower uterine segment: a newly described association with Lynch syndrome. J Clin Oncol 2008;26(36):5965–71.

162. Broaddus RR, Lynch HT, Chen LM, et al. Pathologic features of endometrial carcinoma associated with HNPCC: a comparison with sporadic endometrial carcinoma. Cancer 2006;106(1):87–94.

163. Bruegl AS, Djordjevic B, Batte B, et al. Evaluation of clinical criteria for the identification of Lynch syndrome among unselected patients with endometrial cancer. Cancer Prev Res (Phila) 2014;7(7):686–97.

164. Moline J, Mahdi H, Yang B, et al. Implementation of tumor testing for lynch syndrome in endometrial cancers at a large academic medical center. Gynecol Oncol 2013;130(1):121–6.

165. Ring KL, Connor EV, Atkins KA, et al. Women 50 years or younger with endometrial cancer: the argument for universal mismatch repair screening and potential for targeted therapeutics. Int J Gynecol Cancer 2013;23(5):853–60.

166. Backes FJ, Leon ME, Ivanov I, et al. Prospective evaluation of DNA mismatch repair protein expression in primary endometrial cancer. Gynecol Oncol 2009;114(3):486–90.

167. Batte BA, Bruegl AS, Daniels MS, et al. Consequences of universal MSI/IHC in screening ENDOMETRIAL cancer patients for Lynch syndrome. Gynecol Oncol 2014;134(2):319–25.

168. Bruegl AS, Djordjevic B, Urbauer DL, et al. Utility of MLH1 methylation analysis in the clinical evaluation of Lynch syndrome in women with endometrial cancer. Curr Pharm Des 2014;20(11):1655–63.

169. Tafe LJ, Riggs ER, Tsongalis GJ. Lynch syndrome presenting as endometrial cancer. Clin Chem 2014;60(1):111–21.

170. Clarke BA, Cooper K. Identifying Lynch syndrome in patients with endometrial carcinoma: shortcomings of morphologic and clinical schemas. Adv Anat Pathol 2012;19(4):231–8.

171. Buchanan DD, Tan YY, Walsh MD, et al. Tumor mismatch repair immunohistochemistry and DNA MLH1 methylation testing of patients with endometrial cancer diagnosed at age younger than 60 years optimizes triage for population-level germline mismatch repair gene mutation testing. J Clin Oncol 2014;32(2):90–100.

172. Leenen CH, van Lier MG, van Doorn HC, et al. Prospective evaluation of molecular screening for Lynch syndrome in patients with endometrial cancer ≤70 years. Gynecol Oncol 2012;125(2): 414–20.

173. Mojtahed A, Schrijver I, Ford JM, et al. A two-antibody mismatch repair protein immunohistochemistry screening approach for colorectal carcinomas, skin sebaceous tumors, and gynecologic tract carcinomas. Mod Pathol 2011;24(7): 1004–14.

174. Shia J, Tang LH, Vakiani E, et al. Immunohistochemistry as first-line screening for detecting colorectal cancer patients at risk for hereditary nonpolyposis colorectal cancer syndrome: a 2-antibody panel may be as predictive as a 4-antibody panel. Am J Surg Pathol 2009;33(11):1639–45.

175. Pai RK, Plesec TP, Abdul-Karim FW, et al. Abrupt loss of MLH1 and PMS2 expression in endometrial

carcinoma: molecular and morphologic analysis of 6 cases. Am J Surg Pathol 2015;39(7):993–9.

176. Mittal KR, Chen F, Wei JJ, et al. Molecular and immunohistochemical evidence for the origin of uterine leiomyosarcomas from associated leiomyoma and symplastic leiomyoma-like areas. Mod Pathol 2009;22(10):1303–11.

177. Mittal K, Popiolek D, Demopoulos RI. Uterine myxoid leiomyosarcoma within a leiomyoma. Hum Pathol 2000;31(3):398–400.

178. Quade BJ, Pinto AP, Howard DR, et al. Frequent loss of heterozygosity for chromosome 10 in uterine leiomyosarcoma in contrast to leiomyoma. Am J Pathol 1999;154(3):945–50.

179. Levy B, Mukherjee T, Hirschhorn K. Molecular cytogenetic analysis of uterine leiomyoma and leiomyosarcoma by comparative genomic hybridization. Cancer Genet Cytogenet 2000; 121(1):1–8.

180. Hu J, Khanna V, Jones M, et al. Genomic alterations in uterine leiomyosarcomas: potential markers for clinical diagnosis and prognosis. Genes Chromosomes Cancer 2001;31(2):117–24.

181. Packenham JP, du Manoir S, Schrock E, et al. Analysis of genetic alterations in uterine leiomyomas and leiomyosarcomas by comparative genomic hybridization. Mol Carcinog 1997;19(4):273–9.

182. Zhang Q, Ubago J, Li L, et al. Molecular analyses of 6 different types of uterine smooth muscle tumors: emphasis in atypical leiomyoma. Cancer 2014;120(20):3165–77.

183. Agaram NP, Zhang L, LeLoarer F, et al. Targeted exome sequencing profiles genetic alterations in leiomyosarcoma. Genes Chromosomes Cancer 2016;55(2):124–30.

184. Liang Y, Zhang X, Chen X, et al. Diagnostic value of progesterone receptor, p16, p53 and pHH3 expression in uterine atypical leiomyoma. Int J Clin Exp Pathol 2015;8(6):7196–202.

185. Mills AM, Ly A, Balzer BL, et al. Cell cycle regulatory markers in uterine atypical leiomyoma and leiomyosarcoma: immunohistochemical study of 68 cases with clinical follow-up. Am J Surg Pathol 2013;37(5):634–42.

186. Bertsch E, Qiang W, Zhang Q, et al. MED12 and HMGA2 mutations: two independent genetic events in uterine leiomyoma and leiomyosarcoma. Mod Pathol 2014;27(8):1144–53.

187. Ravegnini G, Marino-Enriquez A, Slater J, et al. MED12 mutations in leiomyosarcoma and extrauterine leiomyoma. Mod Pathol 2013;26(5):743–9.

188. Liau JY, Tsai JH, Jeng YM, et al. Leiomyosarcoma with alternative lengthening of telomeres is associated with aggressive histologic features, loss of ATRX expression, and poor clinical outcome. Am J Surg Pathol 2015;39(2):236–44.

189. Perot G, Croce S, Ribeiro A, et al. MED12 alterations in both human benign and malignant uterine soft tissue tumors. PLoS One 2012;7(6):e40015.

190. Kampjarvi K, Makinen N, Kilpivaara O, et al. Somatic MED12 mutations in uterine leiomyosarcoma and colorectal cancer. Br J Cancer 2012;107(10):1761–5.

191. Matsubara A, Sekine S, Yoshida M, et al. Prevalence of MED12 mutations in uterine and extrauterine smooth muscle tumours. Histopathology 2013;62(4):657–61.

192. Markowski DN, Huhle S, Nimzyk R, et al. MED12 mutations occurring in benign and malignant mammalian smooth muscle tumors. Genes Chromosomes Cancer 2013;52(3):297–304.

193. Liau JY, Lee JC, Tsai JH, et al. Comprehensive screening of alternative lengthening of telomeres phenotype and loss of ATRX expression in sarcomas. Mod Pathol 2015;28(12):1545–54.

194. Yang CY, Liau JY, Huang WJ, et al. Targeted next-generation sequencing of cancer genes identified frequent TP53 and ATRX mutations in leiomyosarcoma. Am J Transl Res 2015;7(10):2072–81.

195. Hrzenjak A, Moinfar F, Tavassoli FA, et al. JAZF1/JJAZ1 gene fusion in endometrial stromal sarcomas: molecular analysis by reverse transcriptase-polymerase chain reaction optimized for paraffin-embedded tissue. J Mol Diagn 2005; 7(3):388–95.

196. Nucci MR, Harburger D, Koontz J, et al. Molecular analysis of the JAZF1-JJAZ1 gene fusion by RT-PCR and fluorescence in situ hybridization in endometrial stromal neoplasms. Am J Surg Pathol 2007; 31(1):65–70.

197. Koontz JI, Soreng AL, Nucci M, et al. Frequent fusion of the JAZF1 and JJAZ1 genes in endometrial stromal tumors. Proc Natl Acad Sci U S A 2001;98(11):6348–53.

198. Micci F, Walter CU, Teixeira MR, et al. Cytogenetic and molecular genetic analyses of endometrial stromal sarcoma: nonrandom involvement of chromosome arms 6p and 7p and confirmation of JAZF1/JJAZ1 gene fusion in t(7;17). Cancer Genet Cytogenet 2003;144(2):119–24.

199. Huang HY, Ladanyi M, Soslow RA. Molecular detection of JAZF1-JJAZ1 gene fusion in endometrial stromal neoplasms with classic and variant histology: evidence for genetic heterogeneity. Am J Surg Pathol 2004;28(2):224–32.

200. Oliva E, de Leval L, Soslow RA, et al. High frequency of JAZF1-JJAZ1 gene fusion in endometrial stromal tumors with smooth muscle differentiation by interphase FISH detection. Am J Surg Pathol 2007;31(8):1277–84.

201. Chiang S, Ali R, Melnyk N, et al. Frequency of known gene rearrangements in endometrial stromal tumors. Am J Surg Pathol 2011;35(9):1364–72.

202. Stewart CJ, Leung YC, Murch A, et al. Evaluation of fluorescence in-situ hybridization in monomorphic endometrial stromal neoplasms and their histological mimics: a review of 49 cases. Histopathology 2014;65(4):473–82.

203. Sreekantaiah C, Li FP, Weidner N, et al. An endometrial stromal sarcoma with clonal cytogenetic abnormalities. Cancer Genet Cytogenet 1991; 55(2):163–6.

204. Dal Cin P, Aly MS, De Wever I, et al. Endometrial stromal sarcoma t(7;17)(p15-21;q12-21) is a nonrandom chromosome change. Cancer Genet Cytogenet 1992;63(1):43–6.

205. Micci F, Panagopoulos I, Bjerkehagen B, et al. Consistent rearrangement of chromosomal band 6p21 with generation of fusion genes JAZF1/ PHF1 and EPC1/PHF1 in endometrial stromal sarcoma. Cancer Res 2006;66(1):107–12.

206. Panagopoulos I, Mertens F, Griffin CA. An endometrial stromal sarcoma cell line with the JAZF1/PHF1 chimera. Cancer Genet Cytogenet 2008;185(2):74–7.

207. D'Angelo E, Ali RH, Espinosa I, et al. Endometrial stromal sarcomas with sex cord differentiation are associated with PHF1 rearrangement. Am J Surg Pathol 2013;37(4):514–21.

208. Panagopoulos I, Micci F, Thorsen J, et al. Novel fusion of MYST/Esa1-associated factor 6 and PHF1 in endometrial stromal sarcoma. PLoS One 2012;7(6):e39354.

209. Micci F, Gorunova L, Gatius S, et al. MEAF6/PHF1 is a recurrent gene fusion in endometrial stromal sarcoma. Cancer Lett 2014;347(1):75–8.

210. Ali RH, Al-Safi R, Al-Waheeb S, et al. Molecular characterization of a population-based series of endometrial stromal sarcomas in Kuwait. Hum Pathol 2014;45(12):2453–62.

211. Panagopoulos I, Thorsen J, Gorunova L, et al. Fusion of the ZC3H7B and BCOR genes in endometrial stromal sarcomas carrying an X;22-translocation. Genes Chromosomes Cancer 2013;52(7):610–8.

212. Dewaele B, Przybyl J, Quattrone A, et al. Identification of a novel, recurrent MBTD1-CXorf67 fusion in low-grade endometrial stromal sarcoma. Int J Cancer 2014;134(5):1112–22.

213. Chiang S, Oliva E. Cytogenetic and molecular aberrations in endometrial stromal tumors. Hum Pathol 2011;42(5):609–17.

214. Lee CH, Nucci MR. Endometrial stromal sarcoma– the new genetic paradigm. Histopathology 2015; 67(1):1–19.

215. Lee CH, Ou WB, Marino-Enriquez A, et al. 14-3-3 fusion oncogenes in high-grade endometrial stromal sarcoma. Proc Natl Acad Sci U S A 2012; 109(3):929–34.

216. Sciallis AP, Bedroske PP, Schoolmeester JK, et al. High-grade endometrial stromal sarcomas: a clinicopathologic study of a group of tumors with heterogenous morphologic and genetic features. Am J Surg Pathol 2014;38(9):1161–72.

217. Lee CH, Ali RH, Rouzbahman M, et al. Cyclin D1 as a diagnostic immunomarker for endometrial stromal sarcoma with YWHAE-FAM22 rearrangement. Am J Surg Pathol 2012;36(10):1562–70.

218. Isphording A, Ali RH, Irving J, et al. YWHAE-FAM22 endometrial stromal sarcoma: diagnosis by reverse transcription-polymerase chain reaction in formalin-fixed, paraffin-embedded tumor. Hum Pathol 2013;44(5):837–43.

219. Croce S, Hostein I, Ribeiro A, et al. YWHAE rearrangement identified by FISH and RT-PCR in endometrial stromal sarcomas: genetic and pathological correlations. Mod Pathol 2013;26(10): 1390–400.

220. O'Meara E, Stack D, Lee CH, et al. Characterization of the chromosomal translocation t(10;17)(q22;p13) in clear cell sarcoma of kidney. J Pathol 2012;227(1):72–80.

221. Fehr A, Hansson MC, Kindblom LG, et al. YWHAE-FAM22 gene fusion in clear cell sarcoma of the kidney. J Pathol 2012;227(4):e5–7.

222. Mirkovic J, Calicchio M, Fletcher CD, et al. Diffuse and strong cyclin D1 immunoreactivity in clear cell sarcoma of the kidney. Histopathology 2015;67(3): 306–12.

223. Shah VI, McCluggage WG. Cyclin D1 does not distinguish YWHAE-NUTM2 high-grade endometrial stromal sarcoma from undifferentiated endometrial carcinoma. Am J Surg Pathol 2015;39(5): 722–4.

224. Howitt BE, Sholl LM, Dal Cin P, et al. Targeted genomic analysis of Mullerian adenosarcoma. J Pathol 2015;235(1):37–49.

225. Piscuoglio S, Burke KA, Ng CK, et al. Uterine adenosarcomas are mesenchymal neoplasms. J Pathol 2016;238(3):381–8.

226. Kurihara S, Oda Y, Ohishi Y, et al. Endometrial stromal sarcomas and related high-grade sarcomas: immunohistochemical and molecular genetic study of 31 cases. Am J Surg Pathol 2008;32(8):1228–38.

Molecular Evaluation of Colorectal Adenocarcinoma
Current Practice and Emerging Concepts

Jonathan A. Nowak, MD, PhD*, Jason L. Hornick, MD, PhD

KEYWORDS

- Colorectal cancer • Targeted therapy • Lynch syndrome • Microsatellite instability
- Mismatch repair • EGFR pathway • Immune checkpoint blockade • PD-L1

Key points

- Molecular testing is a standard component of the routine pathologic evaluation of colorectal carcinoma.
- Assessment of mismatch repair pathway status in colorectal carcinoma provides prognostic information, can guide therapeutic decisions, and serves as an effective method of identifying patients with Lynch syndrome.
- Multiple assays, including immunohistochemistry, microsatellite instability testing, promoter methylation, and sequencing are used to assess mismatch repair pathway status.
- Evaluation of mutations in downstream components of the epidermal growth factor receptor (EGFR) signaling pathway is required to determine which patients with metastatic disease will benefit from targeted anti-EGFR therapy.
- Advances in colorectal carcinoma molecular diagnostics will help refine patient selection for targeted therapies and may enable better disease monitoring.

ABSTRACT

Molecular testing in colorectal cancer helps to address multiple clinical needs. Evaluating the mismatch repair pathway status is the most common use for molecular diagnostics and this testing provides prognostic information, guides therapeutic decisions and helps identify Lynch syndrome patients. For patients with metastatic colorectal cancer, testing for activating mutations in downstream components of the EGFR signaling pathway can identify patients who will benefit from anti-EGFR therapy. Emerging molecular tests for colorectal cancer will help further refine patient selection for targeted therapies and may provide new options for monitoring disease recurrence and the development of treatment resistance.

OVERVIEW

Colorectal carcinoma (CRC) is one of the single most common cancers in both men and women, and is also one of the largest overall contributors to cancer-associated mortality.[1] After lung cancer, it is the single most common solid tumor for which molecular testing is routinely used in clinical practice. CRC has one of the best understood mechanisms of molecular pathogenesis of all solid tumors. In the 25 years since the original stepwise model of CRC pathogenesis was proposed, the

The authors have no financial interests to disclose.

Department of Pathology, Brigham and Women's Hospital, 75 Francis Street, Boston, MA 02115, USA

* Corresponding author.

E-mail address: janowak@partners.org

comprehensive genetic landscape of colorectal cancer has come into focus, greatly aided by modern molecular techniques.[2–4] In this context, molecular evaluation of CRC has been a part of routine pathology practice for nearly 15 years, and addresses multiple clinical needs, including prognostication, therapeutic guidance, and identification of inherited cancer predisposition syndromes. This review focuses on currently accepted molecular testing for CRC, while also briefly highlighting future areas of testing.

As a framework for understanding molecular testing in CRC, it is helpful to recognize that there are 2 generally separate genetic pathways to colorectal carcinogenesis.[5] The classic stepwise sequence, known as the chromosomal instability pathway, is characterized by frequent mutations in the KRAS, APC, and TP53 genes, and is typically associated with chromosomal aneuploidy. In contrast, tumors that arise from the hypermutable pathway, characterized by deficient mismatch repair (MMR) activity, do not have gross chromosomal rearrangements and losses, but rather accumulate many missense mutations and relatively small deletions and insertions in specific types of DNA sequences. Determining which mechanism of genetic instability gave rise to a CRC has inherent prognostic value and also provides the necessary context for interpreting additional molecular alterations.

MISMATCH REPAIR–DEFICIENT TUMORS AND LYNCH SYNDROME

Overall, approximately 15% of CRCs arise from the MMR-deficient hypermutable pathway. These cancers arise in 2 different settings. Approximately 3% to 4% of CRCs arise in the context of Lynch syndrome, an autosomal dominant inherited cancer predisposition syndrome, whereas the remaining 12% arise through sporadic inactivation of the MMR pathway via somatic mutations or epigenetic mechanisms.[6] The MMR pathway primarily functions to repair errors introduced during DNA replication. Four core proteins, MSH2, MSH6, MLH1, and PMS2, are essential for the proper function of this pathway, and germline mutations in any of the 4 corresponding genes can give rise to Lynch syndrome. Additionally, germline deletions involving the EPCAM gene, which is located upstream of the MSH2 gene, can lead to methylation of the MSH2 promoter and subsequent silencing of gene expression. Patients with Lynch syndrome inherit only a single functional copy of 1 of the 4 core MMR genes. Subsequent inactivation of the remaining functional allele can therefore

lead to aberrant accumulation of somatic pathogenic mutations and drive carcinogenesis.

Although Lynch syndrome is most commonly associated with a predisposition for developing colorectal carcinoma, it actually represents a predisposition to develop a wide variety of cancers, including endometrial and ovarian, gastric, small bowel, and urinary tract cancers.[7] Patients with Lynch syndrome also have an elevated risk of developing glioblastomas, an association recognized as Turcot syndrome, and sebaceous neoplasms, referred to as Muir-Torre syndrome. Less robust associations have also been reported with pancreatic, prostate, and breast cancer. Because of these elevated risks, patients with Lynch syndrome should undergo enhanced cancer screening and the relatives of patients with Lynch syndrome also should be tested to determine if they are germline mutation carriers as well.

Rarely, patients may inherit 2 defective copies of a single MMR gene, resulting in Constitutional Mismatch Repair Deficiency syndrome.[8,9] Individuals with this condition are predisposed to develop CRC at a much younger age than patients with Lynch syndrome, frequently develop café-au-lait macules, and are at risk for a slightly different spectrum of cancers than patients with Lynch syndrome.

Since the molecular basis for Lynch syndrome was elucidated in the early 1990s, several systems have been developed to identify patients with Lynch syndrome. The Amsterdam criteria were developed primarily for research use and are based solely on clinical criteria.[10] Although assessment of these criteria may help raise suspicion for Lynch syndrome, they are insufficiently sensitive to serve as a reliable method for identifying all patients with Lynch syndrome. The Bethesda guidelines were developed to help identify which cases of CRC should be evaluated for MMR deficiency and rely on both clinical and histologic data.[11] Notably, cancers that arise due to MMR pathway deficiency often have characteristic pathologic features (Fig. 1) and often induce a prominent lymphocytic inflammatory response that occurs in several distinctive patterns (Fig. 2). Although the presence of these histologic features may be suggestive of MMR deficiency, histology alone is neither specific nor sensitive enough to be used for Lynch syndrome identification. Furthermore, reliance on clinical criteria to identify patients with Lynch syndrome can be challenging in everyday practice due to incomplete knowledge of family pedigrees. Even when the criteria are appropriately evaluated using complete data, the Revised Bethesda Guidelines may fail to detect up to 25% of patients with Lynch syndrome. In

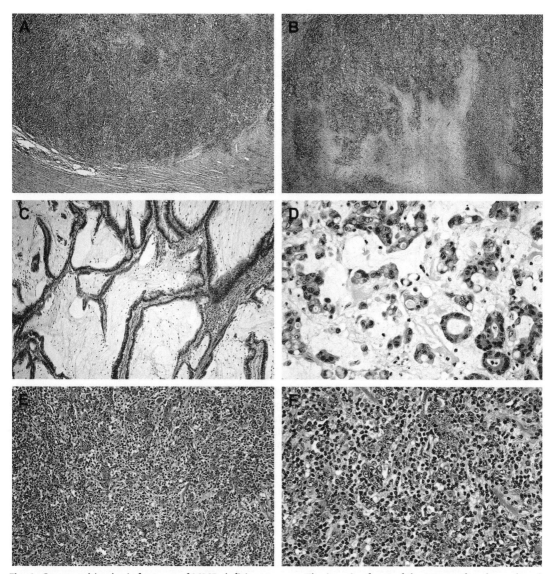

Fig. 1. Common histologic features of MMR-deficient tumors. The invasive front of the tumor often has a sharply delineated, broad "pushing" border as shown here extending into the muscularis propria (*A*). Often grossly visible, tumor necrosis is common in MMR-deficient tumors (*B*). Some MMR-deficient tumors display a mucinous morphology, which may variably manifest as dilated, mucin-filled glandular spaces (*C*) or a prominent signet-ring cell component (*D*). Other MMR-deficient tumors display a medullary morphology, with tumor cells arranged in sheets and cords and minimal gland formation (*E*). Medullary tumors are typified by a prominent intraepithelial lymphocytic infiltrate (*F*).

2009, a study from the Centers for Disease Control and Prevention genomics working group determined that there was sufficient evidence to warrant testing all new CRCs for MMR deficiency so as to identify all patients with Lynch syndrome.[12] Although all health care institutions have not yet adopted universal testing, it has gained broad acceptance and is emerging as a standard of care because of the additional prognostic and therapeutic guidance it provides.

MISMATCH REPAIR PROTEIN EXPRESSION

The mainstay of Lynch syndrome testing at most institutions is immunohistochemical evaluation of MMR protein expression. The 4 core MMR proteins are normally highly expressed in replicating cells, and their presence is usually indicative of a functionally proficient MMR system. Conversely, the absence of MMR protein expression is highly

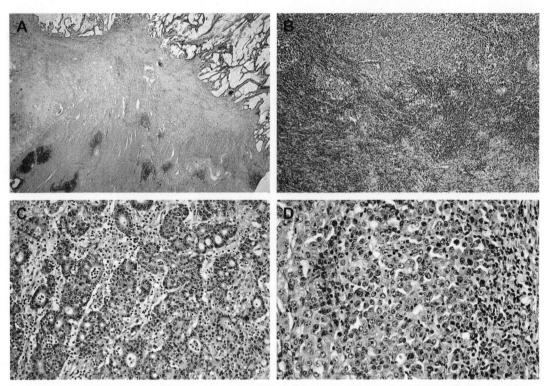

Fig. 2. Common inflammatory infiltrate patterns in MMR-deficient tumors. A Crohn's disease–like lymphoid reaction, characterized by discrete, well-formed lymphoid aggregates surrounding the tumor but separated from the invasive front (*A*). Peritumoral lymphoid reaction, characterized by a lymphocytic infiltrate directly associated with the invasive tumor front (*B*). Intratumoral periglandular reaction, characterized by lymphocytes within the tumor stroma surrounding neoplastic glands (*C*). Tumor-infiltrating lymphocytes, defined as lymphocytes present within neoplastic tumor epithelium (*D*).

correlated with a defective MMR system. When loss of expression is detected, the pattern of MMR protein loss itself can indicate whether a tumor is likely to be sporadic or Lynch-associated (Table 1). Notably, the 4 core MMR proteins function as heterodimers, with pairing of MSH2 with MSH6 and MLH1 with PMS2. In the absence of appropriate partners, the MSH6 and PMS2

Table 1
Common immunohistochemical patterns of mismatch repair (MMR) protein expression and their interpretation

MLH1	PMS2	MSH2	MSH6	Interpretation
+	+	+	+	Intact MMR system • May rarely have germline point mutations or mutations in other genes
−	−	+	+	Defective MMR system, likely sporadic • Usually due to *MLH1* promoter methylation if sporadic • Usually due to *MLH1* germline mutation if Lynch-associated
+	+	−	−	Defective MMR system, likely Lynch-associated • Usually *MSH2* germline mutation
+	−	+	+	Defective MMR system, likely Lynch-associated • Usually *PMS2* germline mutation
+	+	+	−	Defective MMR system, likely Lynch-associated • Usually *MSH6* germline mutation

+, indicates intact expression; −, indicates lost expression.

proteins are unstable and rapidly degraded, leading to the apparent absence of protein expression. In contrast, expression of MSH2 and MLH1 is usually maintained even when their partners are inactivated. Concomitant loss of MLH1 and PMS2 expression, as shown in Fig. 3, is the most common pattern of MMR protein loss and is almost always associated with sporadic inactivation of the MMR system via methylation of the *MLH1* promoter, repressing transcription. Other patterns of MMR protein loss are less common and are generally indicative of Lynch syndrome, with germline inactivating mutations. Although the evaluation of MMR immunohistochemistry is usually straightforward, it is subject to a variety of technical and interpretive pitfalls (Box 1).

MICROSATELLITE INSTABILITY TESTING

In addition to evaluating MMR pathway status by immunohistochemistry for MMR proteins, the function of the repair pathway can be more directly assessed by measuring microsatellite instability. Microsatellites are short stretches of genomic DNA that contain multiple tandem copies of a mononucleotide, dinucleotide, or trinucleotide repeat motif. Due to the highly repetitive nature of these sequences, replication of these stretches during cell division is particularly error-prone. When the MMR pathway is defective, replication errors accumulate in these areas, altering the number of repeats and making their size unstable. Although microsatellites are not the only region of genomic DNA that accumulates errors when the MMR pathway is deficient, these stretches of DNA are very sensitive to MMR deficiency and are reproducibly affected when pathway functionality is lost.

Although recommendations for the number and type of microsatellites to test have varied over the years, it is now clear that mononucleotide repeat microsatellites offer the greatest sensitivity for MMR pathway deficiency.[13,14] Paired germline DNA is also tested to provide a baseline size comparison for each microsatellite (Fig. 5). Testing a panel of 5 microsatellites is currently recommended, with each microsatellite scored as either unstable or stable. By convention, if all 5

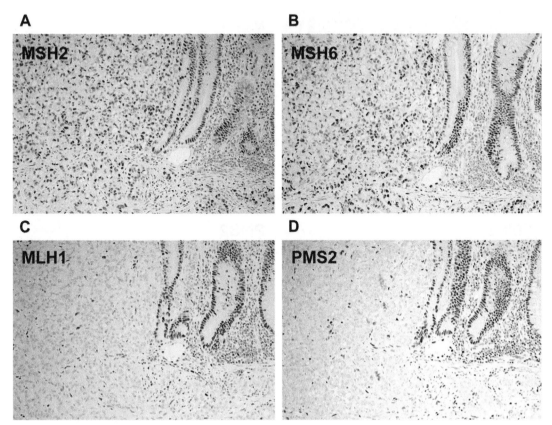

Fig. 3. Immunohistochemical evaluation of MMR protein expression showing the most common pattern of protein loss. MSH2 and MSH6 protein expression is intact in invasive adenocarcinoma (*A, B*), whereas expression of MLH1 and PMS2 proteins is lost (*C, D*). Normal colonic epithelium (*upper right* in all 4 panels), stromal cells, and inflammatory cells serve as internal controls necessary to ensure the technical adequacy of staining.

> **Box 1**
> **Technical and interpretive pitfalls in the evaluation of mismatch repair (MMR) protein immunohistochemistry**
>
> - Inadequate validation of immunohistochemistry methods and failure to maintain rigorous quality control.
> - Suboptimal specimen processing leading to a loss of MMR protein antigenicity and producing a false impression of protein expression loss.
> - Variability in MMR protein staining intensity despite optimal processing with patchy areas of faint staining that may be misinterpreted as protein loss.
> - True subclonal loss of expression, likely representing intratumoral genetic heterogeneity (Fig. 4), which may confound testing on small samples.
> - Missense mutations that occasionally give rise to intact expression of a functionally inactive protein.
> - Reduced expression of MMR proteins, particularly MSH6, after chemoradiation.

microsatellites show an identical size in both tumor and normal tissue, the tumor is classified as microsatellite stable (MSS). If 2 or more microsatellites have an altered size, the tumor is said to demonstrate high-frequency microsatellite instability and is classified as microsatellite instability-high (MSI-H). Finally, tumors that show instability in 1 of the 5 microsatellites are classified as microsatellite instability low (MSI-L). Clinically, MSI-L tumors behave like MSS tumors.

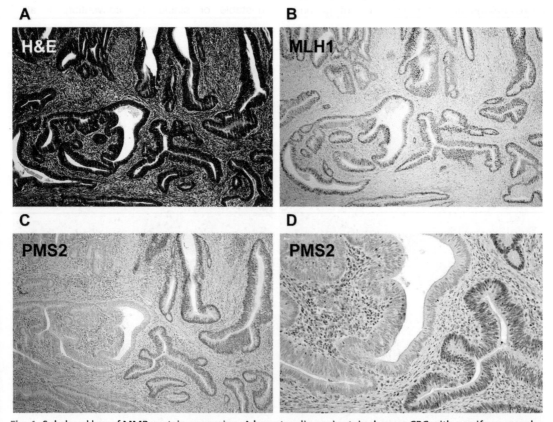

Fig. 4. Subclonal loss of MMR protein expression. A hematoxylin-eosin stain shows a CRC with a uniform morphologic appearance (*A*). The tumor had intact expression of MSH2 and MSH6 (not shown), as well as MLH1 (*B*). However, expression of PMS2 was lost in a distinct subset of tumor cells (*C*), whereas positive staining of adjacent inflammatory cells and stromal cells confirmed technical adequacy of staining (*D*). Such staining patterns likely represent subclonal loss of expression and intratumoral genetic or epigenetic heterogeneity and may not be reflective of a germline mutation.

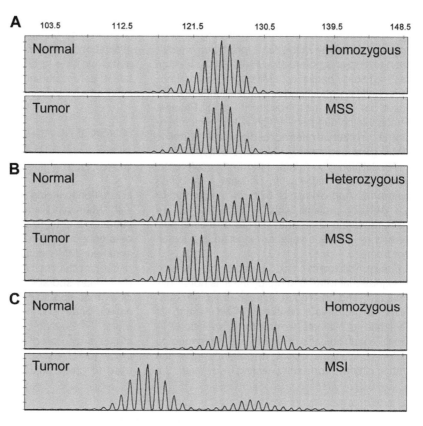

Fig. 5. Example of capillary electrophoresis tracings from polymerase chain reaction–based microsatellite instability testing. An MSS tumor from a patient homozygous at this locus displays a single peak distribution that matches the distribution from normal tissue (*A*). Similarly, an MSS tumor from a patient heterozygous at this locus displays a double-peaked distribution that matches normal tissue (*B*). In contrast, the peak distribution from a microsatellite unstable (MSI-H) tumor is shifted to the left, exposing a small peak distribution derived from admixed non-neoplastic cells and reflecting a shortening of the microsatellite region within tumor cells due to a defective MMR pathway (*C*).

MLH1 PROMOTER METHYLATION AND *BRAF* MUTATION STATUS IN MISMATCH REPAIR–DEFICIENT COLORECTAL CARCINOMA

Although MMR protein immunohistochemistry can suggest whether an MMR-deficient tumor is Lynch-associated or not, several additional molecular tests can help classify an MMR-deficient tumor as sporadic or Lynch-associated. Because MMR deficiency in sporadic tumors is nearly always due to methylation of the *MLH1* promoter, detection of *MLH1* promoter methylation in a tumor with loss of MLH1 and PMS2 protein expression virtually excludes the possibility of a Lynch-associated tumor.

Notably, the *MLH1* promoter methylation that occurs in sporadic MMR-deficient tumors is typically a manifestation of a more general process known at the CpG island methylator phenotype (CIMP). CpG islands are CG-rich stretches of DNA that occur throughout the genome and are present in approximately half of all gene promoters. In non-neoplastic cells, CpG islands located within promoters are usually unmethylated, allowing for appropriate gene expression. In CIMP-positive tumors, however, an imbalance in epigenetic regulators leads to aberrant CpG island methylation within promoters, silencing expression of key genes such as *MLH1*. Although various systems have been put forth for assessing the CIMP status of a tumor, this information does not currently have enough independent clinical utility to merit routine testing.

However, studies of additional molecular alterations in CIMP-positive tumors have revealed a very strong correlation between CIMP positivity and the presence of a V600E mutation in *BRAF*.[15] Because *MLH1* promoter methylation is a frequent manifestation of CIMP positivity, the presence of a *BRAF* V600E mutation can therefore serve as a surrogate for *MLH1* promoter methylation testing.[16] Almost all MSI-H tumors that harbor *BRAF* V600E mutations are sporadic tumors with *MLH1* promoter methylation. In contrast, MSI-H tumors with wild-type *BRAF* status have an approximately 40% chance of being associated with Lynch syndrome.

TEST INTEGRATION FOR ASSESSING MISMATCH REPAIR PATHWAY STATUS

Although both MMR immunohistochemistry and MSI testing offer excellent sensitivity for identifying patients with Lynch syndrome, neither technique is perfect, and optimal identification of patients with Lynch syndrome requires the availability of both approaches, along with *MLH1* promoter methylation testing and potentially *BRAF* V600E testing. Testing usually begins with MMR protein immunohistochemistry and incorporates additional assays based on these initial results (Fig. 6). Patients whose testing suggests a high risk of Lynch syndrome can then be referred for diagnostic germline MMR gene sequencing. Notably, acquisition of MMR deficiency is a later step in the colorectal carcinogenesis pathway in patients with Lynch syndrome. Approximately 50% of the precursor adenomas in patients with Lynch syndrome are MSS, rendering MMR pathway status testing performed on polyps unreliable for excluding Lynch syndrome.[17]

Finally, clinicians and pathologists should recognize that MMR testing, by itself, is not a "genetic test" and does not require any specific counseling or informed consent. MMR testing is designed to evaluate tumors for a specific DNA repair capability and it would be erroneous to make conclusions about a patient's germline status directly from such testing. However, the identification of MMR deficiency in a tumor that is unlikely to be sporadic should prompt further evaluation by an individual trained in medical genetics with the appropriate informed consent obtained before pursuing germline testing for Lynch syndrome.

PROGNOSTIC AND THERAPEUTIC SIGNIFICANCE OF MISMATCH REPAIR DEFICIENCY

In addition to aiding in the identification of patients with Lynch syndrome, knowledge about the MMR status of a CRC has independent prognostic and therapeutic significance. It is known that stage-matched MMR-deficient cancers that arise in patients with Lynch syndrome have a better prognosis than sporadic cancers. Similarly, it is now clear that sporadic MMR-deficient tumors also have a better prognosis than MMR-proficient tumors.[18,19] Beyond prognostic significance, MMR status is also predictive of response to 5-fluorouracil (5-FU)-based chemotherapy and can be used to guide adjuvant chemotherapy decisions

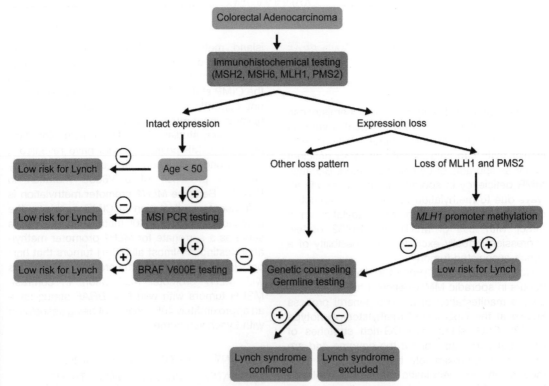

Fig. 6. An example strategy for assessing MMR pathway status in CRC and identifying patients with Lynch syndrome. Testing typically begins with MMR protein immunohistochemistry. The pattern of expression can then direct additional molecular testing if needed. Patients classified as having a high risk for Lynch syndrome are referred for genetic counseling and diagnostic germline testing.

for patients with stage II to III CRC. In contrast to MMR-proficient CRCs, MMR-deficient CRCs do not respond to 5-FU, and treating these patients with 5-FU–based therapies may actually lower their overall survival.[20,21]

MISMATCH REPAIR DEFICIENCY AND PROGRAMMED DEATH-1 INHIBITOR THERAPY

An association between MMR deficiency and high levels of tumor-associated inflammation is well established, and many studies have investigated the composition and significance of the inflammatory infiltrate in CRC.[22] More recently, this robust immune response has been postulated to be due to high levels of "non-self" neoantigens that are produced by the hypermutable phenotype of these tumors. Perhaps as an adaptive response, many tumors express high levels of the programmed death-1 (PD-1) pathway ligands in an attempt to repress the cytotoxic T-cell response that these neoantigens would normally incite (Fig. 7). Therapeutic manipulation of the PD-1 pathway may therefore represent a new treatment modality for MMR-deficient tumors. Although current data are still limited, a small trial that treated 32 patients with advanced CRC with the PD-1 inhibitor pembrolizumab demonstrated a dramatic improvement in survival for patients with MMR-deficient tumors and essentially no response in MMR-proficient tumors.[23] Additional studies will be needed to confirm the efficacy of PD-1 inhibitor therapy for CRC and to determine the best method of selecting patients for this therapy. Notably, expression of PD-1 ligands on the surface of tumor cells is an important but not definitive marker of response to PD-1 blockade.[24,25] It may be the case that the overall neoantigen load itself rather than MMR status is the best predictor of response. If this proves to be the case, the therapeutic benefit of PD-1 blockade may extend to other CRCs with high mutational burdens, such as those with POLD1, POLE, and MYH mutations.

THE EPIDERMAL GROWTH FACTOR RECEPTOR PATHWAY AND INHIBITORS

The epidermal growth factor receptor (EGFR) pathway is well established as a key molecular driver of CRC pathogenesis and also represents the main focus of targeted therapeutic intervention in CRC. Two monoclonal antibodies directed against EGFR, cetuximab and panitumumab, are currently approved for the treatment of metastatic CRC (Fig. 8). These antibodies work by binding to the extracellular domain of the EGFR protein, inhibiting ligand binding and preventing subsequent activation of the downstream signaling pathway. Although both drugs were initially approved for metastatic cancers expressing EGFR, it is now clear that EGFR expression itself is not a strong predictor of response.[26] However, mutations in downstream genes in the EGFR signaling pathway do affect whether a tumor will respond to anti-EGFR therapy, and testing for these mutations is required to identify patients who will benefit from therapy. Interestingly, unlike lung cancer, mutations in EGFR itself are relatively uncommon in CRC and are not a major mechanism of EGFR pathway activation.[4]

EPIDERMAL GROWTH FACTOR RECEPTOR AND THE RAS/RAF/MEK/ERK PATHWAY

One major branch of the EGFR signaling pathway functions through the Ras/Raf/MEK/ERK group of proteins. Resistance to anti-EGFR therapy in this

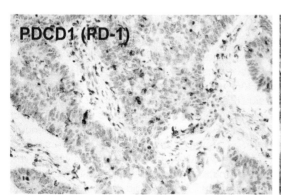
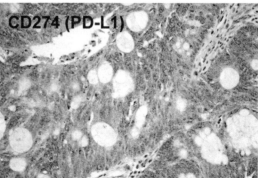

Fig. 7. Expression of immune checkpoint proteins in CRC. PDCD1 (PD-1) protein is expressed on T lymphocytes that infiltrate tumor epithelium and that are present in the stroma surrounding the tumor. The ligand for PDCD1, CD274 (PD-L1) is expressed by the tumor cells and serves to suppress the T-cell cytotoxic response.

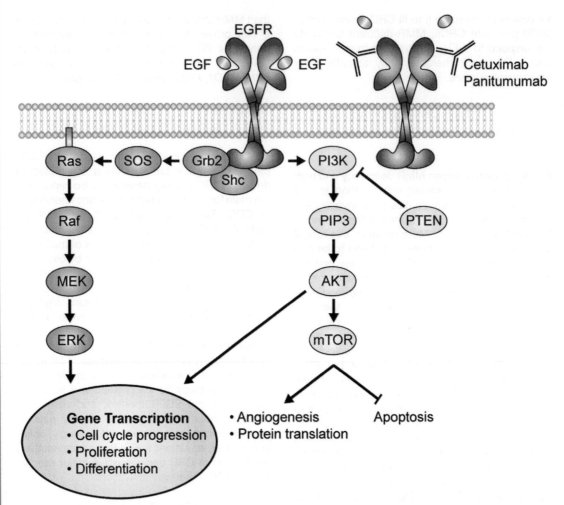

Fig. 8. The EGFR signaling pathway. Signaling occurs when EGF or other ligands bind to the extracellular portion of the EGFR protein. Two main downstream pathways can be activated, including the Ras/Raf/MEK/ERK pathway, leading to altered gene transcription and the PI3K/AKT/mTOR pathway, leading not only to transcriptional changes but alterations in protein translation, angiogenesis, and inhibition of apoptosis as well. Treatment with monoclonal antibodies that are directed against the extracellular domain of EGFR can inhibit ligand binding and downstream signaling. Mutations that affect proteins downstream of EGFR frequently result in constitutive activation of the pathway and resistance to anti-EGFR therapy.

branch of the pathway predominantly occurs due to mutations in the *RAS* genes. Although activating mutations in *KRAS* codons 12 and 13 were the first alterations reported to confer resistance to anti-EGFR therapy, it is now clear that mutations in multiple hotspots in *KRAS* and *NRAS* can confer resistance.[27–30] Mutations in both genes are almost always mutually exclusive, suggesting that they play interchangeable roles in activating EGFR signaling. Although the general mechanism of *RAS* mutation-mediated resistance to anti-EGFR therapy is understood, there are still unresolved questions about testing (Box 2).

In contrast to activating mutations in the *RAS* genes, activating mutations in *BRAF*, which is

farther downstream in the pathway, less clearly predict responsiveness to anti-EGFR therapy. Although a number of studies have reported a slight benefit for using anti-EGFR therapy in *BRAF*-mutant tumors, this effect is not statistically significant when the results of all trials are pooled.[27,31–33] Regardless of its value for predicting anti-EGFR therapeutic benefit, the presence of a *BRAF* mutation carries independent prognostic information. MMR-proficient tumors that have a *BRAF* mutation have a substantially worse prognosis than *BRAF* wild-type MMR-proficient tumors.[34] Furthermore, in the subset of MMR-deficient tumors that are metastatic, *BRAF* mutant status is also associated with a worse prognosis.[35]

Box 2
Unresolved questions involving *RAS* gene testing in colorectal carcinoma

- How functionally similar are mutations within each gene? Most evidence suggests that different activating mutations in a single gene are equivalent, although some evidence suggests differential effects, particularly for *KRAS*.[45]

- How functionally similar are mutations in *KRAS* versus those in *NRAS*?

- What is the significance of rare tumors with co-occurring activating mutations in the 2 different *RAS* genes?

- How does intratumoral molecular heterogeneity influence *RAS* gene mutations, and how should cases with discordant gene status between primary and metastatic disease be treated?[46–48]

Finally, in the near future, it also may be possible for *BRAF* mutation status to determine whether patients would benefit from BRAF inhibitor therapy.[36] Use of single-agent BRAF inhibitors for patients with *BRAF*-mutant CRC has not proven successful thus far. Although tumors may exhibit an initial response, they invariably develop resistance by activating various alternative signaling pathways that induce cell proliferation and survival. Novel BRAF inhibitors are being clinically evaluated and early data on combination therapy that targets BRAF in addition to EGFR or the PIK3CA branch of the pathway appear promising.

EPIDERMAL GROWTH FACTOR RECEPTOR AND THE PHOSPHATIDYLINOSITOL-3-KINASE/AKT/MAMMALIAN TARGET OF RAPAMYCIN PATHWAY

The other major branch of the EGFR signaling pathway operates through the phosphatidylinositol-3-kinase (PI3K)/AKT/mammalian target of rapamycin (mTOR) group of proteins. Although the significance of mutations in this branch of the pathway is not as well defined as those in the Ras/Raf/MEK/ERK branch, this arm of the pathway does play an important role in CRC. In particular, mutations in *PIK3CA* exon 20 are associated with a lack of response to anti-EGFR therapy, whereas coexisting mutations in exons 9 and 20 are associated with a poor prognosis.[37,38] Intriguingly, several studies have reported that patients with *PIK3CA*-mutant tumors derive a greater mortality reduction from adjuvant aspirin treatment than patients with *PIK3CA*–wild-type tumors.[39,40] This effect appears to be due to inhibition of PTGS2 (cyclooxygenase-2), leading to decreased prostaglandin synthesis and lowered PI3K signaling activity. Confirmation of these findings in larger studies may lead to routine *PIK3CA* testing to guide adjuvant therapy.

The Phosphatase and tensin homolog (PTEN) protein functions to inhibit activity in the PIK3CA branch of the EGFR signaling pathway. Loss of PTEN function, due to promoter methylation, inactivating mutations, or deletion, is associated with poor response to anti-EGFR therapy.[29] Outside of predicting anti-EGFR therapy response, however, the significance of PTEN loss is unclear. Some studies have reported that loss is associated with increased rate of relapse and worse overall prognosis, whereas other studies have not found an independent association between PTEN loss and outcome.[41,42] Currently, the routine evaluation of PTEN status has been limited due to poor reproducibility of immunohistochemistry for PTEN protein expression.

SUMMARY AND FUTURE DIRECTIONS

Routine molecular classification of CRC over the past 15 years has focused on evaluating the status of the MMR system and the EGFR signaling pathway. In 2016, the first consensus guidelines for molecular testing in CRC will be jointly released by a group of pathology and oncology professional organizations. These guidelines will provide an evidence-based, uniform strategy for CRC molecular testing and should help maximize the information that can be provided to each patient regarding the biology and expected behavior of their cancer.

Moving forward, it is likely that additional molecular interrogation of EGFR pathway status will have an expanded role in determining which patients may benefit from not only the existing anti-EGFR drugs, but also downstream targeted inhibitors of the pathway. Evaluation of tumor neoantigen load, likely by next-generation sequencing, may also be routinely used to determine which patients should receive immune checkpoint blockade inhibitors. In addition, transcriptional profiling of CRCs may yield additional information that is not captured by mutational testing.[43]

In the more distant future, evaluation of a patient's tumor-associated intestinal microbiome may provide insight into the tumor microenvironment, including the inflammatory milieu and metabolic state.[22] Finally, evaluation of circulating tumor cells or cell-free tumor DNA may provide a noninvasive method of monitoring for disease recurrence or the development of resistance to therapy. Regardless of which new biomarkers prove to have clinical utility, it is increasingly clear that the molecular alterations in CRC will need to be evaluated in the larger context of the patient's environmental, lifestyle, and genetic factors.[44] Dissecting these complex interrelationships will likely increase the value of classically recognized genotype-phenotype molecular alterations in CRC.

REFERENCES

1. Siegel RL, Miller KD, Jemal A. Cancer statistics, 2015. CA Cancer J Clin 2015;65(1):5–29.
2. Fearon ER, Vogelstein B. A genetic model for colorectal tumorigenesis. Cell 1990;61(5):759–67.
3. Vogelstein B, Fearon ER, Hamilton SR, et al. Genetic alterations during colorectal-tumor development. N Engl J Med 1988;319(9):525–32.
4. Cancer Genome Atlas Network. Comprehensive molecular characterization of human colon and rectal cancer. Nature 2012;487(7407):330–7.
5. Grady WM, Markowitz SD. The molecular pathogenesis of colorectal cancer and its potential application to colorectal cancer screening. Dig Dis Sci 2015; 60(3):762–72.
6. Giardiello FM, Allen JI, Axilbund JE, et al. Guidelines on genetic evaluation and management of Lynch syndrome: a consensus statement by the US Multisociety Task Force on colorectal cancer. Am J Gastroenterol 2014;109(8):1159–79.
7. Tiwari AK, Roy HK, Lynch HT. Lynch syndrome in the 21st century: clinical perspectives. QJM 2016; 109(3):151–8.
8. Wimmer K, Etzler J. Constitutional mismatch repair-deficiency syndrome: have we so far seen only the tip of an iceberg? Hum Genet 2008; 124(2):105–22.
9. Durno CA, Holter S, Sherman PM, et al. The gastrointestinal phenotype of germline biallelic mismatch repair gene mutations. Am J Gastroenterol 2010; 105(11):2449–56.
10. Vasen HF, Watson P, Mecklin JP, et al. New clinical criteria for hereditary nonpolyposis colorectal cancer (HNPCC, Lynch syndrome) proposed by the International Collaborative group on HNPCC. Gastroenterology 1999;116(6):1453–6.
11. Umar A, Boland CR, Terdiman JP, et al. Revised Bethesda Guidelines for hereditary nonpolyposis colorectal cancer (Lynch syndrome) and microsatellite instability. J Natl Cancer Inst 2004;96(4):261–8.
12. Evaluation of Genomic Applications in Practice and Prevention Working Group. Recommendations from the EGAPP Working Group: genetic testing strategies in newly diagnosed individuals with colorectal cancer aimed at reducing morbidity and mortality from Lynch syndrome in relatives. Genet Med 2009;11(1):35–41.
13. Cicek MS, Lindor NM, Gallinger S, et al. Quality assessment and correlation of microsatellite instability and immunohistochemical markers among population- and clinic-based colorectal tumors results from the Colon Cancer Family Registry. J Mol Diagn 2011;13(3):271–81.
14. Murphy KM, Zhang S, Geiger T, et al. Comparison of the microsatellite instability analysis system and the Bethesda panel for the determination of microsatellite instability in colorectal cancers. J Mol Diagn 2006;8(3):305–11.
15. Weisenberger DJ, Siegmund KD, Campan M, et al. CpG island methylator phenotype underlies sporadic microsatellite instability and is tightly associated with BRAF mutation in colorectal cancer. Nat Genet 2006;38(7):787–93.
16. Deng G, Bell I, Crawley S, et al. BRAF mutation is frequently present in sporadic colorectal cancer with methylated hMLH1, but not in hereditary nonpolyposis colorectal cancer. Clin Cancer Res 2004; 10(1 Pt 1):191–5.
17. Yurgelun MB, Goel A, Hornick JL, et al. Microsatellite instability and DNA mismatch repair protein deficiency in Lynch syndrome colorectal polyps. Cancer Prev Res 2012;5(4):574–82.
18. Gryfe R, Kim H, Hsieh ET, et al. Tumor microsatellite instability and clinical outcome in young patients with colorectal cancer. N Engl J Med 2000;342(2): 69–77.
19. Guastadisegni C, Colafranceschi M, Ottini L, et al. Microsatellite instability as a marker of prognosis and response to therapy: a meta-analysis of colorectal cancer survival data. Eur J Cancer 2010; 46(15):2788–98.
20. Sargent DJ, Marsoni S, Monges G, et al. Defective mismatch repair as a predictive marker for lack of efficacy of fluorouracil-based adjuvant therapy in colon cancer. J Clin Oncol 2010;28(20):3219–26.
21. Ribic CM, Sargent DJ, Moore MJ, et al. Tumor microsatellite-instability status as a predictor of benefit from fluorouracil-based adjuvant chemotherapy for colon cancer. N Engl J Med 2003;349(3):247–57.
22. Lasry A, Zinger A, Ben-Neriah Y. Inflammatory networks underlying colorectal cancer. Nat Immunol 2016;17(3):230–40.
23. Le DT, Uram JN, Wang H, et al. PD-1 blockade in tumors with mismatch-repair deficiency. N Engl J Med 2015;372(26):2509–20.
24. Herbst RS, Soria JC, Kowanetz M, et al. Predictive correlates of response to the anti-PD-L1 antibody

MPDL3280A in cancer patients. Nature 2014; 515(7528):563–7.

25. Taube JM, Klein A, Brahmer JR, et al. Association of PD-1, PD-1 ligands, and other features of the tumor immune microenvironment with response to anti-PD-1 therapy. Clin Cancer Res 2014;20(19):5064–74.

26. Hebbar M, Wacrenier A, Desauw C, et al. Lack of usefulness of epidermal growth factor receptor expression determination for cetuximab therapy in patients with colorectal cancer. Anticancer Drugs 2006;17(7):855–7.

27. Van Cutsem E, Kohne CH, Hitre E, et al. Cetuximab and chemotherapy as initial treatment for metastatic colorectal cancer. N Engl J Med 2009;360(14): 1408–17.

28. Tol J, Koopman M, Cats A, et al. Chemotherapy, bevacizumab, and cetuximab in metastatic colorectal cancer. N Engl J Med 2009;360(6):563–72.

29. Therkildsen C, Bergmann TK, Henrichsen-Schnack T, et al. The predictive value of KRAS, NRAS, BRAF, PIK3CA and PTEN for anti-EGFR treatment in metastatic colorectal cancer: a systematic review and meta-analysis. Acta Oncol 2014;53(7): 852–64.

30. De Roock W, Claes B, Bernasconi D, et al. Effects of KRAS, BRAF, NRAS, and PIK3CA mutations on the efficacy of cetuximab plus chemotherapy in chemotherapy-refractory metastatic colorectal cancer: a retrospective consortium analysis. Lancet 2010;11(8):753–62.

31. Pietrantonio F, Petrelli F, Coinu A, et al. Predictive role of BRAF mutations in patients with advanced colorectal cancer receiving cetuximab and panitumumab: a meta-analysis. Eur J Cancer 2015;51(5): 587–94.

32. Bokemeyer C, Van Cutsem E, Rougier P, et al. Addition of cetuximab to chemotherapy as first-line treatment for KRAS wild-type metastatic colorectal cancer: pooled analysis of the CRYSTAL and OPUS randomised clinical trials. Eur J Cancer 2012;48(10):1466–75.

33. Rowland A, Dias MM, Wiese MD, et al. Meta-analysis of BRAF mutation as a predictive biomarker of benefit from anti-EGFR monoclonal antibody therapy for RAS wild-type metastatic colorectal cancer. Br J Cancer 2015;112(12):1888–94.

34. Lochhead P, Kuchiba A, Imamura Y, et al. Microsatellite instability and BRAF mutation testing in colorectal cancer prognostication. J Natl Cancer Inst 2013;105(15):1151–6.

35. Russo AL, Borger DR, Szymonifka J, et al. Mutational analysis and clinical correlation of metastatic colorectal cancer. Cancer 2014;120(10):1482–90.

36. Temraz S, Alameddine R, Shamseddine A. B-type proto-oncogene-mutated tumors of colon cancer:

promising therapeutic approaches. Curr Opin Oncol 2015;27(3):276–81.

37. Liao X, Morikawa T, Lochhead P, et al. Prognostic role of PIK3CA mutation in colorectal cancer: cohort study and literature review. Clin Cancer Res 2012; 18(8):2257–68.

38. Mao C, Yang ZY, Hu XF, et al. PIK3CA exon 20 mutations as a potential biomarker for resistance to anti-EGFR monoclonal antibodies in KRAS wild-type metastatic colorectal cancer: a systematic review and meta-analysis. Ann Oncol 2012;23(6): 1518–25.

39. Paleari L, Puntoni M, Clavarezza M, et al. PIK3CA mutation, aspirin use after diagnosis and survival of colorectal cancer. A systematic review and meta-analysis of epidemiological studies. Clin Oncol 2016;28(5):317–26.

40. Liao X, Lochhead P, Nishihara R, et al. Aspirin use, tumor PIK3CA mutation, and colorectal-cancer survival. N Engl J Med 2012;367(17):1596–606.

41. Lin PC, Lin JK, Lin HH, et al. A comprehensive analysis of phosphatase and tensin homolog deleted on chromosome 10 (PTEN) loss in colorectal cancer. World J Surg Oncol 2015;13:186.

42. Jang KS, Song YS, Jang SH, et al. Clinicopathological significance of nuclear PTEN expression in colorectal adenocarcinoma. Histopathology 2010;56(2): 229–39.

43. Sadanandam A, Wang X, de Sousa EMF, et al. Reconciliation of classification systems defining molecular subtypes of colorectal cancer: interrelationships and clinical implications. Cell Cycle 2014; 13(3):353–7.

44. Ogino S, Chan AT, Fuchs CS, et al. Molecular pathological epidemiology of colorectal neoplasia: an emerging transdisciplinary and interdisciplinary field. Gut 2011;60(3):397–411.

45. Imamura Y, Morikawa T, Liao X, et al. Specific mutations in KRAS codons 12 and 13, and patient prognosis in 1075 BRAF wild-type colorectal cancers. Clin Cancer Res 2012;18(17):4753–63.

46. Siyar Ekinci A, Demirci U, Cakmak Oksuzoglu B, et al. KRAS discordance between primary and metastatic tumor in patients with metastatic colorectal carcinoma. J BUON 2015;20(1):128–35.

47. Dono M, Massucco C, Chiara S, et al. Low percentage of KRAS mutations revealed by locked nucleic acid polymerase chain reaction: implications for treatment of metastatic colorectal cancer. Mol Med 2012;18:1519–26.

48. Italiano A, Hostein I, Soubeyran I, et al. KRAS and BRAF mutational status in primary colorectal tumors and related metastatic sites: biological and clinical implications. Ann Surg Oncol 2010;17(5): 1429–34.

Molecular Diagnostics in the Evaluation of Pancreatic Cysts

 CrossMark

Brian K. Theisen, MD, Abigail I. Wald, PhD,
Aatur D. Singhi, MD, PhD*

KEYWORDS

- Pancreas • Pancreatic ductal adenocarcinoma • Intraductal papillary mucinous neoplasm
- Mucinous cystic neoplasm • Serous cystadenoma • Solid-pseudopapillary neoplasm • Genetics
- Diagnostics

Key points

- Due to the increased use and improvements in abdominal imaging, incidental pancreatic cysts are becoming frequently encountered.

- Although many cysts, such as serous cystadenomas (SCAs) and non-neoplastic cysts have a benign clinical course, others, such as intraductal papillary mucinous neoplasms (IPMN) and mucinous cystic neoplasms (MCN), represent precursor lesions to invasive pancreatic ductal adenocarcinoma (PDAC).

- Despite a multidisciplinary approach, preoperatively distinguishing pancreatic cysts from one another can be challenging and, if incorrect, can pose a significant health risk to the patient. The application of molecular techniques has emerged as a promising adjunct to the evaluation of pancreatic cysts.

- DNA obtained from preoperative endoscopic ultrasound fine-needle aspiration can be analyzed for genetic abnormalities that are specific for cyst type and likelihood of progression to high-grade dysplasia and/or PDAC.

- Mutations in *KRAS*, *GNAS*, and/or *RNF43* are highly specific for IPMNs and MCNs. Moreover, IPMNs and MCNs with high-grade dysplasia and/or PDAC often harbor alterations in *TP53*, *PIK3CA*, and/or *PTEN*. In contrast, *VHL* mutations and/or deletions are present in SCAs. Non-neoplastic cysts are devoid of any alterations in the aforementioned genes.

ABSTRACT

Within the past few decades, there has been a dramatic increase in the detection of incidental pancreatic cysts. It is reported a pancreatic cyst is identified in up to 2.6% of abdominal scans. Many of these cysts, including serous cystadenomas and pseudocysts, are benign and can be monitored clinically. In contrast, mucinous cysts, which include intraductal papillary mucinous neoplasms and mucinous cystic neoplasms, have the potential to progress to pancreatic adenocarcinoma. In this review, we discuss the current management guidelines for pancreatic cysts, their underlying genetics, and the integration of molecular testing in cyst classification and prognostication.

OVERVIEW

With the rapid utilization and ongoing advancements in cross-sectional abdominal imaging, the detection of pancreatic cysts has become increasingly frequent. It is estimated that more than 2% of the general population in the United States harbors a pancreatic cyst.[1,2] Many of these cysts, including serous cystadenomas (SCA) and pseudocysts, are benign and can be monitored clinically. In contrast, mucinous cysts, which include intraductal papillary mucinous neoplasms (IPMNs)

Funding Support: None.
Disclosure/Conflict of Interest: The authors have no conflicts of interest to declare.
Department of Pathology, University of Pittsburgh Medical Center, Pittsburgh, PA, USA
* Corresponding author. UPMC Presbyterian Hospital, 200 Lothrop Street, Room A616.2, Pittsburgh, PA 15213.
E-mail address: singhiad@upmc.edu

Surgical Pathology 9 (2016) 441–456
http://dx.doi.org/10.1016/j.path.2016.04.008
1875-9181/16/$ – see front matter © 2016 Elsevier Inc. All rights reserved.

surgpath.theclinics.com

and mucinous cystic neoplasms (MCNs), have the potential to progress to pancreatic adenocarcinoma (PDAC).[3–6]

Currently, a multidisciplinary approach is recommended for the assessment of pancreatic cysts. This includes clinical and radiographic evaluation, endoscopic ultrasound-guided fine-needle aspiration (EUS-FNA), cytology, cyst fluid analysis (eg, viscosity), and tumor markers (eg, carcinoembryonic antigen [CEA]). Despite a combination of methodologies, the distinction between mucinous cysts from other pancreatic cysts and those that will progress to PDAC can be challenging. Recently, the application of molecular techniques has emerged as a promising adjunct to the evaluation of pancreatic cysts.[7–10] Although the cellular content of pancreatic cyst aspirates is often suboptimal, DNA from lysed or exfoliated epithelial cells shed into the fluid from the cyst lining can be analyzed for genetic abnormalities. Moreover, whole-exome sequencing of the most common pancreatic cysts has identified distinct mutational profiles for each cyst type as well as genetic alterations that coincide with the development of PDAC. Within this review, we discuss the current management guidelines for pancreatic cysts, their underlying genetics, and the integration of molecular testing within this field.

PANCREATIC CANCER AND PANCREATIC CYSTS

In the United States, PDAC is the fourth leading cause of cancer deaths in both men and women. In 2015, an estimated 48,960 individuals were diagnosed with PDAC, and approximately 40,560 died from this deadly disease. Although surgical resection offers the only possibility of cure, more than 85% of patients present with inoperable disease at the time of diagnosis. Therefore, chemotherapy and radiation are the mainstay of treatment in most patients. Despite aggressive combined modality treatment approaches, the 5-year survival of PDAC is a dismal 6% and has remained unchanged for the past 40 years.

Fundamentally, PDAC is a genetic disease and arises from noninvasive intraductal precursor lesions that accumulate genetic alterations. Among these precursor lesions are microscopic pancreatic intraepithelial neoplasia (PanIN), and macroscopic mucinous pancreatic cysts that include IPMNs and MCNs. Although PanINs are too small to identify radiographically, mucinous cysts are often found by routine imaging and represent ideal lesions for early detection strategies. However, most IPMNs and MCNs are indolent, with a

minority undergoing malignant transformation. In addition, as surgical intervention remains the preferred treatment option for mucinous cysts, patients must consider the operative mortality and morbidity of these procedures, which range from 2% to 4% and 40% to 50%, respectively. Consequently, treatment guidelines by the International Association of Pancreatology and American Gastroenterological Association (AGA) were established.

SENDAI AND FUKUOKA GUIDELINES

During the 11th Congress of the International Association of Pancreatology held in Sendai, Japan, in 2004, a multidisciplinary group of physicians held a consensus meeting on the management of IPMNs and MCNs. Published in 2006, the "Sendai" guidelines recommended that patients with main duct or mixed main and branch duct IPMN should undergo surgical resection assuming good surgical candidacy and reasonable life expectancy. Similarly, it was recommended that all MCNs should be resected. However, for branch duct IPMNs, the Sendai guidelines advocated serial imaging yearly for mucinous cysts smaller than 1 cm, twice a year for cysts between 1 and 2 cm, 2 to 4 times a year for cysts between 2 and 3 cm, and surgical resection for cysts larger than 3 cm (Fig. 1).[11] Further, surgical resection was recommended for any cyst with at least 1 of the following: associated clinical symptoms, the presence of a mural nodule, a dilated pancreatic duct (≥6 mm), and/or "positive cytology." Subsequently, several studies found the Sendai guidelines were highly sensitive for mucinous cysts harboring advanced neoplasia (high-grade dysplasia and PDAC) but suffered from low specificity. In a review of 147 patients, Pelaez-Luna and colleagues[12] reported a sensitivity of 100%, but a specificity of 23% in detecting advanced neoplasia. Similarly, Tang and colleagues[13] found implementation of the consensus guidelines within their cohort of 204 patients would have 100% sensitivity, but 31% specificity.

In retrospect, both cyst size and nonspecific patient symptoms are considered poor predictors of advanced neoplasia. Although a mucinous cyst 3 cm or larger is more likely to have advanced neoplasia, smaller cysts also can have significant malignant potential. Buscaglia and colleagues[14] published a cyst size of larger than 1.5 cm as the optimal cutoff as a predictor of advanced neoplasia. In comparison, Gomez and colleagues[15] proposed a cyst size of larger than 2.5 cm. Moreover, many benign, nonmucinous cysts can be quite large. Clinical symptoms as a

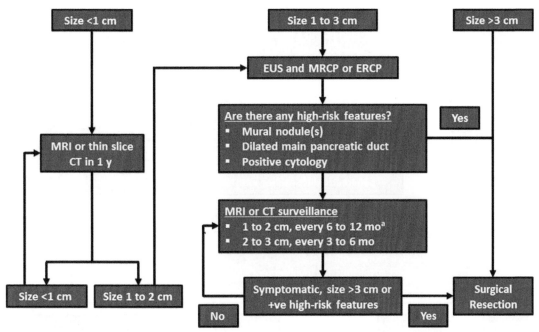

Fig. 1. Sendia guidelines. An international consensus algorithmic approach to the management of branch duct IPMNs established at the 11th Congress of the International Association of Pancreatology held in Sendai, Japan. ᵃ The interval of follow-up examination can be lengthened after 2 years without change in cyst characteristics. CT, computed tomography; ERCP, endoscopic retrograde cholangiopancreatography; MRCP, magnetic resonance cholangiopancreatography. (*Adapted from* Tanaka M, Chari S, Adsay V, et al. International consensus guidelines for management of intraductal papillary mucinous neoplasms and mucinous cystic neoplasms of the pancreas. Pancreatology 2006;6:17–32; with permission.)

predictor of advanced neoplasia are also controversial. Although advanced neoplasia is more likely to occur in a symptomatic patient, an asymptomatic patient is equally likely to have advanced neoplasia.[16] In addition, determining whether symptoms are due to the pancreatic cyst or are nonspecific in nature can be difficult. Although the most common complaint at presentation may be nonspecific epigastric/abdominal pain, 20% to 28% of patients with advanced neoplasia are asymptomatic.[16–20]

Therefore, in 2010, a new multidisciplinary group of physicians met in Fukuoka, Japan, to revisit the Sendai guidelines and discuss current advances within the field.[21] The Fukuoka guidelines were once again labeled as consensus guidelines and aimed to decrease the frequency of imaging and surgical intervention. Like the Sendai guidelines, surgical intervention was recommended for patients with main duct or mixed main and branch duct IPMNs and for MCNs. Criteria for surgical resection of branch duct IPMNs was at least one of the following: obstructive jaundice in a patient with a pancreatic head cyst, an enhancing solid component, and a dilated main pancreatic duct of 10 mm or larger (**Fig. 2**).

Surveillance in asymptomatic patients changed to every 2 to 3 years for cysts smaller than 1 cm; yearly for cysts 1 to 2 cm; twice a year for cysts 2 to 3 cm, which can be lengthened as appropriate; and close surveillance for cysts larger than 3 cm. These guidelines also recommended EUS-FNA for cysts 3 cm or larger, and smaller than 3 cm with worrisome features to further risk stratify patients before considering surgery. Worrisome features included the presence of a nonenhancing mural nodule, thickened cyst wall, main pancreatic duct size of 5 to 9 mm, an abrupt caliber change in the main pancreatic duct with distal parenchymal atrophy, and regional lymphadenopathy. Cytology that was suspicious or positive for malignancy also was an indication for surgery.

Although the Fukuoka guidelines demonstrated increased specificity for advanced neoplasia, this was at a loss in sensitivity. Kaimakliotis and colleagues[22] identified no statistically significant differences between the Sendai and the Fukuoka guidelines in detecting advanced neoplasia. The central issue with both guidelines is they address only IPMNs and MCNs, and are not informative for other cystic lesions. In other words, they

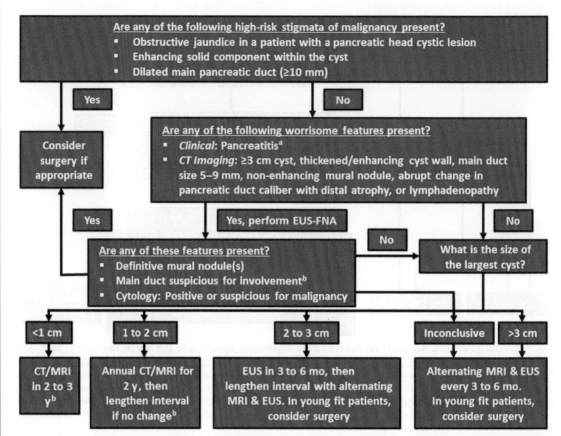

Fig. 2. Fukuoka guidelines. As a follow-up to the Sendai guidelines, the Fukuoka guidelines were established in 2012. [a] Pancreatitis may be an indication for surgery for the relief of symptoms. [b] The presence of thickened walls, intraductal mucin, or mural nodules is suggestive of main duct involvement. In their absence, main duct involvement is inconclusive. (*Adapted from* Tanaka M, Fernandez-del Castillo C, Adsay V, et al. International consensus guidelines 2012 for the management of IPMN and MCN of the pancreas. Pancreatology 2012;12:183–97; with permission.)

assume that mucinous cysts can be diagnosed correctly based on standard clinical, imaging, and laboratory criteria, before following their respective management algorithm. However, as mentioned previously, preoperatively distinguishing pancreatic cysts from one another can be challenging and contribute to the limitations in applying the Sendai and Fukuoka guidelines.

AMERICAN GASTROENTEROLOGICAL ASSOCIATION GUIDELINES

In 2015, the AGA presented its own guidelines, which were labeled as evidence-based, rather than consensus, and were accompanied by an extensive technical review of the literature.[23,24] The investigators acknowledged that all evidence related to the management of pancreatic cysts was graded as very low quality or with "great

uncertainty regarding the estimate of effect." Analogous to the Sendai and Fukuoka guidelines, resection was recommended for patients with main duct or mixed main and branch duct IPMNs and MCNs. However, in contrast to the Sendai and Fukuoka guidelines, the AGA guidelines recommended a uniform approach to management for all other neoplastic cysts, including branch duct IPMNs (Fig. 3). This controversial aspect of the AGA guidelines also advocated a higher threshold for surgical intervention, requiring more than 2 high-risk features (size >3 cm, dilated main pancreatic duct, and the presence of a mural nodule) and/or positive cytology. Considering that the AGA guidelines pertain to the management of all asymptomatic neoplastic cysts, rather than just mucinous cysts alone, imaging surveillance is recommended for a much larger population of patients with pancreatic cysts. In addition, the AGA recommends less surveillance for cysts with

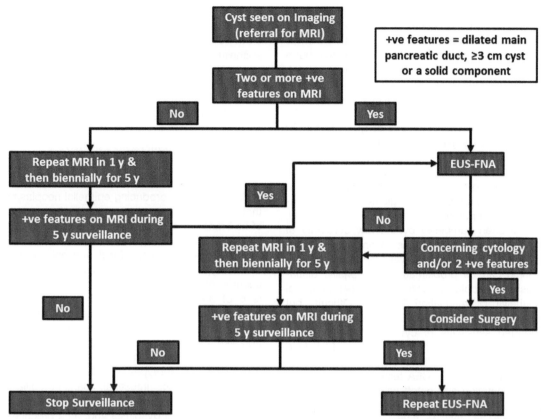

Fig. 3. AGA guidelines. In 2015, the AGA established evidence-based guidelines for the management of asymptomatic neoplastic pancreatic cysts. Of note, these guidelines are based on no or very low quality of evidence. (*Adapted from* Vege SS, Ziring B, Jain R, et al. American Gastroenterological Association institute guideline on the diagnosis and management of asymptomatic neoplastic pancreatic cysts. Gastroenterology 2015;148:819–22.)

1 or no high-risk features, defined as biennial imaging after 1 year follow-up and discontinuing surveillance within 5 years if the cyst is stable, regardless of size. Last, EUS-FNA is considered only when more than 2 high-risk features are present or if additional characterization of the cyst would change management. EUS-FNA offers the highest imaging resolution and the opportunity to sample the cyst for cytology and ancillary studies. As expected, publication of the AGA guidelines resulted in significant concern that adoption will lead to inaccuracies in identifying advanced neoplasia and questions over whether the surveillance of all pancreatic cysts is warranted.[25–27]

Analyzing a cohort of 225 patients, Singhi and colleagues[10] reported that the AGA guidelines had a sensitivity and specificity of 62% and 79%, respectively, in detecting advanced neoplasia. Among IPMNs, the application of the AGA guidelines would have missed 45% with adenocarcinoma or high-grade dysplasia. As the AGA manages all pancreatic cysts similarly, 27 SCAs within their cohort would continue radiographic surveillance. SCAs are nearly always benign and most investigators would agree that these neoplasms could be monitored clinically rather than by radiographic imaging. Considering the cost of abdominal MRI at the current national average for 27 patients, this would amount to more than $200,000 over a 5-year time frame. Equally concerning is that continuing to survey patients with SCAs also exposes them to the possibility of unnecessary surgical resection, as SCAs may mimic high-risk mucinous cysts.

The limitations of the AGA guidelines can be attributed to the inherent performance characteristics of cross-sectional imaging, EUS, and cytologic evaluation. Cross-sectional imaging cannot reliably distinguish neoplastic cysts with malignant potential from other types of pancreatic cysts. Although EUS allows for high-resolution imaging of a cyst and main pancreatic duct, morphologic features alone are still poor predictors of advanced neoplasia. Within a technical review published by the AGA, cyst size of 3 cm or larger by EUS has a pooled sensitivity of 74% for malignancy, but a poor pooled specificity of 49%.[24] In contrast, main pancreatic duct dilatation and the presence of a mural nodule have specificities of 80% and 91%, but poor sensitivities of 32% and 48%, respectively.[24] The limitations of EUS are further compounded when factoring interobserver variability and operator dependence.[28] The utility of EUS is, however, enhanced when coupled with FNA, cytologic evaluation, and pancreatic cyst fluid ancillary studies. Cytologic evaluation for malignancy is reported to approach 100% specificity, but is often hampered by the low cellularity of pancreatic cyst fluid.[29] In the absence of frank malignancy, differentiating advanced neoplasia from benign can be problematic. Additionally, distinguishing neoplastic cells from gastrointestinal epithelial contamination is often difficult, but imperative to establishing a diagnosis. Thus, the reported sensitivity of cytology for malignancy varies widely from 25% to 88%.[24,30,31] EUS-FNA also allows for ancillary studies that include pancreatic cyst fluid CEA testing. Depending on instrumentation and cutoff values, elevated CEA has a sensitivity ranging from 70% to 82% and a specificity of 79% to 84%.[24,31] However, it is important to note, CEA will not differentiate between advanced neoplasia and benign cyst lining. Hence, more reliable biomarkers than those used in the AGA, Sendai, and Fukuoka guidelines are needed for optimal management of patients with pancreatic cysts.

PANCREATIC CYSTS AND THEIR GENETICS

Within the past decade, deep sequencing technologies have uncovered recurrent mutations within the major cystic lesions of the pancreas. Several studies have sequenced DNA isolated from preoperative cyst fluid, postoperative cyst fluid, and cyst epithelium to reveal genetic alterations specific for pancreatic cyst type and likelihood of progression to PDAC (Table 1). Moreover, these genetic alterations have the potential to be used diagnostically and aid in the management of pancreatic cysts.

INTRADUCTAL PAPILLARY MUCINOUS NEOPLASM

IPMNs are mucin-producing epithelial neoplasms that arise from the main pancreatic duct or one of its branches (Fig. 4). These lesions are usually larger than 1 cm, frequently arise in the head of the pancreas, can be multifocal, and lack a dense ovarian-type stroma. These neoplasms more often occur in men than women with a male-to-female ratio of 3:2. The mean age of presentation is approximately 65 years old. Microscopically, the neoplastic epithelium often adopts a papillary or frondlike architecture and is composed of tall columnar mucin-producing cells. On the basis of morphology, 4 subtypes of IPMNs have been described: gastric, intestinal, pancreatobiliary, and oncocytic. The neoplastic cells may show varying degrees of dysplasia and, as a precursor of PDAC, can be associated with invasive adenocarcinoma. However, the overall incidence of an associated invasive adenocarcinoma is relatively low. For IPMNs involving the main pancreatic duct, the prevalence of invasive adenocarcinoma is 30%; whereas branch duct IPMNs have a lower prevalence of 15% or lower.[4,32]

The most frequent genetic alteration in IPMNs is an oncogenic KRAS mutation, which has a prevalence of more than 80%. KRAS encodes for a G-protein, or a guanosine-nucleotide-binding protein, that functions as a small GTPase and

Table 1
Key genetic mutations and/or deletions in pancreatic cysts

Pancreatic Cyst Type	KRAS	GNAS	RNF43	VHL	CTNNB1	TP53	PIK3CA	PTEN	CDKN2A	SMAD4
Intraductal papillary mucinous neoplasm	+	+	+	−	−	+[a]	+[a]	+[a]	+[a]	+[a]
Mucinous cystic neoplasm	+	−	−	−	−	+[a]	+[a]	+[a]	+[a]	+[a]
Serous cystadenoma	−	−	−	+	−	−	−	−	−	−
Solid-pseudopapillary neoplasm	−	−	−	−	+	+[b]	+[b]	−	−	−
Non-neoplastic cysts	−	−	−	−	−	−	−	−	−	−

[a] Alterations in these genes are associated with advanced neoplasia.
[b] Although mutations in these genes have been described, they are rare findings.
+, presence; −, absence.

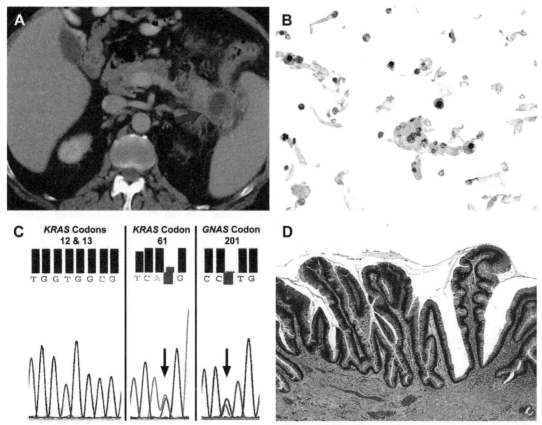

Fig. 4. DNA-based molecular testing of an IPMN. (A) Contrast-enhanced computed tomography imaging demonstrated a 3.8-cm cystic lesion (red arrow) involving the tail of the pancreas that was interpreted to be walled off necrosis in a patient with a history of chronic pancreatitis. (B) An EUS-FNA was performed and cytology revealed a hypocellular specimen that consisted of macrophages, red blood cells, and debris. Overall, the cytologic and imaging findings were not concerning for a neoplasm (Pap stain, original magnification ×600). (C) However, molecular analysis of EUS-FNA pancreatic cyst fluid identified missense mutations (black arrows) in codon 61 of KRAS and codon 201 GNAS (electropherograms). (D) Consistent with the molecular results, resection of the pancreatic cyst showed a mixed branch duct and main duct IPMN with low-grade dysplasia and of gastric histologic subtype (hematoxylin-eosin [H&E], original magnification ×200).

mediates downstream MAPK/ERK signaling from growth factor receptors.[8,33] Missense mutations result in constitutive activation of KRAS and occurs primarily in codon 12 and, to a lesser extent, codons 13 and 61.[8] KRAS mutations are detected in all histologic subtypes of IPMNs, but are more likely present in the gastric and pancreatobiliary types. Further, Nikiforova and colleagues[8] found KRAS mutations in IPMNs are associated with a branch duct location. In addition to KRAS, 65% of IPMNs harbor somatic mutations in the oncogene GNAS, which encodes for the G-protein stimulating α subunit (Gsα). Mutations in GNAS at either codon 201 or 227 result in constitutive activation of Gsα and its effector adenylate cyclase, leading to autonomous synthesis of cAMP and uncontrolled growth signaling.[33,34] GNAS mutations are typically present in IPMNs

involving the main pancreatic duct and of an intestinal histologic subtype. Collectively, activating mutations in KRAS and/or GNAS are present in more than 96% of IPMNs and are considered early events during tumorigenesis.[33]

Inactivating mutations in the tumor suppressor gene RNF43 occur in 14% to 38% of IPMNs, with frequent loss of heterozygosity at the RNF43 locus on chromosome 17q.[35,36] RNF43 encodes for an E3 ubiquitin ligase that regulates the Wnt signaling pathway. Other potential genes mutated in IPMNs include TP53, PIK3CA, PTEN, CDKN2A, and SMAD4. TP53 mutations are late events in the neoplastic progression of IPMNs and are frequently seen in advanced neoplasia.[37] Similarly, Garcia-Carracedo and colleagues[38] found PIK3CA mutations and deletions in PTEN are strongly associated with high-grade IPMNs and

PDAC. Losses in *CDKN2A* are an uncommon finding, but more prevalent in IPMNs with high-grade dysplasia than low-grade dysplasia.[39,40] *SMAD4* is also rarely inactivated in low-grade IPMNs, but mutation with corresponding loss of heterozygosity typically occurs in advanced neoplasia.[39,40]

MUCINOUS CYSTIC NEOPLASM

Similar to IPMNs, MCNs are mucin-producing epithelial neoplasms, but arise outside of the large ducts of the pancreas (Fig. 5). These cystic neoplasms occur almost exclusively in the pancreatic body and tail, and are far more common in women than men.[41] The average age at diagnosis is approximately 50 years. MCNs are solitary and do not communicate with the larger pancreatic ducts. They are composed of multiloculated

thick-walled cysts that may contain papillary excrescences and/or mural nodules. Microscopically, MCNs are lined by tall columnar mucinous epithelium with varying degrees of dysplasia and are supported by a dense ovarian-type stroma, the cells of which may express hormone receptors. Similar to IPMNs, MCNs are also precursor lesions to pancreatic ductal adenocarcinoma with a reported incidence ranging from 4% to 34%.[42,43]

Genetic alterations in *KRAS*, *RNF43*, *TP53*, *PIK3CA*, *PTEN*, *CDKN2A*, and *SMAD4* are reported to be present in MCNs. *KRAS* mutations are the most common finding and their prevalence increases with the degree of dysplasia. Jimenez and colleagues[44] detected *KRAS* mutations in 26% of low-grade MCNs and in 89% of MCNs with advanced neoplasia. Analogous to IPMNs, between 8% and 35% of MCNs harbor somatic

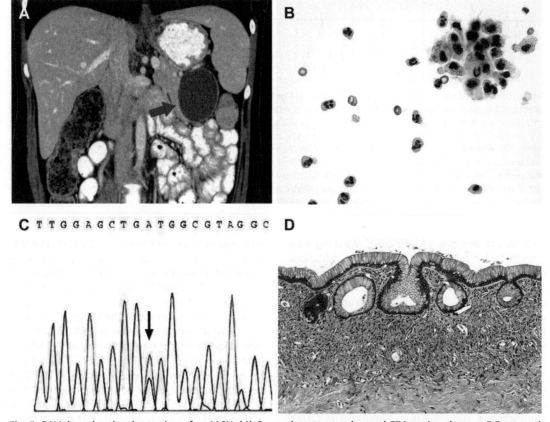

C TTGGAGCTGATGGCGTAGGC

Fig. 5. DNA-based molecular testing of an MCN. (*A*) Coronal contrast-enhanced CT imaging shows a 5.5-cm cystic lesion (*red arrow*) involving the pancreatic body and tail. (*B*) A corresponding cytology specimen consisted of histiocytes and acute inflammation with a differential diagnosis of a pseudocyst (Pap stain, original magnification ×600). (*C*) Mutational testing for *KRAS* revealed a missense mutation (*black arrow*) in codon 12 (electropherogram). (*D*) On resection, the pancreatic cyst correlated with the molecular findings and was consistent with a mucinous cystic neoplasm with low-grade dysplasia (H&E, original magnification ×200).

mutations in *RNF43*, suggesting a common role for this gene in the development of mucin-producing cystic neoplasms of the pancreas.[35,36] In addition, mutations and/or deletions in *TP53*, *PIK3CA*, *PTEN*, *CDKN2A*, and *SMAD4* appear in MCNs with advanced neoplasia. However, in contrast to IPMNs, *GNAS* mutations are distinctly absent in MCNs.[35,36]

SEROUS CYSTADENOMA

SCAs are benign cystic epithelial neoplasms characterized by a glycogen-rich cellular lining and serous fluid (**Fig. 6**). There is a slight female predominance and the mean patient age at diagnosis is 65 years.[45–47] Radiographically, these cystic neoplasms are often described with a central stellate scar and sunburst pattern of calcifications. SCAs are typically solitary and well-demarcated with a slight predilection for the pancreatic body and tail. On cross-section, most serous cystadenomas are composed of small cysts containing watery serous fluid, but oligocystic variants do occur and represent a clinical challenge in their distinction from mucinous cysts. The cysts are lined by single-to-multiple layers of cuboidal epithelial cells containing abundant glycogen and a centrally placed, round nucleus. Occasionally, the epithelium of SCAs may form small projections, reminiscent of an IPMN. Although a neoplasm, SCAs have no malignant potential and do not warrant surgical intervention, unless they become large, resulting in pancreatitis and, rarely, jaundice.

Molecularly, SCAs do not harbor the mutations found in mucinous cysts, such as *KRAS*, *GNAS*, *RNF43*, *TP53*, *PIK3CA*, *PTEN*, *CDKN2A*, and *SMAD4*.[9,10,35,36] Instead, alterations in the tumor

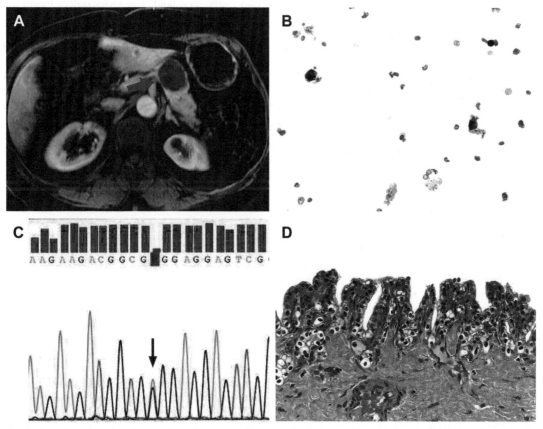

Fig. 6. DNA-based molecular testing of an SCA. (*A*) MRI demonstrated a 4.5-cm cyst (*red arrow*) within the pancreatic body. Comparison with the patient's prior scan 1 year ago showed an interval increase in cyst size by 1 cm. These findings were concerning for an IPMN. (*B*) Pancreatic cyst cytology revealed a hypocellular specimen with predominantly blood and macrophages (Pap stain, original magnification ×600). (*C*) Molecular analysis of the pancreatic cyst fluid detected a missense mutation in codon 30 (*black arrow*) of *VHL* (electropherogram). (*D*) Consistent with the molecular results, the pancreatectomy specimen showed a serous cystadenoma (H&E, original magnification ×200). An IPMN was not identified.

suppressor gene *VHL* have been described in 89% to 100% of SCAs and include inactivating mutations, loss of heterozygosity, and aneuploidy of chromosome 3p.[35,36] The finding of *VHL* alterations in SCAs is not surprising. A subset of patients with von Hippel-Lindau disease, which is associated with germline mutations in *VHL*, develop multiple serous cystadenomas throughout the pancreas. *VHL* encodes for an E3 ubiquitin ligase that inhibits the transactivation function of the oncogene hypoxia-inducible factor 1-α (HIF1-α) and targets it for degradation. Thus, inactivation of VHL results in the transactivation of HIF1-α targets that promote cell survival, angiogenesis, metabolism, and growth.

SOLID-PSEUDOPAPILLARY NEOPLASM

Solid-pseudopapillary neoplasms (SPNs) are rare, low-grade malignant epithelial neoplasms of enigmatic origin. Patients are typically young women with a mean age of 25 years and often present with nonspecific abdominal symptoms.[48,49] SPNs can occur throughout the pancreas and generally form a solitary, well-demarcated solid and cystic lesion with an average diameter of 7.5 cm. As their name would imply, SPNs are histologically composed of poorly cohesive, monomorphic cells forming a heterogeneous growth pattern of solid, pseudopapillary and hemorrhagic, pseudocystic structures.

Whole-exome sequencing has detected a relative paucity of genetic alterations in SPNs.[35] The only recurrent mutations identified were those located in the oncogene *CTNNB1*. CTNNB1 is a key mediator of the Wnt signaling pathway. Under normal conditions, CTNNB1 is primarily located at the plasma cell membrane and targeted for degradation by phosphorylation. Missense mutations within exon 3 of *CTNNB1*, as those found in SPNs, inhibit phosphorylation, resulting in cytoplasmic and eventual nuclear accumulation of CTNNB1.[50,51] On nuclear translocation, mutant CTNNB1 interacts with the DNA-bound lymphoid enhancer-binding protein 1/T-cell factor (LEF1/TCF) resulting in transcriptional activation of Wnt-responsive genes including those that regulate cell cycle and survival.[52] Mutations in *TP53* and *PIK3CA* have been described in SPNs; however, these are rare findings.[36] In addition, SPNs do not harbor mutations in *KRAS*, *GNAS*, *RNF43*, *PTEN*, *CDKN2A*, *SMAD4*, and *VHL*.[35,36]

NON-NEOPLASTIC CYSTS

Genetic alterations in non-neoplastic cysts, such as retention cysts (**Fig. 7**), pseudocysts,

lymphoepithelial cysts, and squamoid cysts of pancreatic ducts have not been reported.[8–10] Interestingly, targeted next-generation sequencing has also failed to reveal mutations in acinar cell cystadenomas, consistent with the notion that these cysts represent the benign cystic transformation of both acinar and ductal epithelium.[10,53]

DIAGNOSTIC DNA TESTING OF PANCREATIC CYST FLUID

Considering the differences in genetic profiles of the major pancreatic cysts, it is not surprising that there have been many attempts to assess DNA-based markers in cyst fluid as a means of diagnostic classification and prognostication. In a pilot study, Khalid and colleagues[7,54] prospectively evaluated the presence of mutations in *KRAS* and allelic imbalance in 7 tumor suppressor genes by Sanger sequencing in preoperative pancreatic cyst fluid. The investigators found the combination of *KRAS* mutations and allelic loss was predictive of advanced neoplasia in a pancreatic cyst with a sensitivity of 91% and specificity of 93%.[7] These results were expanded into a multicenter prospective study (Pancreatic Cyst DNA Analysis Study or PANDA study) of 113 patients with pancreatic cysts who underwent surgical resection or had diagnostic cytology.[55] Pancreatic cyst fluid was collected preoperatively by EUS-FNA and assessed for *KRAS* mutations and the overall fraction of alleles lost compared with germ-line DNA (mean allelic loss amplitude or MALA).

In the PANDA study, the presence of mutant *KRAS* alone had a sensitivity and specificity of 45% and 96%, respectively, for a mucinous pancreatic cyst, but was not predictive of advanced neoplasia. In contrast, a high MALA (>82%) had 90% sensitivity and 67% specificity for advanced neoplasia. But there were a number of weaknesses in the study design that diminished the overall significance of these results. Notably, it was unclear if DNA analysis would add value to established management guidelines (eg, Sendai, Fukuoka, and AGA). Furthermore, there was concern that MALA may be confounded by DNA degradation, gastrointestinal contamination during EUS-FNA, and other variables due to the underlying heterogeneity in cyst fluid composition. Indeed, follow-up studies demonstrated broad variability in agreement between molecular and clinical diagnoses. Shen and colleagues[56] reported an 89% concordance between molecular and clinical consensus diagnoses. Conversely, Panarelli and colleagues[57] and Toll and

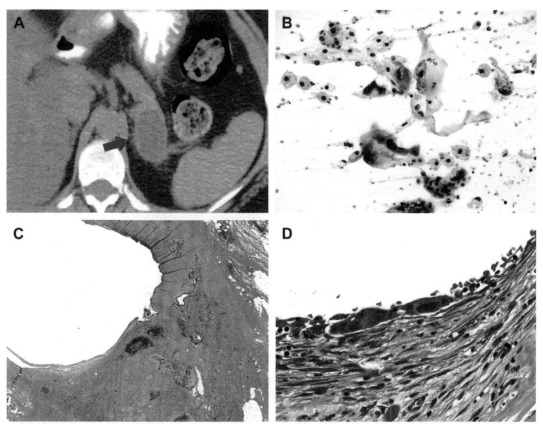

Fig. 7. DNA-based molecular testing of a retention cyst. (*A*) Contrast-enhanced CT imaging demonstrated a 4.2-cm cyst (*red arrow*) within the pancreatic tail of a patient with a history of chronic pancreatitis. (*B*) The subsequent cytology specimen revealed epithelial cells with marked nuclear enlargement, anisonucleosis and prominent nucleoli that was interpreted as positive for adenocarcinoma (Pap stain, original magnification ×600). However, molecular analysis detected no mutations in *KRAS*, *GNAS*, *VHL*, *TP53*, *PIK3CA*, or *PTEN*. (*C*) The surgical resection specimen was consistent with a retention cyst in a background of marked chronic pancreatitis (H&E, original magnification ×40). (*D*) Within the retention cyst lining, reactive ductal epithelium similar to the cytologic findings was identified (H&E, original magnification ×600).

colleagues[58] reported a concordance rate of 39% and 56%, respectively.

Regardless of the issues with MALA, *KRAS* testing proved to be a cost-effective strategy to identify patients with mucinous pancreatic cysts. In a cohort of more than 600 patients, Nikiforova and colleagues[8] found mutant *KRAS* had 54% sensitivity and 100% specificity for a mucinous cyst. This assay was superior to CEA testing alone and used significantly less pancreatic cyst fluid for analysis. Moreover, the combination of *KRAS* point mutations and elevated CEA improved the sensitivity to 83% and maintained a high specificity of 85%. The sensitivity of molecular analysis for mucinous cysts was further increased by the addition of *GNAS* testing. Singhi and colleagues[9] showed the detection of mutant *KRAS* and/or *GNAS* had a sensitivity and specificity of 65%

and 100%, respectively. However, there was significant discordance between the rates of detection of *KRAS* and *GNAS* mutations between preoperative EUS-FNA and postoperative pancreatic cyst fluid. The investigators underscored the possibility that limitations of their assay may be due to the inherent sensitivity and specimen requirements of conventional Sanger sequencing.

The limit of detection of Sanger sequencing is approximately 15% to 20% of mutant alleles. In comparison, next-generation sequencing has a limit of detection of approximately 3% to 5% of mutant alleles. Recent studies have shown the application of next-generation sequencing technologies to pancreatic cyst fluid ranges from 86% to 90% in sensitivity and 75% to 100% in specificity for mucinous differentiation.[10,59] Other advantages of next-generation sequencing are the small amounts of DNA required for analysis and the ability to assay multiple genes simultaneously. Using a broad panel of genes to include KRAS, GNAS, VHL, TP53, CDKN2A, and SMAD4, among others, Jones and colleagues[59] identified a high concordance rate between molecular and clinical diagnoses. Similarly, Singhi and colleagues[10] showed the detection of VHL alterations to be highly correlative with a diagnosis of serous cystadenoma. In addition, the investigators found mutations in TP53, PIK3CA, and/or PTEN to have 83% sensitivity and 97% specificity in detecting advanced neoplasia.

INTEGRATION OF DIAGNOSTIC DNA TESTING IN THE MANAGEMENT OF PANCREATIC CYSTS

As diagnostic DNA testing of pancreatic cyst fluid continues to evolve, questions remain as to how these alterations will influence patient management. As a first step, the University of Pittsburgh Medical Center proposed a novel pathway of pancreatic cyst evaluation and management that integrates ancillary pancreatic cyst fluid testing to include molecular analysis (Fig. 8).[10] In this approach, an EUS-FNA is recommended for pancreatic cysts not involving the main pancreatic duct and that meet any of the following MRI criteria: 1.5 cm or larger in size, dilated main pancreatic duct (>5 mm), a mural nodule/solid component, clinical symptoms that can be attributed to the pancreatic cyst, and a family history of pancreatic cancer. On EUS-FNA evaluation, pancreatic cyst fluid should be submitted for cytology, CEA analysis, and molecular testing. Molecular testing is to be performed using next-generation sequencing for KRAS, GNAS, VHL, TP53, PIK3CA, and PTEN. Indications for surgery are (1) suspicious or positive cytology; (2) a mucinous cyst 3 cm or larger in size (based on EUS findings, cytology, elevated CEA, or mutant KRAS and/or GNAS), with main duct dilatation (≥5 mm) and/or a definitive mural nodule (uniform echogenic nodule without a lucent center or hyperechoic rim); or (3) the identification of KRAS and/or GNAS mutations in association with mutant TP53 and either a PIK3CA or PTEN mutation. If none of these criteria are met, various management plans are considered based on imaging, cytology, CEA, and molecular findings (Fig. 9).

For a mucinous cyst smaller than 3 cm with main duct dilatation and/or a definitive mural nodule, or the detection of mutant KRAS and/or GNAS in conjunction with a TP53, PIK3CA, or PTEN mutation, a close surveillance plan should be followed that consists of alternating MRI and EUS-FNA every 6 to 12 months. In addition, for young, fit patients, surgery should be strongly considered. Patients with mucinous cysts as assessed by imaging, cytology, elevated CEA, or the presence of KRAS and/or GNAS mutations, but who do not satisfy the aforementioned criteria should undergo biennial MRI surveillance for up to 5 years. If at any time during these 5 years an interval change in cyst features occurs, the cyst should be reevaluated by EUS-FNA with ancillary studies. For young to middle-age patients, continuation of MRI surveillance is suggested after 5 years based on a discussion between the patient and his or her physician. Management of pancreatic cysts that are not consistent with mucinous differentiation by imaging, cytology, or CEA, and lack any genetic alterations should be discussed with the patient's physician. As these cases most likely represent non-neoplastic cysts, such as pseudocysts, retention cysts, and lymphoepithelial cysts, among others, no surveillance should be considered. However, as there is a small margin of error in detecting a mucinous cyst using the described diagnostic modalities, biennial MRI surveillance for 5 years may be warranted and should be based on a discussion with the patient and his or her physician. Last, imaging, cytology, and/or the presence of VHL alterations that are consistent with a serous cystadenoma should require no further imaging surveillance unless the patient becomes clinically symptomatic due to the cyst.

This pathway bears resemblance to the Sendai, Fukuoka, and AGA guidelines, but key differences include a lower threshold for initial EUS-FNA evaluation, reliance on ancillary studies, and individualization of patient management based on imaging, cytology, and fluid analysis. Although this approach has not been prospectively or independently validated, it is reported to detect advanced neoplasia with 100% sensitivity, 90% specificity, 79% positive predictive value, and 100% negative predictive value.[10] The authors admit that on validation, additional criteria or changes within this algorithm may be necessary, such as the inclusion of acute pancreatitis or new-onset diabetes.

FUTURE TRENDS

The growing use and development of sophisticated abdominal imaging has been a double-edged sword. There has been a dramatic increase in the detection of incidental pancreatic cysts, which results in significant patient and physician anxiety with regard to management decisions. DNA-based diagnostic testing offers a reliable

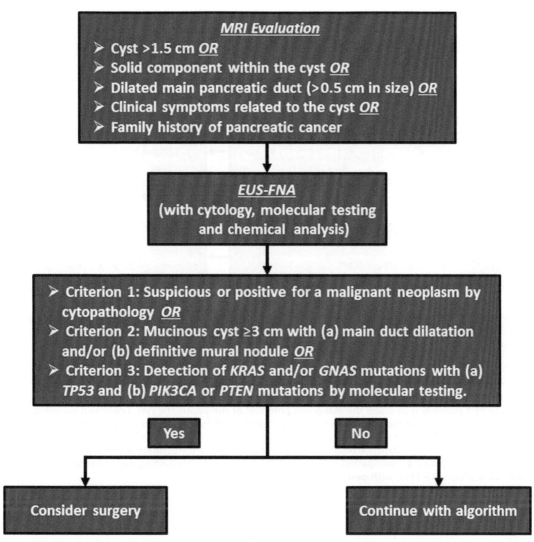

Fig. 8. UPMC pancreatic cyst pathway. The pancreatic cyst pathway is an algorithmic approach to the management of pancreatic cysts with integration of DNA analysis. Referral for EUS-FNA is based on the presence of specific MRI criteria. In addition, EUS-FNA includes cytopathologic evaluation, CEA analysis, and molecular testing for *KRAS*, *GNAS*, *VHL*, *TP53*, *PIK3CA*, and *PTEN*. Selection criteria for surgery should be demonstration of either a malignant neoplasm (eg, adenocarcinoma, cystic pancreatic neuroendocrine tumor) or a mucinous cyst 3 cm or larger with concerning features. A mucinous cyst is defined by mucinous cytopathology, elevated CEA, and/or detection of *KRAS* and/or *GNAS* mutations by molecular testing. Concerning features include an associated main duct dilatation, the presence of a definitive mural nodule, and/or the combination of *TP53* and either *PIK3CA* or *PTEN* mutations. (*Adapted from* Singhi AD, Zeh HJ, Brand RE, et al. American Gastroenterological Association guidelines are inaccurate in detecting pancreatic cysts with advanced neoplasia: a clinicopathologic study of 225 patients with supporting molecular data. Gastrointest Endosc 2016;83(6):1107–1117; with permission.)

methodology to not only classify a pancreatic cyst, but the potential to identify the likelihood of progression to PDAC. However, the utility of DNA testing is not limited to the evaluation of pancreatic cysts alone, but could be extended to other types of pancreatobiliary specimens, such as biliary duct brushings and FNAs of solid lesions. As research continues to unravel the underlying genetics and pathogenesis of neoplasms arising within the pancreatobiliary system, these studies should lead to diagnostic targets, which, in time, may prove to be invaluable for early detection strategies and treatment stratification.

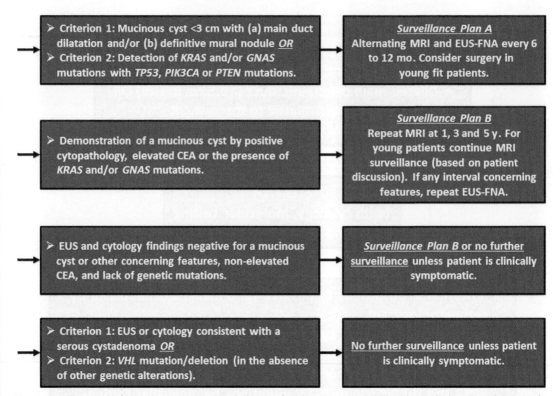

> Criterion 1: Mucinous cyst <3 cm with (a) main duct dilatation and/or (b) definitive mural nodule _OR_
> Criterion 2: Detection of _KRAS_ and/or _GNAS_ mutations with _TP53_, _PIK3CA_ or _PTEN_ mutations.

Surveillance Plan A
Alternating MRI and EUS-FNA every 6 to 12 mo. Consider surgery in young fit patients.

> Demonstration of a mucinous cyst by positive cytopathology, elevated CEA or the presence of _KRAS_ and/or _GNAS_ mutations.

Surveillance Plan B
Repeat MRI at 1, 3 and 5 y. For young patients continue MRI surveillance (based on patient discussion). If any interval concerning features, repeat EUS-FNA.

> EUS and cytology findings negative for a mucinous cyst or other concerning features, non-elevated CEA, and lack of genetic mutations.

Surveillance Plan B or no further surveillance unless patient is clinically symptomatic.

> Criterion 1: EUS or cytology consistent with a serous cystadenoma _OR_
> Criterion 2: _VHL_ mutation/deletion (in the absence of other genetic alterations).

No further surveillance unless patient is clinically symptomatic.

Fig. 9. Continuation of UPMC pancreatic cyst pathway. If surgery is not initially indicated, the second step in the pancreatic cyst pathway is to assess the malignant potential of the pancreatic cyst. Continued surveillance and frequency of surveillance is based on both EUS findings and ancillary studies. Concerning features include cyst size 3 cm or larger, main duct dilatation, or the presence of a mural nodule. (_Adapted from_ Singhi AD, Zeh HJ, Brand RE, et al. American Gastroenterological Association guidelines are inaccurate in detecting pancreatic cysts with advanced neoplasia: a clinicopathologic study of 225 patients with supporting molecular data. Gastrointest Endosc 2016;83(6):1107–1117; with permission.)

REFERENCES

1. Spinelli KS, Fromwiller TE, Daniel RA, et al. Cystic pancreatic neoplasms: observe or operate. Ann Surg 2004;239:651–7, [discussion: 657–9].
2. Laffan TA, Horton KM, Klein AP, et al. Prevalence of unsuspected pancreatic cysts on MDCT. AJR Am J Roentgenol 2008;191:802–7.
3. Sohn TA, Yeo CJ, Cameron JL, et al. Intraductal papillary mucinous neoplasms of the pancreas: an updated experience. Ann Surg 2004;239:788–97, [discussion: 797–9.
4. Schnelldorfer T, Sarr MG, Nagorney DM, et al. Experience with 208 resections for intraductal papillary mucinous neoplasm of the pancreas. Arch Surg 2008;143:639–46, [discussion: 646].
5. Hruban RH, Maitra A, Kern SE, et al. Precursors to pancreatic cancer. Gastroenterol Clin North Am 2007;36:831–49, vi.
6. Matthaei H, Schulick RD, Hruban RH, et al. Cystic precursors to invasive pancreatic cancer. Nat Rev Gastroenterol Hepatol 2011;8:141–50.
7. Khalid A, McGrath KM, Zahid M, et al. The role of pancreatic cyst fluid molecular analysis in predicting cyst pathology. Clin Gastroenterol Hepatol 2005;3:967–73.
8. Nikiforova MN, Khalid A, Fasanella KE, et al. Integration of KRAS testing in the diagnosis of pancreatic cystic lesions: a clinical experience of 618 pancreatic cysts. Mod Pathol 2013;26:1478–87.
9. Singhi AD, Nikiforova MN, Fasanella KE, et al. Preoperative GNAS and KRAS testing in the diagnosis of pancreatic mucinous cysts. Clin Cancer Res 2014;20:4381–9.
10. Singhi AD, Zeh HJ, Brand RE, et al. American Gastroenterological Association guidelines are inaccurate in detecting pancreatic cysts with advanced neoplasia: a clinicopathologic study of 225 patients with supporting molecular data. Gastrointest Endosc 2015, [Epub ahead of print].
11. Tanaka M, Chari S, Adsay V, et al. International consensus guidelines for management of intraductal papillary mucinous neoplasms and mucinous cystic neoplasms of the pancreas. Pancreatology 2006;6: 17–32.

12. Pelaez-Luna M, Chari ST, Smyrk TC, et al. Do consensus indications for resection in branch duct intraductal papillary mucinous neoplasm predict malignancy? A study of 147 patients. Am J Gastroenterol 2007;102:1759–64.

13. Tang RS, Weinberg B, Dawson DW, et al. Evaluation of the guidelines for management of pancreatic branch-duct intraductal papillary mucinous neoplasm. Clin Gastroenterol Hepatol 2008;6: 815–9, [quiz: 719].

14. Buscaglia JM, Giday SA, Kantsevoy SV, et al. Patient- and cyst-related factors for improved prediction of malignancy within cystic lesions of the pancreas. Pancreatology 2009;9:631–8.

15. Gomez D, Rahman SH, Wong LF, et al. Predictors of malignant potential of cystic lesions of the pancreas. Eur J Surg Oncol 2008;34:876–82.

16. Mimura T, Masuda A, Matsumoto I, et al. Predictors of malignant intraductal papillary mucinous neoplasm of the pancreas. J Clin Gastroenterol 2010;44:e224–9.

17. Genevay M, Mino-Kenudson M, Yaeger K, et al. Cytology adds value to imaging studies for risk assessment of malignancy in pancreatic mucinous cysts. Ann Surg 2011;254:977–83.

18. Schmidt CM, White PB, Waters JA, et al. Intraductal papillary mucinous neoplasms: predictors of malignant and invasive pathology. Ann Surg 2007;246: 644–51, [discussion: 651–4].

19. Wiesenauer CA, Schmidt CM, Cummings OW, et al. Preoperative predictors of malignancy in pancreatic intraductal papillary mucinous neoplasms. Arch Surg 2003;138:610–7, discussion: 617–8.

20. Salvia R, Fernandez-del Castillo C, Bassi C, et al. Main-duct intraductal papillary mucinous neoplasms of the pancreas: clinical predictors of malignancy and long-term survival following resection. Ann Surg 2004;239:678–85, [discussion: 685–7].

21. Tanaka M, Fernandez-del Castillo C, Adsay V, et al. International consensus guidelines 2012 for the management of IPMN and MCN of the pancreas. Pancreatology 2012;12:183–97.

22. Kaimakliotis P, Riff B, Pourmand K, et al. Sendai and Fukuoka consensus guidelines identify advanced neoplasia in patients with suspected mucinous cystic neoplasms of the pancreas. Clin Gastroenterol Hepatol 2015;13:1808–15.

23. Vege SS, Ziring B, Jain R, et al. American Gastroenterological Association institute guidelines on the diagnosis and management of asymptomatic neoplastic pancreatic cysts. Gastroenterology 2015;148:819–22, [quiz: 12–3].

24. Scheiman JM, Hwang JH, Moayyedi P. American Gastroenterological Association technical review on the diagnosis and management of asymptomatic neoplastic pancreatic cysts. Gastroenterology 2015; 148:824–48.e22.

25. Canto MI, Hruban RH. Managing pancreatic cysts: less is more? Gastroenterology 2015;148:688–91.

26. Fernandez-Del Castillo C, Tanaka M. Management of pancreatic cysts: the evidence is not here yet. Gastroenterology 2015;148:685–7.

27. Lennon AM, Ahuja N, Wolfgang CL. AGA guidelines for the management of pancreatic cysts. Gastroenterology 2015;149:825.

28. Ahmad NA, Kochman ML, Brensinger C, et al. Interobserver agreement among endosonographers for the diagnosis of neoplastic versus non-neoplastic pancreatic cystic lesions. Gastrointest Endosc 2003;58:59–64.

29. Pitman MB, Lewandrowski K, Shen J, et al. Pancreatic cysts: preoperative diagnosis and clinical management. Cancer Cytopathol 2010;118:1–13.

30. Maker AV, Lee LS, Raut CP, et al. Cytology from pancreatic cysts has marginal utility in surgical decision-making. Ann Surg Oncol 2008;15:3187–92.

31. Khalid A, Brugge W. ACG practice guidelines for the diagnosis and management of neoplastic pancreatic cysts. Am J Gastroenterol 2007;102:2339–49.

32. Shimizu Y, Yamaue H, Maguchi H, et al. Predictors of malignancy in intraductal papillary mucinous neoplasm of the pancreas: analysis of 310 pancreatic resection patients at multiple high-volume centers. Pancreas 2013;42:883–8.

33. Wu J, Matthaei H, Maitra A, et al. Recurrent GNAS mutations define an unexpected pathway for pancreatic cyst development. Sci Transl Med 2011;3:92ra66.

34. Tan MC, Basturk O, Brannon AR, et al. GNAS and KRAS mutations define separate progression pathways in intraductal papillary mucinous neoplasm-associated carcinoma. J Am Coll Surg 2015;220: 845–54.e1.

35. Wu J, Jiao Y, Dal Molin M, et al. Whole-exome sequencing of neoplastic cysts of the pancreas reveals recurrent mutations in components of ubiquitin-dependent pathways. Proc Natl Acad Sci U S A 2011;108:21188–93.

36. Springer S, Wang Y, Dal Molin M, et al. A combination of molecular markers and clinical features improve the classification of pancreatic cysts. Gastroenterology 2015;149:1501–10.

37. Kanda M, Sadakari Y, Borges M, et al. Mutant TP53 in duodenal samples of pancreatic juice from patients with pancreatic cancer or high-grade dysplasia. Clin Gastroenterol Hepatol 2013;11:719–30.e5.

38. Garcia-Carracedo D, Chen ZM, Qiu W, et al. PIK3CA mutations in mucinous cystic neoplasms of the pancreas. Pancreas 2014;43:245–9.

39. Sasaki S, Yamamoto H, Kaneto H, et al. Differential roles of alterations of p53, p16, and SMAD4 expression in the progression of intraductal papillary-mucinous tumors of the pancreas. Oncol Rep 2003;10:21–5.

40. Biankin AV, Biankin SA, Kench JG, et al. Aberrant p16(INK4A) and DPC4/Smad4 expression in intraductal papillary mucinous tumours of the pancreas is associated with invasive ductal adenocarcinoma. Gut 2002;50:861–8.

41. Zamboni G, Scarpa A, Bogina G, et al. Mucinous cystic tumors of the pancreas: clinicopathological features, prognosis, and relationship to other mucinous cystic tumors. Am J Surg Pathol 1999;23:410–22.

42. Gil E, Choi SH, Choi DW, et al. Mucinous cystic neoplasms of the pancreas with ovarian stroma. ANZ J Surg 2013;83:985–90.

43. Crippa S, Salvia R, Warshaw AL, et al. Mucinous cystic neoplasm of the pancreas is not an aggressive entity: lessons from 163 resected patients. Ann Surg 2008;247:571–9.

44. Jimenez RE, Warshaw AL, Z'Graggen K, et al. Sequential accumulation of K-ras mutations and p53 overexpression in the progression of pancreatic mucinous cystic neoplasms to malignancy. Ann Surg 1999;230:501–9, [discussion: 509–11].

45. Kimura W, Moriya T, Hirai I, et al. Multicenter study of serous cystic neoplasm of the Japan Pancreas Society. Pancreas 2012;41:380–7.

46. Valsangkar NP, Morales-Oyarvide V, Thayer SP, et al. 851 resected cystic tumors of the pancreas: a 33-year experience at the Massachusetts General Hospital. Surgery 2012;152:S4–12.

47. Jais B, Rebours V, Malleo G, et al. Serous cystic neoplasm of the pancreas: a multinational study of 2622 patients under the auspices of the International Association of Pancreatology and European Pancreatic Club (European Study Group on Cystic Tumors of the Pancreas). Gut 2016;65:305–12.

48. Estrella JS, Li L, Rashid A, et al. Solid pseudopapillary neoplasm of the pancreas: clinicopathologic and survival analyses of 64 cases from a single institution. Am J Surg Pathol 2014;38:147–57.

49. Law JK, Ahmed A, Singh VK, et al. A systematic review of solid-pseudopapillary neoplasms: are these rare lesions? Pancreas 2014;43:331–7.

50. Abraham SC, Klimstra DS, Wilentz RE, et al. Solid-pseudopapillary tumors of the pancreas are genetically distinct from pancreatic ductal adenocarcinomas and almost always harbor beta-catenin mutations. Am J Pathol 2002;160:1361–9.

51. Tanaka Y, Kato K, Notohara K, et al. Frequent beta-catenin mutation and cytoplasmic/nuclear accumulation in pancreatic solid-pseudopapillary neoplasm. Cancer Res 2001;61:8401–4.

52. Singhi AD, Lilo M, Hruban RH, et al. Overexpression of lymphoid enhancer-binding factor 1 (LEF1) in solid-pseudopapillary neoplasms of the pancreas. Mod Pathol 2014;27:1355–63.

53. Singhi AD, Norwood S, Liu TC, et al. Acinar cell cystadenoma of the pancreas: a benign neoplasm or non-neoplastic ballooning of acinar and ductal epithelium? Am J Surg Pathol 2013;37:1329–35.

54. Khalid A, Pal R, Sasatomi E, et al. Use of microsatellite marker loss of heterozygosity in accurate diagnosis of pancreaticobiliary malignancy from brush cytology samples. Gut 2004;53:1860–5.

55. Khalid A, Zahid M, Finkelstein SD, et al. Pancreatic cyst fluid DNA analysis in evaluating pancreatic cysts: a report of the PANDA study. Gastrointest Endosc 2009;69:1095–102.

56. Shen J, Brugge WR, Dimaio CJ, et al. Molecular analysis of pancreatic cyst fluid: a comparative analysis with current practice of diagnosis. Cancer 2009;117:217–27.

57. Panarelli NC, Sela R, Schreiner AM, et al. Commercial molecular panels are of limited utility in the classification of pancreatic cystic lesions. Am J Surg Pathol 2012;36:1434–43.

58. Toll AD, Kowalski T, Loren D, et al. The added value of molecular testing in small pancreatic cysts. JOP 2010;11:582–6.

59. Jones M, Zheng Z, Wang J, et al. Impact of next-generation sequencing on the clinical diagnosis of pancreatic cysts. Gastrointest Endosc 2016;83:140–8.

Molecular Pathogenesis and Diagnostic, Prognostic and Predictive Molecular Markers in Sarcoma

CrossMark

Adrián Mariño-Enríquez, MD, PhD[a],*,
Judith V.M.G. Bovée, MD, PhD[b]

KEYWORDS

- Molecular diagnostics • Molecular markers • Soft tissue tumor • Bone • Sarcoma • GIST

Key points

- Sarcomas are characterized by notable morphologic and molecular heterogeneity. Molecular studies in the clinical setting provide refinements to morphologic classification, and contribute diagnostic and predictive information.

- Sarcomas with simple genome can be driven by transcriptional deregulation, abnormal kinase signaling, or epigenetic reprogramming. This group of sarcomas can be identified with specific molecular markers.

- Sarcomas with complex genome show multiple, nonrecurrent molecular alterations. There are no specific molecular diagnostic markers for these tumors. Some prognostic information may be derived from loss of tumor suppressor genes. The high mutational load may make these tumors good candidates for immunotherapies relying on neoantigens.

ABSTRACT

Sarcomas are infrequent mesenchymal neoplasms characterized by notable morphological and molecular heterogeneity. Molecular studies in sarcoma provide refinements to morphologic classification, and contribute diagnostic information (frequently), prognostic stratification (rarely) and predict therapeutic response (occasionally). Herein, we summarize the major molecular mechanisms underlying sarcoma pathogenesis and present clinically useful diagnostic, prognostic and predictive molecular markers for sarcoma. Five major molecular alterations are discussed, illustrated with representative sarcoma types, including 1. the presence of chimeric transcription factors, in vascular tumors; 2. abnormal kinase signaling, in gastrointestinal stromal tumor; 3. epigenetic deregulation, in chondrosarcoma, chondroblastoma, and other tumors; 4. deregulated cell survival and proliferation, due to focal copy number alterations, in dedifferentiated liposarcoma; 5. extreme genomic instability, in conventional osteosarcoma as a representative example of sarcomas with highly complex karyotype.

OVERVIEW

Sarcomas are infrequent malignant mesenchymal neoplasms characterized by notable morphologic and molecular heterogeneity. The current World Health Organization classification recognizes more than 100 soft tissue and bone tumor types,

Financial Support: A. Mariño-Enríquez receives research support from The Sarcoma Alliance for Research through Collaboration (SARC sarcoma SPORE - U54 CA168512).

Conflict of Interest: The authors have nothing to disclose.

[a] Department of Pathology, Brigham and Women's Hospital, Harvard Medical School, 75 Francis Street, Boston, MA 02115, USA; [b] Department of Pathology, Leiden University Medical Center, Albinusdreef 2, Leiden 2333 ZA, The Netherlands

* Corresponding author.

E-mail address: admarino@partners.org

http://dx.doi.org/10.1016/j.path.2016.04.009

surgpath.theclinics.com

more than 70 of which are sarcomas,[1] illustrating a nosologic complexity that reflects biologic complexity and leads to substantial challenges in diagnosis and clinical management. Sarcoma diagnosis is based on morphology, immunohisto-chemistry, and clinicopathological correlation. In addition, molecular studies in the clinical setting provide refinements to morphologic sarcoma clas-sification, and contribute diagnostic information (frequently), prognostic stratification (rarely), and predictive information concerning specific thera-pies (only occasionally, but most excitingly). Much of the current molecular understanding of sarcomas derives from conventional karyotypic analysis, which has been extremely fruitful in this field over the past 25 years.[2] At the cytogenetic level, a binary distinction between sarcomas with simple karyotype versus those with complex kar-yotype has been long recognized, and provides a simple conceptual framework of some academic value but limited clinicopathological significance.[3] The molecular correlates of these cytogenetic pre-sentations are recurrent genomic rearrangements and activating gene mutations for sarcomas with relatively simple karyotype, and multiple, diverse genomic events including gene amplifications and nonrecurrent rearrangements, for those with complex karyotype. Biologically, oncogenic mechanisms are better understood for sarcomas with simple karyotype, and fall typically into 2 broad categories: transcriptional deregulation or deregulated signaling. This is in contrast to sar-comas with highly complex karyotypes, which typically do not harbor single "driver" genetic alter-ations, and rather display nonspecific molecular changes that promote oncogenic traits, such as cell cycle deregulation or genomic instability.

In this review, we summarize the major molecular mechanisms that underlie sarcoma pathogenesis, highlighting those alterations that provide diag-nostic, prognostic, or predictive information (the so-called "clinically actionable" alterations). The discussion focuses on representative mesen-chymal tumor types, following a pathogenic classification combining cytogenetic/genomic in-formation and molecular biologic features, as sum-marized in **Box 1**. We address 5 major molecular alterations, including the presence of chimeric tran-scription factors, in vascular tumors; deregulated kinase signaling, in gastrointestinal stromal tumor (GIST); epigenetic deregulation by oncometabo-lites, as a result of metabolic enzyme mutations in chondrosarcoma and other tumor types; deregu-lated cell survival and proliferation, due to extreme copy number alterations, in dedifferentiated lipo-sarcoma (DDLPS); and extreme genomic instability in conventional osteosarcoma, as a representative example of sarcomas with highly complex karyo-type. **Table 1** provides a noncomprehensive list of clinically actionable genetic alterations commonly encountered in soft tissue and bone tumors.

SARCOMAS WITH SIMPLE GENOME

Sarcomas with simple genomic profiles usually harbor a recurrent molecular aberration, either a balanced chromosomal rearrangement or a muta-tion in a known oncogene or tumor suppressor gene, which is critical for tumorigenesis and is considered the main oncogenic driver. These al-terations are usually present in the context of a relatively stable genome, with a low mutational load and a (near) diploid karyotype; additional point mutations or copy number alterations may occur during tumor progression, frequently following reproducible patterns, in contrast with the remarkable variability observed in genomically complex sarcomas (**Fig. 1**).

TUMORS WITH CHIMERIC TRANSCRIPTION FACTORS: VASCULAR TUMORS

An expanding list of mesenchymal tumors contain recurrent balanced chromosomal rearrangements,

---

Box 1
Molecular genetic categories of soft tissue and bone tumors, and representative tumor types

1. Sarcomas with simple genome
   a. Tumors with chimeric transcription factors and transcriptional deregulation; eg, vascular tumors
   b. Tumors with deregulated kinase signaling; eg, gastrointestinal stromal tumor
   c. Tumors driven by oncometabolites (via epigenetic deregulation); eg, chondrosarcoma
   d. Tumors driven by primary epigenetic deregulation; eg, chondroblastoma
2. Sarcomas with complex genome
   a. Tumors with characteristic copy number alterations; eg, dedifferentiated liposarcoma
   b. Tumors with highly complex karyotypes; eg, osteosarcoma

Table 1
Diagnostic, prognostic and predictive molecular markers in sarcoma

Type of Alteration	Genes	Entities	Clinical Value
Recurrent rearrangements	Fusion oncogenes	Sarcoma-type specific (reviewed by Mertens et al,[4] 2015)	Diagnostic/ Prognostic
Point mutations or small indels	KIT/PDGFRA, SDHA/B	GIST	Predictive/ Diagnostic
	CTNNB1	Desmoid tumor	Diagnostic
	IDH1, IDH2	Enchondroma/chondrosarcoma	Diagnostic
	SUZ12, EED	Malignant peripheral nerve sheath tumor	Diagnostic
	PIK3CA	Myxoid liposarcoma	Predictive
	KDR	Angiosarcoma	Diagnostic
	NRAS, KRAS, HRAS, FGFR4	Embryonal rhabdomyosarcoma	Diagnostic/ Predictive
	MYOD1	Spindle-cell rhabdomyosarcoma	Diagnostic
	MED12	Leiomyoma (and small subset of leiomyosarcoma)	Diagnostic
	NF1	MPNST and others	Diagnostic
Copy number gain/ Amplification	MDM2, CDK4	WD/DDLPS	Diagnostic
	CDK4		Predictive
	MYC	Postradiation sarcoma	Diagnostic
	MYOCD	Leiomyosarcoma	Diagnostic
Copy number loss/ Deletion	TP53	Osteosarcoma, leiomyosarcoma, and others	Prognostic
	SMARCB1	Rhabdoid tumor, epithelioid sarcoma	Diagnostic
	SMARCA4	SMARCA4-deficient thoracic sarcomas	Diagnostic
	CDKN2A	MPNST, fibrosarcomatous DFSP, advanced GIST	Predictive
	RB1	Spindle cell lipoma, and others	Diagnostic
		High-grade sarcomas with complex karyotype	Predictive
	NF1	MPNST and others	Diagnostic/ Predictive

Abbreviations: DDLPS, dedifferentiated liposarcoma; DFSP, dermatofibrosarcoma protuberans; GIST, gastrointestinal stromal tumor; MPNST, malignant peripheral nerve sheath tumor; WD, well-differentiated.

most often translocations, and most fusion genes produced by these rearrangements encode chimeric transcription factors that cause transcriptional deregulation. The best studied example is the *EWSR1-FLI1* and *EWSR1-ERG* fusions in Ewing sarcoma. Chimeric transcription factors are thought to deregulate the expression of specific repertoires of target genes, thereby orchestrating several of the defined "hallmarks of cancer."[5] The group of sarcomas carrying a specific translocation constituted approximately 15% to 20% of all sarcomas in 2007.[6] However, using next-generation sequencing (NGS) approaches that can detect cryptic alterations not detectable by conventional approaches, this group is rapidly expanding. Examples of the increased resolution of current methods include the detection of fusions by a paracentric inversion in solitary fibrous tumor,[7] and mesenchymal chondrosarcoma,[8] both of which have provided useful diagnostic immunohistochemical or molecular markers for the diagnosis of these entities. More recently, novel gene fusions have been discovered by transcriptome analysis in several vascular tumor types, a group of lesions characteristically difficult to classify because of overlapping morphology, which display a range of biological behaviors, including clinically benign, intermediate, and overtly malignant tumors.

Key Features
OF SARCOMAS WITH SIMPLE GENOME

- Simple karyotype, low mutational rate.

- Often show relatively monomorphic morphology.

- Wide range of clinical behavior.

- Clinically useful diagnostic molecular markers (amenable to detection by FISH, RT-PCR, sequencing, and occasionally immunohistochemistry).

- No therapies targeting chimeric transcription factors or epigenetic alterations (yet).

- Effective targeted therapies against most deregulated kinases.

- Prototypical examples: Ewing sarcoma, rhabdomyosarcoma, synovial sarcoma.

Vascular tumors of bone and soft tissue constitute a heterogeneous group of tumors displaying endothelial differentiation. Tumors range from benign (hemangioma) to intermediate (various types of hemangioendothelioma) to malignant (epithelioid hemangioendothelioma and angiosarcoma). Over the past decade, with the advance of NGS techniques, the molecular background of some of these lesions has been elucidated.

Epithelioid hemangioma (previously known as angiolymphoid hyperplasia with eosinophilia or histiocytoid hemangioma) is a benign (in soft tissue) or locally aggressive (in bone) neoplasm composed of cells that have an endothelial phenotype and epithelioid morphology.[1] Transcriptome sequencing of epithelioid hemangioma revealed a recurrent translocation breakpoint involving the *FOS* gene fused to different partners.[9,10] The break was observed in exon 4 of the *FOS* gene and the fusion event led to the introduction of a Stop codon, truncating the FOS protein and resulting in loss of the Trans-Activation Domain.[9] Atypical variants of epithelioid hemangioma were shown to harbor *ZFP36-FOSB* fusions.[11] The distinction between epithelioid hemangioma and angiosarcoma can be challenging, and detection of *FOS* rearrangements may assist in the differential diagnosis.[9,10,12] Treatment with curettage or marginal en bloc excision is usually sufficient for epithelioid hemangioma, whereas patients with angiosarcoma need more aggressive treatment.[13]

Interestingly, *FOSB* rearrangements are also the hallmark of pseudomyogenic hemangioendothelioma (a.k.a. epithelioid sarcomalike hemangioendothelioma), which is a rare, distinctive entity frequently presenting as multiple discontiguous nodules in different tissue planes of a limb in young adult men.[14,15] In epithelioid hemangioendothelioma (EHE), considered a low-grade angiosarcoma, cords of epithelioid endothelial cells are seen in a distinctive myxohyaline stroma.[1] EHE is characterized by the recurrent t(1;3) translocation, resulting in a *WWTR1-CAMTA1* fusion gene.[16,17] Interestingly, like epithelioid hemangioma, EHE also often presents as multifocal lesions, which are all monoclonal.[9,16] Immunohistochemistry for CAMTA1 can be used as a surrogate marker for the translocation (**Fig. 2**).[18,19] A specific entity has been described, focally resembling EHE, with more solid architecture admixed with the formation of well-formed vascular channels, genetically characterized by *YAP1-TFE3* fusions.[20] These tumors have strong immunoreactivity for TFE3 (see **Fig. 2D**).[20] Future studies should reveal if there is a final common pathway affected by these fusion genes that is involved in the development of these vascular tumors.

Unlike epithelioid hemangioma and the different types of hemangioendothelioma, in angiosarcomas, translocations are infrequent. Instead, angiosarcomas often show more complex genomic findings. Most radiation-induced angiosarcomas show *MYC* gene amplification,[21] which can be a helpful diagnostic tool, either using fluorescence in situ hybridization (FISH) or immunohistochemistry, in the differential diagnosis of atypical radiation-induced vascular proliferation versus angiosarcoma.[22] However, *MYC* amplification is not restricted to radiation-induced angiosarcomas, as it also can be found in a subset of primary angiosarcomas.[23,24] In addition to *MYC* amplification, co-amplification of *FLT4* or *PTPRB* and/or *PLCG1* mutations can be found in secondary angiosarcomas.[25] Angiosarcomas of the breast can present with *KDR* mutations (10%).[26] Many of these genes involve angiogenic signaling. Recently, *CIC* rearrangements or mutations have been found in 9% of primary angiosarcomas, which were predominantly epithelioid with solid growth and affected younger patients, with an inferior disease-free survival.[27]

TUMORS WITH DEREGULATED KINASE SIGNALING: GASTROINTESTINAL STROMAL TUMOR

Deregulation of cellular signaling driving sustained proliferation is a major hallmark of cancer[5] and, as

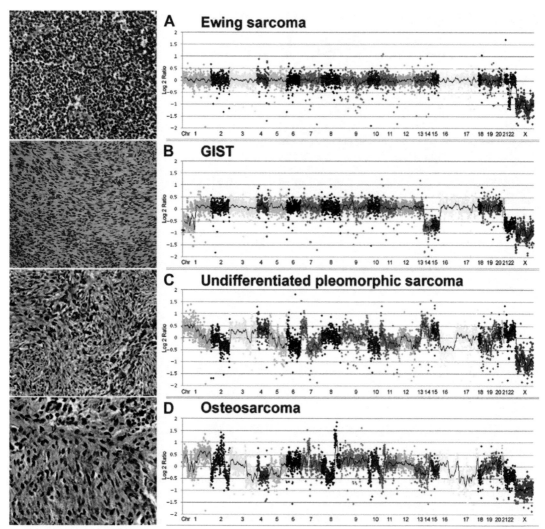

Fig. 1. Copy number profile of sarcomas with simple genome (*top*) in comparison with sarcomas with complex genome (*bottom*), as determined by a next generation sequencing platform.[28] These graphs represents the ratio of sequence coverage on the y axis (log 2 scale) plotted along the chromosomes, on the x axis. (*A*) Ewing sarcoma affecting a 9-year-old boy. Note the simple genomic profile. This tumor harbored a *EWSR1-FLI1* fusion, identified by this assay. (*B*) High-risk, spindle-cell intestinal GIST in a 60-year-old man. The tumor harbored a KIT K642E mutation detected by the assay. Note a relatively simple genomic profile, with near-diploid karyotype and loss of chromosomes 1p, 14q, 15q, and 22q, characteristic of advanced GIST. (*C*) Undifferentiated pleomorphic sarcoma arising in the deltoid of a 55-year-old man and (*D*) conventional osteosarcoma in the femur of a 7-year-old boy: multiple chromosomal gains and losses in a nonrecurrent pattern. Both these tumors showed alterations in *TP53* (copy number loss and truncating mutations).

such, contributes to the biology of most sarcoma types. In certain sarcoma types, however, signaling alteration due to kinase activation is the main oncogenic driver and, most likely, the initiating oncogenic event. Prominent examples among sarcoma include mutationally activated receptor tyrosine kinases, such as KIT in GIST, recurrent chimeric kinase fusions like anaplastic lymphoma kinase oncoproteins in inflammatory myofibroblastic tumor and *ETV6-NTRK3* in

infantile fibrosarcoma, or kinase receptors activated by autocrine mechanisms, such as PDGFR, in dermatofibrosarcoma protuberans. Advances in pharmacology have generated a collection of kinase inhibitors with variable potency that provide tremendous clinical benefit to patients affected by this group of sarcomas. The most remarkable example of clinically effective targeted inhibition of oncogenic kinase mutations in sarcoma is inhibition of KIT/PDGFRA in GIST.

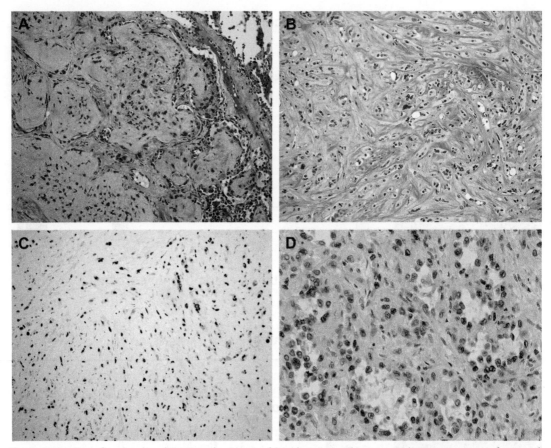

Fig. 2. Detection of molecular alterations by immunohistochemistry. Chromosomal rearrangements frequently result in overexpression of transcription factors that can be detected by immunohistochemistry. (*A*) Pulmonary epithelioid hemangioendothelioma with *WWTR1-CAMTA1* fusion, composed of small tumor nodules growing along the preexistent alveolar spaces. (*B*) Cords and strands of endothelial epithelioid cells, with intracytoplasmic lumina, embedded in a myxohyaline stroma characteristic of epithelioid hemangioendothelioma. (*C*) CAMTA1 expression in epithelioid hemangioendothelioma with *WWTR1-CAMTA1* gene fusion. (*D*) *YAP1-TFE3*-rearranged epithelioid hemangioendothelioma; TFE3 overexpression can be detected by immunohistochemistry in this unusually vasoformative variant of epithelioid hemangioendothelioma.

GIST is a model of oncogenic addiction: GIST cell viability and proliferation is absolutely dependent on signaling from the receptor tyrosine kinase KIT, which is constitutively active due to gain-of-function mutations in approximately 70% to 80% of cases.[29] An additional approximately 10% of GISTs are driven by analogous activating mutations in the receptor tyrosine kinase PDGFRA. PDGFRA-driven GIST show predilection for gastric location and an epithelioid phenotype.[30–32] KIT and PDGFRA mutations, which are mutually exclusive, lead to ligand-independent activation, which in turn activates intracellular signaling pathways controlling cell differentiation, survival, and proliferation.[33] *KIT* primary mutations usually affect exon 11 (70%), exon 9 (10%), exon 13 (1%), or exon 17 (1%), whereas *PDGFRA* mutations affect exons 18 (5%), 12 (1%), or 14

(<1%).[29] All of them are activating mutations, but the activation results from different alterations of various functional domains of the protein, which tolerate different kinds of mutational mechanisms (**Fig. 3**): the juxtamembrane domain, encoded by exons 11 in *KIT* and 12 in *PDGFRA*, is an autoinhibitory domain that can be disrupted by in-frame deletions, in-frame insertion-deletions, or point mutations. The 2 domains that form the split kinase, namely the ATP-binding pocket (exons 13–15) and the activation loop (exons 17 and 18), are usually activated by point mutations.

The diagnostic value of KIT mutations is limited in GIST, as the diagnosis is often achievable by the detection of KIT and DOG1 expression by immunohistochemistry, in the appropriate morphologic and clinical context. An infrequent exception would be rare cases of KIT-negative

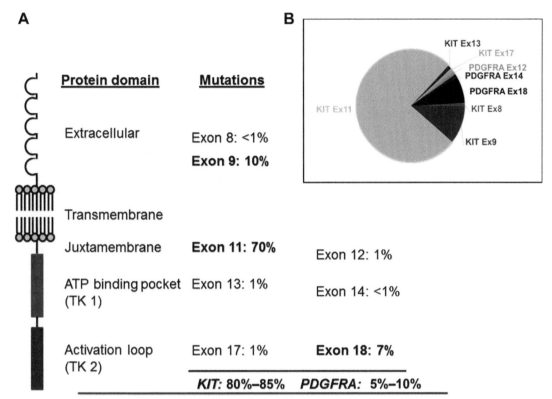

A

Protein domain   **Mutations**

Extracellular   Exon 8: <1%

                        Exon 9: 10%

Transmembrane

Juxtamembrane   **Exon 11: 70%**   Exon 12: 1%

ATP binding pocket   Exon 13: 1%   Exon 14: <1%
(TK 1)

Activation loop   Exon 17: 1%   **Exon 18: 7%**
(TK 2)

                        KIT: 80%–85%   *PDGFRA:* 5%–10%

B

KIT Ex13
KIT Ex17
PDGFRA Ex12
PDGFRA Ex14
PDGFRA Ex18
KIT Ex11
KIT Ex8
KIT Ex9

Fig. 3. (*A*) *KIT* mutations in untreated GISTs involve exons 11, 9, 13, and 17, encoding regions of the extracellular, juxtamembrane, ATP-binding pocket, and activation-loop domains, respectively. *PDGFRA* mutations, found in fewer than 10% of GISTs, involve analogous domains. (*B*) Relative frequency of the most common KIT and PDGFRA primary mutations in GIST.

GIST, which may lose KIT expression during tumor progression.[34,35] Regarding prognosis, some *KIT/PDGFRA* mutations are associated with poor clinical outcomes in untreated populations, such as in-frame deletions of codons 557 to 558 in exon 11.[36] However, the prognostic differences between mutations are minor, in comparison to their variable sensitivity to drug inhibitors, so that the most important clinical value of mutational analysis in GIST is prediction of response to therapy.

Most mutant KIT and PDGFRA oncoproteins can be effectively inhibited by small molecule kinase inhibitors, such as imatinib. The degree of inhibition, and hence the potential clinical benefit, correlates tightly with the specific mutation (**Box 2**): in general, exon 11 *KIT* mutations are extremely sensitive to imatinib, whereas exon 9 mutations, typically a 6-nucleotide duplication affecting codons 502 and 503, are less sensitive and require a higher dose of imatinib (usually 800 mg, double the standard dose).[37] Approximately 80% of patients with metastatic GIST initially respond to imatinib, 50% showing partial responses, 30% with stable disease, resulting in a 3-year survival rate of 69% to 74% and a median

overall survival of 5 years, compared with only 19 months in the pre-imatinib era.[38–40] Primary drug resistance to imatinib results mostly from PDGFRA D842V point mutation or KIT/PDGFRA wild-type status.[41] Of note, PDGFRA D842V mutation is cross-resistant to most tyrosine kinase inhibitors, with the potential exception of crenolanib that has shown activity in vitro.[42] Primary resistance to imatinib due to hyperactivation of signaling effectors downstream of KIT is possible, but much less common, and which reflects the need of activation of several signaling pathways by independent mechanisms for sustained proliferation in GIST.[43]

Even in patients with near-complete initial response to imatinib, secondary resistance develops due to the invariable presence of residual viable GIST cells, including drug-resistant subclones, which subsequently manifest as clinical progression.[44] Up to 50% of patients with GIST who initially respond to imatinib develop secondary resistance within 2 years of therapy. Resistance results in most cases from secondary mutations that affect nonrandom residues in KIT, typically in either the ATP-binding pocket (exons

13–15), or the kinase activation loop (exons 17 or 18 of *KIT* and exon 18 of *PDGFRA*), which occur *in cis* with the primary mutation (ie, in the same allele).[45–47] These secondary KIT mutations are almost never detectable in untreated GIST, likely reflecting their negative effect in cellular fitness in the absence of pharmacologic inhibitors. In the presence of an active inhibitor, however, activating mutations in these domains are selected, and clonal expansion of tumor cells may result in the presence of different secondary mutations in different tumor cell subpopulations in a single patient. A mutation assay with high sensitivity of detection is critical in the setting of therapeutic resistance to appropriately detect heterogeneous secondary mutations, which can be missed by Sanger sequencing.

The predictive nature of KIT mutations in relation to imatinib treatment response extends to sunitinib and regorafenib, and, essentially, to every tyrosine kinase inhibitor. Sunitinib is effective against mutations in exons 13 to 15 of *KIT* (ATP-binding pocket), but is ineffective against mutations in exons 17 to 18 (activation loop),[41] whereas the opposite is true for regorafenib.[48,49] Therefore, the sequence of treatment for advanced GIST determined historically, first-line imatinib, followed

by sunitinib and third-line regorafenib, is not surprisingly supported by the biology and the natural history of the various KIT/PDGFRA mutations in GIST. It is worth noting that sunitinib and regorafenib are less specific inhibitors than imatinib, with activity against a substantially wider range of kinases; this may explain part of their pharmacologic activity, and determines a higher incidence of significant side effects.

The 3 major signaling pathways activated by constitutive KIT and PDGFRA activation are the PI3K/AKT/mTOR pathway, the RAS/RAF/MAPK pathway, and the JAK/STAT pathway.[33,50] The latter pathway is known to be relevant in mast cell disease harboring *KIT* mutations, but plays only a limited role in GIST. The PI3K/AKT/mTOR and the RAS/RAF/MAPK pathways, on the other hand, are crucial for proliferation in GIST and may be further activated and contribute to disease progression in high-risk GIST or advanced tumors.[51,52] Although primary mutations in effectors downstream of KIT are not common, they can occur at the time of progression in the setting of therapeutic resistance[43,53,54]; activating mutations and loss of negative regulators of these pathways can be easily detected given the low mutation rate and relatively quiet copy number variation profile of GIST (see **Fig. 1**B).

TUMORS DRIVEN BY ONCOMETABOLITES (VIA EPIGENETIC DEREGULATION): CHONDROSARCOMA

Mutations in metabolic enzymes lead to deregulated cellular energetics in cancer cells and, more importantly, result in the production of metabolites that may alter tightly regulated physiologic processes such as gene expression and epigenetic regulation. Several metabolic enzymes are frequently mutated in particular tumor types. An illustrative example is the metabolic enzyme isocitrate dehydrogenase (IDH), in which somatic mutations were first described in gliomas,[55] followed by other tumors[56] including approximately 50% of chondrosarcomas.[57] Heterozygous somatic mosaic mutations in *IDH* were later found in up to 81% of patients with multiple enchondromas (Ollier disease/Maffucci syndrome).[58,59]

Mutations in *IDH* cause epigenetic changes[60–62] by the formation of a neoenzyme that catalyzes the reduction of α-ketoglutarate to D-2-hydroxyglutarate (D2HG).[63–65] D2HG is considered an oncometabolite and inhibits α-ketoglutarate–dependent oxygenases like TET2.[66,67] This results in inhibition of DNA demethylation, causing hypermethylation. Indeed, *IDH1* mutations are associated with a hypermethylated phenotype in cartilage tumors.[59] D2HG also inhibits other α-ketoglutarate–dependent oxygenases,[68,69] such as the Jumonji domain histone demethylases, thereby increasing histone methylation as well.[61] These epigenetic changes are thought to affect differentiation. Indeed, when mesenchymal stem cells are treated with D2HG, or when an *IDH* mutation is introduced, this results in inhibition of osteogenic differentiation and stimulation of chondrogenic differentiation, explaining the development of enchondromas during bone development.[70,71]

Chondrosarcoma can arise secondarily within a benign enchondroma, or as a primary tumor. It is the second most frequent primary bone malignancy, predominantly affecting adults.[1] The development of chondrosarcoma occurs through the acquisition of additional genetic alterations (multistep genetic progression model),[72] involving among others the pRb pathway.[73] In high-grade

chondrosarcomas, the *IDH* mutation is no longer essential for tumor growth.[74,75]

Detection of hotspot mutations in *IDH1* or *IDH2* can be useful in the differential diagnosis of chondrosarcoma versus chordoma or chondroblastic osteosarcoma, which can sometimes be challenging. A specific antibody against the IDH R132H mutation, widely used for the diagnosis of gliomas, permits detection of this specific mutation by immunohistochemistry, although it is a rare mutation in chondrosarcoma (**Fig. 4**A). *IDH* mutations are present in 87% of Ollier-associated enchondromas, 86% of secondary central chondrosarcoma, 38% to 70% of primary central chondrosarcoma, approximately 15% of periosteal chondrosarcoma, and 54% of dedifferentiated chondrosarcoma,[57–59,76] and are absent in peripheral chondrosarcoma, osteosarcoma, and chordoma.[57,59,77,78]

Other metabolic enzymes, including succinate dehydrogenase (SDH) and fumarate hydratase

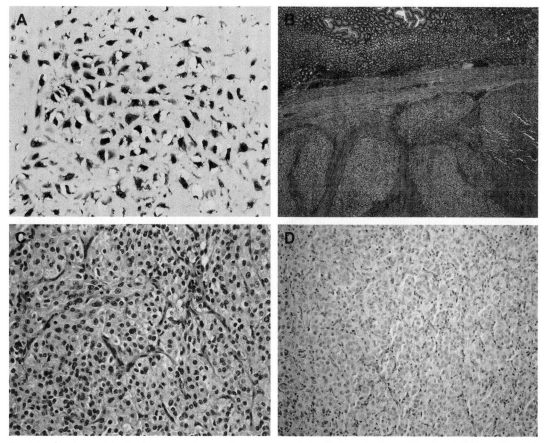

Fig. 4. Molecular metabolic aberrations leading to epigenetic deregulation. (*A*) Detection of R132H mutant IDH1 in chondrosarcoma. Note that this antibody only detects the specific R132H mutation, which is infrequent in chondrosarcoma, in contrast to gliomas in which it is the most common mutation. Thus, negative immunohistochemistry in chondrosarcoma does not rule out a mutation in IDH1. (*B*) Low-magnification view of SDH-deficient gastric GIST, demonstrating its characteristic multinodular growth pattern. (*C*) Epithelioid cytomorphology and (*D*) loss of SDHB expression in SDH-deficient gastric GIST. (*B–D, Courtesy of* Leona Doyle, Brigham and Women's Hospital, Boston, MA.)

(FH) are also known to be mutated in cancer and to cause defective energy metabolism as well as epigenetic deregulation in cancer cells. Inactivating mutations in subunits of mitochondrial complex II including the SDH subunit D (SDHD), C (SDHC), and B (SDHB) genes, are found in patients with head and neck paragangliomas and pheochromocytomas.[79] A subset of gastrointestinal stromal tumors, lacking mutations in KIT or PDGFRA, also carry mutations in one of the SDH genes[80] or an SDHC epimutation,[81] both of which are associated with global hypermethylation.[82] These gastric GISTs tend to affect young patients, and are morphologically distinct, with a multinodular architecture and epithelioid cytomorphology (Fig. 4B, C). Mutations in one of the SDH subunits destabilize the SDH complex, causing degradation and loss of SDHB. Immunohistochemistry for SDHB is therefore a surrogate marker for mutations in one of the SDH subunits (Fig. 4D).[83]

Inactivating germline mutations of FH cause autosomal dominant HLRCC syndrome (hereditary leiomyomas and type 2 papillary renal cell carcinoma), including benign cutaneous and uterine leiomyomas and renal cell cancer,[84] whereas somatic mutations are rare. The accumulation of fumarate, caused by mutations in FH, leads to aberrant succination of proteins. Positive staining for (S)-2-succinocysteine (2SC) can be used as a robust biomarker for mutations in FH.[85,86] Similar to mutations in IDH, FH as well as SDH mutations affect epigenetic signaling, by inhibition of histone demethylases and the TET family of 5 hydroxymethylcytosine (5mC)-hydroxylases by accumulated fumarate and succinate, respectively.[87–89] Loss of 5hmC and increased H3K9me3 can be demonstrated by immunohistochemistry in SDH and FH-mutant tumor cells.[90] In SDH-mutant GIST, 5-hmC staining is also low to absent.[91]

TUMORS DRIVEN BY PRIMARY EPIGENETIC DEREGULATION: CHONDROBLASTOMA

Epigenetic deregulation is emerging as a very prevalent oncogenic mechanism in a wide variety of tumors, beyond the effects of metabolic enzymes and oncometabolites. Molecular alterations of components of the Polycomb group, the SWI/SNF complex, and other genes involved in chromatin structure and regulation are being increasingly identified in many cancer types.[92] Frequently, epigenetic deregulation is an additional feature in a cancer cell, contributing to a complex genomic environment in which several other oncogenic mechanisms are already in place (eg, mutations in members of the PRC2 complex in malignant peripheral nerve sheath tumors).[93,94] In some tumor types, however, mutations causing epigenetic deregulation seem to occur as early events, in a background of low mutational rate, and may serve as primary drivers of tumorigenesis.[95] Examples include histone mutations in giant cell tumor of bone (GCTB) and chondroblastoma, SMARCB1 homozygous deletions in rhabdoid tumor,[96] and SMARCA4 inactivation in SMARCA4-deficient thoracic sarcomas.[97] It has recently become apparent that the main oncogenic effect of some chimeric transcription factors, specifically the SS18-SSX fusions in synovial sarcoma, is epigenetic reprograming by mechanisms such as disruption of SWI/SNF complexes[98]; these 2 pathogenetic categories are therefore not mutually exclusive, and as our biological understanding improves, it is to be expected that other sarcomas with chimeric transcription factors may fit better in more specific pathogenetic categories.

Mutations affecting epigenetic signaling in bone tumors include histone H3.3 mutations in GCTB and chondroblastoma,[99] both of which are locally aggressive bone tumors predominantly affecting young patients. In 92% of GCTBs, mutations are found in the H3F3A gene, whereas in 95% of chondroblastomas, mutations are found in the H3F3B gene.[99] Both genes encode for histone H3.3. The exact mechanism by which these histone H3.3 mutations cause the formation of these giant cell–containing tumors is currently unknown. The distinction between GCTB and chondroblastoma and their distinction from other giant cell–containing lesions of bone such as aneurysmal bone cyst, telangiectatic osteosarcoma or chondromyxoid fibroma can be a challenge. Mutation analysis may be a useful diagnostic tool, in addition to the possible detection of S100 or DOG1 expression by immunohistochemistry, which would favor chondroblastoma.[100,101] Moreover, in chondromyxoid fibroma, GRM1 rearrangements are found.[102] When using mutation analysis for diagnosis, one should realize that the mutation is present in the mononuclear stromal cells, which constitute only a minority of the cells in the tumor. Sensitivity is therefore highly dependent on the technique used, detecting mutations in H3F3A in 69% of the GCTBs using classical Sanger sequencing[103] compared with 92% using a targeted NGS approach.[99]

SARCOMAS WITH COMPLEX GENOME

Most sarcomas show complex genomic profiles, with inconsistent, nonspecific molecular alterations. These are aggressive tumors that tend to affect older adults (with the exception of some osteosarcomas, and most radiation-associated sarcomas). Morphologically, sarcomas with complex genomes are heterogeneous, usually of high histologic grade, frequently cytologically pleomorphic, and may show signs of differentiation along several mesenchymal lineages or may be undifferentiated.[1] Notable examples include high-grade leiomyosarcoma, pleomorphic and dedifferentiated liposarcoma, high-grade myxofibrosarcoma, angiosarcoma, and undifferentiated pleomorphic sarcoma. Although the genomics of these lesions are highly variable from case to case, there is a predominance of copy number alterations over single nucleotide variants. Such high chromosomal instability occurs in the context of TP53 pathway alterations, very often *TP53* mutation, which is likely an early event in tumorigenesis. Additional molecular alterations include activation of the alternative lengthening of telomeres, often facilitated by loss of the chromatin remodeling factor *ATRX*, and loss of multiple tumor suppressor genes. The Rb/E2F pathway is also critical for tumor development and multiple members of this pathway are frequently mutated by different mechanisms. The extreme heterogeneity and complexity explains the limited number of specific molecular markers available for these sarcomas, which at present only benefit from molecular studies in rare occasions in clinical settings. Nonetheless, the higher mutational load in these tumors potentially makes them good candidates for immunotherapy, with drugs such as immune checkpoint blockers.[104]

Key Features
OF SARCOMAS WITH COMPLEX GENOME

- Complex unbalanced karyotype. Numerous copy number changes reflecting chromosomal instability.

- High mutational load.

- Pleomorphic, high-grade morphology.

- No specific molecular markers.

- Frequent loss of tumor suppressor genes (most often *TP53*).

- No effective targeted therapies.

- Subsets respond to conventional chemotherapy and radiation therapy.

- Prototypical examples: osteosarcoma, leiomyosarcoma, pleomorphic liposarcoma, myxofibrosarcoma, undifferentiated pleomorphic sarcoma.

TUMORS WITH CHARACTERISTIC COPY NUMBER ALTERATIONS: DEDIFFERENTIATED LIPOSARCOMA

Subsets of sarcoma cases with complex genome show somewhat reproducible pattern of chromosomal alterations. One example is well-differentiated/dedifferentiated liposarcoma (WD/DDLPS), whose genome is characterized by multiple copy number changes, mostly gains and amplifications, with multiple intrachromosomal and interchromosomal rearrangements. More than 90% of WD/DDLPS have characteristic neochromosomes, either linear or circular, giant markers and ring chromosomes, respectively, which are dynamic structures composed of genetic material from various distinct chromosomal regions. The composition of WD/DDLPS neochromosomes is highly variable, from case to case and during clonal evolution within an individual WD/DDLPS, but almost invariably includes a core group of genes from chromosome 12q13-15, including multiple copies of the *MDM2* and *CDK4* oncogenes. These represent at least 2 independent 12q amplicons, among approximately 15 to 20 amplicons characteristically present in WD/DDLPS cells. Some of the additional amplicons are relatively consistent, such as 1q25 or 6q21, but their extension and amplitude is highly variable. The mutational mechanisms underlying the formation of WD/DDLPS neochromosomes are the subject of intense investigations[105] and can be systematized as (1) an early initiation phase, in which a single catastrophic event (chromothripsis) results in massive fragmentation, rearrangement, and circularization of chromosome 12; (2) an amplification phase, in which hundreds of repetitive cycles of break-fusion-bridge allow for amplification, loss, and variable incorporation of additional chromosomal regions to the ring neochromosomes; and (3) a linearization phase, in which the neochromosomes are stabilized by capturing new chromothriptic telomeres. Understanding these events is helpful in the interpretation of copy number alterations and chromosomal rearrangements in WD/DDLPS cases.

The highly recurrent nature of *MDM2* and *CDK4* amplification provides a useful diagnostic marker. Detection of *MDM2* amplification by FISH is currently the gold standard for diagnosis of WD/DDLPS,[106] whereas combined immunohistochemical detection of the MDM2 and CDK4 proteins is a very useful diagnostic tool in surgical pathology routine practice.[107,108] *MDM2* and *CDK4* amplification are readily detectable by NGS (**Fig. 5**). These tests are particularly useful in 3 situations: (1) to confirm the diagnosis of WDLPS in an adipocytic

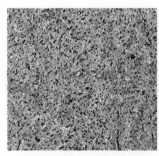
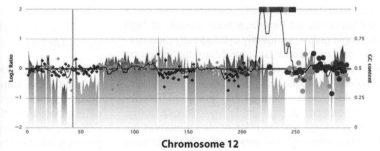

Chromosome 12

Fig. 5. High copy number gain of *MDM2* and *CDK4*, in 2 independent amplicons, in chromosomal region 12q13 - 15. The graph represents the ratio of sequence coverage on the y axis (log 2 scale) plotted along chromosome 12, on the x axis. This lesion was a dedifferentiated liposarcoma with spindle cell and pleomorphic morphologies, arising in a well-differentiated liposarcoma in the inguinal region of a 72-year-old man.

lesion of minimal cytologic atypia; (2) more commonly, to establish the diagnosis of DDLPS in a relatively nondescript spindle-cell or pleomorphic sarcoma in a deep somatic location; and (3) very rarely, to reclassify a high-grade pleomorphic adipocytic sarcoma as DDLPS with homologous lipoblastic differentiation that could be mistaken for pleomorphic liposarcoma.[109]

Biologically, amplification of *MDM2* results in inactivation of p53, whereas *CDK4* amplification leads to cell cycle progression.[110] Both alterations can be pharmacologically targeted with compounds that are at different stages of clinical development. MDM2 inhibitors restore p53 function disrupting the p53-MDM2 interaction. Several classes are being evaluated in clinical trials in DDLPS and other forms of cancer. Despite the strong biologic rationale and proven on-target activity, initial clinical experiences demonstrate that few patients with DDLPS achieve disease stabilization after MDM2 inhibition, at the expense of substantial adverse effects.[111] Such drug toxicities seem to be class-specific, and novel compounds are being evaluated that may overcome the limitations of current drugs. CDK4 inhibition can be achieved with compounds that typically target CDK4 and CDK6, such as palbociclib.[112] In DDLPS, CDK4 inhibition provides limited benefits as a single agent,[113] but may be therapeutically effective in combination regimens, as recently demonstrated in other cancer types.[114] The combination of MDM2 and CDK4 inhibition in DDLPS is an attractive concept supported by a strong biological rationale, but it may not be achievable due to the combined toxicities of these drugs.

TUMORS WITH HIGHLY COMPLEX KARYOTYPES: OSTEOSARCOMA

Most pleomorphic sarcomas have complex karyotypes lacking specific genetic aberrations and recognizable chromosomal patterns. These high-grade sarcomas often harbor aberrations in the Rb or p53 pathway. One such tumor is conventional osteosarcoma, the most frequent primary high-grade bone tumor in humans, which occurs predominantly in children and adolescents.[1,115] It has a high risk of metastasis, and despite intensive treatment strategies, the chance at cure of patients with resectable osteosarcoma has remained approximately 60% to 65% in the past 3 decades.[116] At the genetic level, osteosarcoma is extremely unstable with many translocations, amplifications, mutations and deletions (see **Fig. 1**D). The detection of specific driver genes and pathways is therefore extremely difficult. Recently 2 phenomena were described that reflect this genomic instability. One is chromothripsis,[117] a cataclysmic event in which chromosomes are fragmented and subsequently aberrantly assembled. Chromothripsis was also identified in other tumors, but is most prevalent in bone tumors.[118] The other phenomenon is kataegis, reflected by a localized hypermutation area, which also occurs at a high frequency in osteosarcoma (~50%).[118]

Both chromothrypsis and kataegis can conceivably generate numerous neoantigens, which may predict response to certain immunotherapies. Despite an improved mechanistic understanding of these chromosomal alterations, they provide no clinically-useful information beyond the presence of genomic instability, which is itself a predictor of poor outcome. At present, molecular markers are not routinely used for the management of sarcomas with highly complex karyotype.

REFERENCES

1. Fletcher CDM, World Health Organization, International Agency for Research on Cancer. WHO classification of tumours of soft tissue and bone. 4th edition. Lyon (France): IARC Press; 2013. p. 468.

2. Fletcher JA, Kozakewich HP, Hoffer FA, et al. Diagnostic relevance of clonal cytogenetic aberrations in malignant soft-tissue tumors. N Engl J Med 1991;324(7):436–42.

3. Bovee JV, Hogendoorn PC. Molecular pathology of sarcomas: concepts and clinical implications. Virchows Arch 2010;456(2):193–9.

4. Mertens F, Johansson B, Fioretos T, et al. The emerging complexity of gene fusions in cancer. Nat Rev Cancer 2015;15(6):371–81.

5. Hanahan D, Weinberg RA. Hallmarks of cancer: the next generation. Cell 2011;144(5):646–74.

6. Mitelman F, Johansson B, Mertens F. The impact of translocations and gene fusions on cancer causation. Nat Rev Cancer 2007;7(4):233–45.

7. Robinson DR, Wu YM, Kalyana-Sundaram S, et al. Identification of recurrent NAB2-STAT6 gene fusions in solitary fibrous tumor by integrative sequencing. Nat Genet 2013;45(2):180–5.

8. Wang L, Motoi T, Khanin R, et al. Identification of a novel, recurrent HEY1-NCOA2 fusion in mesenchymal chondrosarcoma based on a genome-wide screen of exon-level expression data. Genes Chromosomes Cancer 2012;51(2):127–39.

9. van IJzendoorn DG, de Jong D, Romagosa C, et al. Fusion events lead to truncation of FOS in epithelioid hemangioma of bone. Genes Chromosomes Cancer 2015;54(9):565–74.

10. Huang SC, Zhang L, Sung YS, et al. Frequent FOS gene rearrangements in epithelioid hemangioma: a molecular study of 58 cases with morphologic reappraisal. Am J Surg Pathol 2015;39(10):1313–21.

11. Antonescu CR, Chen HW, Zhang L, et al. ZFP36-FOSB fusion defines a subset of epithelioid hemangioma with atypical features. Genes Chromosomes Cancer 2014;53(11):951–9.

12. Errani C, Zhang L, Panicek DM, et al. Epithelioid hemangioma of bone and soft tissue: a reappraisal of a controversial entity. Clin Orthop Relat Res 2012;470(5):1498–506.

13. Nielsen GP, Srivastava A, Kattapuram S, et al. Epithelioid hemangioma of bone revisited: a study of 50 cases. Am J Surg Pathol 2009;33(2):270–7.

14. Walther C, Tayebwa J, Lilljebjorn H, et al. A novel SERPINE1-FOSB fusion gene results in transcriptional up-regulation of FOSB in pseudomyogenic haemangioendothelioma. J Pathol 2014;232(5):534–40.

15. Trombetta D, Magnusson L, von Steyern FV, et al. Translocation t(7;19)(q22;q13)—a recurrent chromosome aberration in pseudomyogenic hemangioendothelioma? Cancer Genet 2011;204(4):211–5.

16. Errani C, Zhang L, Sung YS, et al. A novel WWTR1-CAMTA1 gene fusion is a consistent abnormality in epithelioid hemangioendothelioma of different anatomic sites. Genes Chromosomes Cancer 2011;50(8):644–53.

17. Tanas MR, Sboner A, Oliveira AM, et al. Identification of a disease-defining gene fusion in epithelioid hemangioendothelioma. Sci Transl Med 2011;3(98):98ra82.

18. Doyle LA, Fletcher CD, Hornick JL. Nuclear expression of CAMTA1 distinguishes epithelioid hemangioendothelioma from histologic mimics. Am J Surg Pathol 2016;40(1):94–102.

19. Shibuya R, Matsuyama A, Shiba E, et al. CAMTA1 is a useful immunohistochemical marker for diagnosing epithelioid haemangioendothelioma. Histopathology 2015;67(6):827–35.

20. Antonescu CR, Le Loarer F, Mosquera JM, et al. Novel YAP1-TFE3 fusion defines a distinct subset of epithelioid hemangioendothelioma. Genes Chromosomes Cancer 2013;52(8):775–84.

21. Manner J, Radlwimmer B, Hohenberger P, et al. MYC high level gene amplification is a distinctive feature of angiosarcomas after irradiation or chronic lymphedema. Am J Pathol 2010;176(1):34–9.

22. Mentzel T, Schildhaus HU, Palmedo G, et al. Post-radiation cutaneous angiosarcoma after treatment of breast carcinoma is characterized by MYC amplification in contrast to atypical vascular lesions after radiotherapy and control cases: clinicopathological, immunohistochemical and molecular analysis of 66 cases. Mod Pathol 2012;25(1):75–85.

23. Shon W, Sukov WR, Jenkins SM, et al. MYC amplification and overexpression in primary cutaneous angiosarcoma: a fluorescence in-situ hybridization and immunohistochemical study. Mod Pathol 2014;27(4):509–15.

24. Verbeke SL, de Jong D, Bertoni F, et al. Array CGH analysis identifies two distinct subgroups of primary angiosarcoma of bone. Genes Chromosomes Cancer 2015;54(2):72–81.

25. Behjati S, Tarpey PS, Sheldon H, et al. Recurrent PTPRB and PLCG1 mutations in angiosarcoma. Nat Genet 2014;46(4):376–9.

26. Antonescu CR, Yoshida A, Guo T, et al. KDR activating mutations in human angiosarcomas are sensitive to specific kinase inhibitors. Cancer Res 2009;69(18):7175–9.

27. Huang SC, Zhang L, Sung YS, et al. Recurrent CIC gene abnormalities in angiosarcomas: a molecular study of 120 cases with concurrent investigation of PLCG1, KDR, MYC, and FLT4 gene alterations. Am J Surg Pathol 2016;40(5):645–55.

28. MacConaill LE, Garcia E, Shivdasani P, et al. Prospective enterprise-level molecular genotyping of a cohort of cancer patients. J Mol Diagn 2014;16(6):660–72.

29. Corless CL, Fletcher JA, Heinrich MC. Biology of gastrointestinal stromal tumors. J Clin Oncol 2004;22(18):3813–25.

30. Heinrich MC, Corless CL, Duensing A, et al. PDGFRA activating mutations in gastrointestinal stromal tumors. Science 2003;299(5607):708–10.

31. Wardelmann E, Pauls K, Merkelbach-Bruse S, et al. Gastrointestinal stromal tumors carrying PDGFRalpha mutations occur preferentially in the stomach and exhibit an epithelioid or mixed phenotype. Verh Dtsch Ges Pathol 2004;88:174–83, [in German].

32. Corless CL, Schroeder A, Griffith D, et al. PDGFRA mutations in gastrointestinal stromal tumors: frequency, spectrum and in vitro sensitivity to imatinib. J Clin Oncol 2005;23(23):5357–64.

33. Duensing A, Medeiros F, McConarty B, et al. Mechanisms of oncogenic KIT signal transduction in primary gastrointestinal stromal tumors (GISTs). Oncogene 2004;23(22):3999–4006.

34. Medeiros F, Corless CL, Duensing A, et al. KIT-negative gastrointestinal stromal tumors: proof of concept and therapeutic implications. Am J Surg Pathol 2004;28(7):889–94.

35. Espinosa I, Lee CH, Kim MK, et al. A novel monoclonal antibody against DOG1 is a sensitive and specific marker for gastrointestinal stromal tumors. Am J Surg Pathol 2008;32(2):210–8.

36. Martin J, Poveda A, Llombart-Bosch A, et al. Deletions affecting codons 557-558 of the c-KIT gene indicate a poor prognosis in patients with completely resected gastrointestinal stromal tumors: a study by the Spanish group for sarcoma research (GEIS). J Clin Oncol 2005;23(25):6190–8.

37. Gastrointestinal Stromal Tumor Meta-Analysis Group (MetaGIST). Comparison of two doses of imatinib for the treatment of unresectable or metastatic gastrointestinal stromal tumors: a meta-analysis of 1,640 patients. J Clin Oncol 2010;28(7):1247–53.

38. Dematteo RP, Lewis JJ, Leung D, et al. Two hundred gastrointestinal stromal tumors: recurrence patterns and prognostic factors for survival. Ann Surg 2000;231(1):51–8.

39. Verweij J, Casali PG, Zalcberg J, et al. Progression-free survival in gastrointestinal stromal tumours with high-dose imatinib: randomised trial. Lancet 2004; 364(9440):1127–34.

40. Blanke CD, Demetri GD, von MM, et al. Long-term results from a randomized phase II trial of standard- versus higher-dose imatinib mesylate for patients with unresectable or metastatic gastrointestinal stromal tumors expressing KIT. J Clin Oncol 2008;26(4):620–5.

41. Heinrich MC, Maki RG, Corless CL, et al. Primary and secondary kinase genotypes correlate with the biological and clinical activity of sunitinib in imatinib-resistant gastrointestinal stromal tumor. J Clin Oncol 2008;26(33):5352–9.

42. Heinrich MC, Griffith D, McKinley A, et al. Crenolanib inhibits the drug-resistant PDGFRA D842V mutation associated with imatinib-resistant gastrointestinal stromal tumors. Clin Cancer Res 2012;18(16):4375–84.

43. Serrano C, Wang Y, Marino-Enriquez A, et al. KRAS and KIT gatekeeper mutations confer polyclonal primary imatinib resistance in GI stromal tumors: relevance of concomitant phosphatidylinositol 3-kinase/AKT dysregulation. J Clin Oncol 2015;33(22): e93–6.

44. Heinrich MC, Corless CL, Blanke CD, et al. Molecular correlates of imatinib resistance in gastrointestinal stromal tumors. J Clin Oncol 2006;24(29):4764–74.

45. Fletcher JA, Corless C, Dimitrijevic S. Mechanisms of resistance to imatinib mesylate (IM) in advanced gastrointestinal stromal tumors (GIST). Proc Am Soc Clin Oncol 2003;22:815.

46. Debiec-Rychter M, Cools J, Dumez H, et al. Mechanisms of resistance to imatinib mesylate in gastrointestinal stromal tumors and activity of the PKC412 inhibitor against imatinib-resistant mutants. Gastroenterology 2005;128(2):270–9.

47. Liegl B, Kepten I, Le C, et al. Heterogeneity of kinase inhibitor resistance mechanisms in GIST. J Pathol 2008;216(1):64–74.

48. George S, Wang Q, Heinrich MC, et al. Efficacy and safety of regorafenib in patients with metastatic and/or unresectable GI stromal tumor after failure of imatinib and sunitinib: a multicenter phase II trial. J Clin Oncol 2012;30(19):2401–7.

49. Demetri GD, Reichardt P, Kang YK, et al. Efficacy and safety of regorafenib for advanced gastrointestinal stromal tumours after failure of imatinib and sunitinib (GRID): an international, multicentre, randomised, placebo-controlled, phase 3 trial. Lancet 2013;381(9863):295–302.

50. Bauer S, Duensing A, Demetri GD, et al. KIT oncogenic signaling mechanisms in imatinib-resistant gastrointestinal stromal tumor: PI3-kinase/AKT is a crucial survival pathway. Oncogene 2007;26(54): 7560–8.

51. Floris G, Wozniak A, Sciot R, et al. A potent combination of the novel PI3K inhibitor, GDC-0941, with imatinib in gastrointestinal stromal tumor xenografts: long-lasting responses after treatment withdrawal. Clin Cancer Res 2013;19(3):620–30.

52. Patel S. Exploring novel therapeutic targets in GIST: focus on the PI3K/akt/mTOR pathway. Curr Oncol Rep 2013;15(4):386–95.

53. Yang J, Ikezoe T, Nishioka C, et al. Long-term exposure of gastrointestinal stromal tumor cells to sunitinib induces epigenetic silencing of the PTEN gene. Int J Cancer 2012;130(4):959–66.

54. Quattrone A, Wozniak A, Dewaele B, et al. Frequent mono-allelic loss associated with deficient PTEN expression in imatinib-resistant gastrointestinal stromal tumors. Mod Pathol 2014;27(11):1510–20.

55. Yan H, Parsons DW, Jin G, et al. IDH1 and IDH2 mutations in gliomas. N Engl J Med 2009;360(8): 765–73.

56. Schaap FG, French PJ, Bovee JV. Mutations in the isocitrate dehydrogenase genes IDH1 and IDH2 in tumors. Adv Anat Pathol 2013;20(1):32–8.

57. Amary MF, Bacsi K, Maggiani F, et al. IDH1 and IDH2 mutations are frequent events in central chondrosarcoma and central and periosteal chondromas but not in other mesenchymal tumours. J Pathol 2011;224(3):334–43.

58. Amary MF, Damato S, Halai D, et al. Ollier disease and Maffucci syndrome are caused by somatic mosaic mutations of IDH1 and IDH2. Nat Genet 2011;43(12):1262–5.

59. Pansuriya TC, van Eijk R, d'Adamo P, et al. Somatic mosaic IDH1 and IDH2 mutations are associated with enchondroma and spindle cell hemangioma in Ollier disease and Maffucci syndrome. Nat Genet 2011;43(12):1256–61.

60. Turcan S, Rohle D, Goenka A, et al. IDH1 mutation is sufficient to establish the glioma hypermethylator phenotype. Nature 2012;483(7390):479–83.

61. Lu C, Ward PS, Kapoor GS, et al. IDH mutation impairs histone demethylation and results in a block to cell differentiation. Nature 2012;483(7390):474–8.

62. Sasaki M, Knobbe CB, Munger JC, et al. IDH1(R132H) mutation increases murine haematopoietic progenitors and alters epigenetics. Nature 2012;488(7413):656–9.

63. Ward PS, Patel J, Wise DR, et al. The common feature of leukemia-associated IDH1 and IDH2 mutations is a neomorphic enzyme activity converting alpha-ketoglutarate to 2-hydroxyglutarate. Cancer Cell 2010;17(3):225–34.

64. Gross S, Cairns RA, Minden MD, et al. Cancer-associated metabolite 2-hydroxyglutarate accumulates in acute myelogenous leukemia with isocitrate dehydrogenase 1 and 2 mutations. J Exp Med 2010;207(2):339–44.

65. Dang L, White DW, Gross S, et al. Cancer-associated IDH1 mutations produce 2-hydroxyglutarate. Nature 2010;465(7300):966.

66. Xu W, Yang H, Liu Y, et al. Oncometabolite 2-hydroxyglutarate is a competitive inhibitor of alpha-ketoglutarate-dependent dioxygenases. Cancer Cell 2011;19(1):17–30.

67. Figueroa ME, Abdel-Wahab O, Lu C, et al. Leukemic IDH1 and IDH2 mutations result in a hypermethylation phenotype, disrupt TET2 function, and impair hematopoietic differentiation. Cancer Cell 2010;18(6):553–67.

68. McDonough MA, Loenarz C, Chowdhury R, et al. Structural studies on human 2-oxoglutarate dependent oxygenases. Curr Opin Struct Biol 2010; 20(6):659–72.

69. Chowdhury R, Yeoh KK, Tian YM, et al. The oncometabolite 2-hydroxyglutarate inhibits histone lysine demethylases. EMBO Rep 2011;12(5):463–9.

70. Suijker J, Baelde HJ, Roelofs H, et al. The oncometabolite D-2-hydroxyglutarate induced by mutant IDH1 or -2 blocks osteoblast differentiation in vitro and in vivo. Oncotarget 2015;6(17):14832–42.

71. Jin Y, Elalaf H, Watanabe M, et al. Mutant IDH1 dysregulates the differentiation of mesenchymal stem cells in association with gene-specific histone modifications to cartilage- and bone-related genes. PLoS One 2015;10(7):e0131998.

72. Bovee JV, Hogendoorn PC, Wunder JS, et al. Cartilage tumours and bone development: molecular pathology and possible therapeutic targets. Nat Rev Cancer 2010;10(7):481–8.

73. Schrage YM, Lam S, Jochemsen AG, et al. Central chondrosarcoma progression is associated with pRb pathway alterations: CDK4 down-regulation and p16 overexpression inhibit cell growth in vitro. J Cell Mol Med 2009;13(9A):2843–52.

74. Suijker J, Oosting J, Koornneef A, et al. Inhibition of mutant IDH1 decreases D-2-HG levels without affecting tumorigenic properties of chondrosarcoma cell lines. Oncotarget 2015;6(14):12505–19.

75. Li L, Paz AC, Wilky BA, et al. Treatment with a small molecule mutant IDH1 inhibitor suppresses tumorigenic activity and decreases production of the oncometabolite 2-hydroxyglutarate in human chondrosarcoma cells. PLoS One 2015;10(9): e0133813.

76. Cleven AH, Zwartkruis E, Hogendoorn PC, et al. Periosteal chondrosarcoma: a histopathological and molecular analysis of a rare chondrosarcoma subtype. Histopathology 2015;67(4):483–90.

77. Damato S, Alorjani M, Bonar F, et al. IDH1 mutations are not found in cartilaginous tumours other than central and periosteal chondrosarcomas and enchondromas. Histopathology 2012;60(2):363–5.

78. Arai M, Nobusawa S, Ikota H, et al. Frequent IDH1/2 mutations in intracranial chondrosarcoma: a possible diagnostic clue for its differentiation from chordoma. Brain Tumor Pathol 2012;29(4): 201–6.

79. Eng C, Kiuru M, Fernandez MJ, et al. A role for mitochondrial enzymes in inherited neoplasia and beyond. Nat Rev Cancer 2003;3(3):193–202.

80. Janeway KA, Kim SY, Lodish M, et al. Defects in succinate dehydrogenase in gastrointestinal stromal tumors lacking KIT and PDGFRA mutations. Proc Natl Acad Sci U S A 2011;108(1):314–8.

81. Killian JK, Miettinen M, Walker RL, et al. Recurrent epimutation of SDHC in gastrointestinal stromal tumors. Sci Transl Med 2014;6(268):268ra177.

82. Killian JK, Kim SY, Miettinen M, et al. Succinate dehydrogenase mutation underlies global epigenomic divergence in gastrointestinal stromal tumor. Cancer Discov 2013;3(6):648–57.

83. Kirmani S, Young WF. Hereditary paraganglioma–pheochromocytoma syndromes. In: Pagon RA, Bird TD, Dolan CR, et al, editors. GeneReviews(R). Seattle, WA: University of Washington; 1993-2016. Available online at: http://www.ncbi.nlm.nih.gov/books/NBK1548. Accessed March 22, 2016.

84. Tomlinson IP, Alam NA, Rowan AJ, et al. Germline mutations in FH predispose to dominantly inherited uterine fibroids, skin leiomyomata and papillary renal cell cancer. Nat Genet 2002;30(4):406–10.

85. Bardella C, El-Bahrawy M, Frizzell N, et al. Aberrant succination of proteins in fumarate hydratase-deficient mice and HLRCC patients is a robust biomarker of mutation status. J Pathol 2011; 225(1):4–11.

86. Ternette N, Yang M, Laroyia M, et al. Inhibition of mitochondrial aconitase by succination in fumarate hydratase deficiency. Cell Rep 2013;3(3):689–700.

87. Cervera AM, Bayley JP, Devilee P, et al. Inhibition of succinate dehydrogenase dysregulates histone modification in mammalian cells. Mol Cancer 2009;8:89.

88. Smith EH, Janknecht R, Maher LJ 3rd. Succinate inhibition of alpha-ketoglutarate-dependent enzymes in a yeast model of paraganglioma. Hum Mol Genet 2007;16(24):3136–48.

89. Xiao M, Yang H, Xu W, et al. Inhibition of alpha-KG-dependent histone and DNA demethylases by fumarate and succinate that are accumulated in mutations of FH and SDH tumor suppressors. Genes Dev 2012;26(12):1326–38.

90. Hoekstra AS, de Graaff MA, Briaire-de Bruijn IH, et al. Inactivation of SDH and FH cause loss of 5hmC and increased H3K9me3 in paraganglioma/pheochromocytoma and smooth muscle tumors. Oncotarget 2015;6(36):38777–88.

91. Mason EF, Hornick JL. Succinate dehydrogenase deficiency is associated with decreased 5-hydroxymethylcytosine production in gastrointestinal stromal tumors: implications for mechanisms of tumorigenesis. Mod Pathol 2013;26(11):1492–7.

92. Kadoch C, Hargreaves DC, Hodges C, et al. Proteomic and bioinformatic analysis of mammalian SWI/SNF complexes identifies extensive roles in human malignancy. Nat Genet 2013;45(6): 592–601.

93. Lee W, Teckie S, Wiesner T, et al. PRC2 is recurrently inactivated through EED or SUZ12 loss in malignant peripheral nerve sheath tumors. Nat Genet 2014;46(11):1227–32.

94. De Raedt T, Beert E, Pasmant E, et al. PRC2 loss amplifies ras-driven transcription and confers sensitivity to BRD4-based therapies. Nature 2014; 514(7521):247–51.

95. Wilson BG, Wang X, Shen X, et al. Epigenetic antagonism between polycomb and SWI/SNF complexes during oncogenic transformation. Cancer Cell 2010;18(4):316–28.

96. Lee RS, Stewart C, Carter SL, et al. A remarkably simple genome underlies highly malignant pediatric rhabdoid cancers. J Clin Invest 2012;122(8): 2983–8.

97. Le Loarer F, Watson S, Pierron G, et al. SMARCA4 inactivation defines a group of undifferentiated thoracic malignancies transcriptionally related to BAF-deficient sarcomas. Nat Genet 2015;47(10): 1200–5.

98. Kadoch C, Crabtree GR. Reversible disruption of mSWI/SNF (BAF) complexes by the SS18-SSX oncogenic fusion in synovial sarcoma. Cell 2013; 153(1):71–85.

99. Behjati S, Tarpey PS, Presneau N, et al. Distinct H3F3A and H3F3B driver mutations define chondroblastoma and giant cell tumor of bone. Nat Genet 2013;45(12):1479–82.

100. Cleven AH, Briaire-de Bruijn I, Szuhai K, et al. DOG1 expression in giant cell containing bone tumours. Histopathology 2016;68(6):942–5.

101. Akpalo H, Lange C, Zustin J. Discovered on gastrointestinal stromal tumour 1 (DOG1): a useful immunohistochemical marker for diagnosing chondroblastoma. Histopathology 2012;60(7):1099–106.

102. Nord KH, Lilljebjorn H, Vezzi F, et al. GRM1 is upregulated through gene fusion and promoter swapping in chondromyxoid fibroma. Nat Genet 2014; 46(5):474–7.

103. Cleven AH, Hocker S, Briaire-de Bruijn I, et al. Mutation analysis of H3F3A and H3F3B as a diagnostic tool for giant cell tumor of bone and chondroblastoma. Am J Surg Pathol 2015;39(11): 1576–83.

104. Lim J, Poulin NM, Nielsen TO. New strategies in sarcoma: linking genomic and immunotherapy approaches to molecular subtype. Clin Cancer Res 2015;21(21):4753–9.

105. Garsed DW, Marshall OJ, Corbin VD, et al. The architecture and evolution of cancer neochromosomes. Cancer Cell 2014;26(5):653–67.

106. Sirvent N, Coindre JM, Maire G, et al. Detection of MDM2-CDK4 amplification by fluorescence in situ hybridization in 200 paraffin-embedded tumor samples: utility in diagnosing adipocytic lesions and comparison with immunohistochemistry and real-time PCR. Am J Surg Pathol 2007;31(10): 1476–89.

107. Binh MB, Sastre-Garau X, Guillou L, et al. MDM2 and CDK4 immunostainings are useful adjuncts in diagnosing well-differentiated and dedifferentiated liposarcoma subtypes: a comparative analysis of 559 soft tissue neoplasms with genetic data. Am J Surg Pathol 2005;29(10): 1340–7.

108. Binh MB, Garau XS, Guillou L, et al. Reproducibility of MDM2 and CDK4 staining in soft tissue tumors. Am J Clin Pathol 2006;125(5):693–7.

109. Marino-Enriquez A, Fletcher CD, Dal Cin P, et al. Dedifferentiated liposarcoma with "homologous" lipoblastic (pleomorphic liposarcoma-like)

differentiation: clinicopathologic and molecular analysis of a series suggesting revised diagnostic criteria. Am J Surg Pathol 2010;34(8):1122–31.

110. Conyers R, Young S, Thomas DM. Liposarcoma: molecular genetics and therapeutics. Sarcoma 2011;2011:483154.

111. Ray-Coquard I, Blay JY, Italiano A, et al. Effect of the MDM2 antagonist RG7112 on the P53 pathway in patients with MDM2-amplified, well-differentiated or dedifferentiated liposarcoma: an exploratory proof-of-mechanism study. Lancet Oncol 2012; 13(11):1133–40.

112. Zhang YX, Sicinska E, Czaplinski JT, et al. Antiproliferative effects of CDK4/6 inhibition in CDK4-amplified human liposarcoma in vitro and in vivo. Mol Cancer Ther 2014;13(9):2184–93.

113. Dickson MA, Tap WD, Keohan ML, et al. Phase II trial of the CDK4 inhibitor PD0332991 in patients with advanced CDK4-amplified well-differentiated or dedifferentiated liposarcoma. J Clin Oncol 2013;31(16):2024–8.

114. Turner NC, Ro J, Andre F, et al. Palbociclib in hormone-receptor-positive advanced breast cancer. N Engl J Med 2015;373(3):209–19.

115. Savage SA, Mirabello L. Using epidemiology and genomics to understand osteosarcoma etiology. Sarcoma 2011;2011:548151.

116. Anninga JK, Gelderblom H, Fiocco M, et al. Chemotherapeutic adjuvant treatment for osteosarcoma: where do we stand? Eur J Cancer 2011; 47(16):2431–45.

117. Stephens PJ, Greenman CD, Fu B, et al. Massive genomic rearrangement acquired in a single catastrophic event during cancer development. Cell 2011;144(1):27–40.

118. Chen X, Bahrami A, Pappo A, et al. Recurrent somatic structural variations contribute to tumorigenesis in pediatric osteosarcoma. Cell Rep 2014;7(1):104–12.

Molecular Pathology
Prognostic and Diagnostic Genomic Markers for Myeloid Neoplasms

Frank C. Kuo, MD, PhD

KEYWORDS

- Next-generation sequencing • Myeloid neoplasms • Genomic alterations • Mutations
- Copy number variation

Key points

- The same genetic alterations are shared among diverse types of myeloid neoplasms.
- Complex combination of genetic alterations is the rule, not the exception.
- Genetic alterations in the same pathway tend to be mutually exclusive.
- Clonal heterogeneity is a way of life for myeloid neoplasms.
- When multiple alterations are present, temporal sequence of acquisition of genetic alterations varies among different individuals and may have prognostic significance.

ABSTRACT

Application of next-generation sequencing (NGS) on myeloid neoplasms has expanded our knowledge of genomic alterations in this group of diseases. Genomic alterations in myeloid neoplasms are complex, heterogeneous, and not specific to a disease entity. NGS-based panel testing of myeloid neoplasms can complement existing diagnostic modalities and is gaining acceptance in the clinics and diagnostic laboratories. Prospective, randomized trials to evaluate the prognostic significance of genomic markers in myeloid neoplasms are under way in academic medical centers.

OVERVIEW

The term myeloid neoplasm refers to a group of neoplastic diseases that include acute myeloid leukemia (AML) and its frequent predecessors, myelodysplastic syndromes (MDS) and myeloproliferative neoplasms (MPN). Myeloid neoplasms originate from the hematopoietic stem/progenitor cells (HSCs).[1] Multistep accumulation of genetic alterations in HSCs progressively confer growth advantages to certain clones and lead to a state of "clonal hematopoiesis" that can either exist transiently or last for many years. Further acquisition of "driver" type of genetic alterations, coupled with epigenetic and environmental changes, eventually enables the uncontrolled outgrowth of cells with reduced capacity to differentiate into more mature hematopoietic elements.[2]

KNOWLEDGE DERIVED FROM NEXT-GENERATION SEQUENCING TESTING OF MYELOID NEOPLASMS

With the advance in next-generation sequencing (NGS)[3] and its application to analyzing patient samples, our understanding of the genetic

---

No commercial or financial conflicts of interest to disclose.
No external funding source to disclose.
Center for Advanced Molecular Diagnostics, Brigham and Women's Hospital, Harvard Medical School, 75 Francis Street, Boston, MA 02115, USA
E-mail address: fkuo@partners.org

Surgical Pathology 9 (2016) 475–488
http://dx.doi.org/10.1016/j.path.2016.04.010
1875-9181/16/$ – see front matter © 2016 Elsevier Inc. All rights reserved.

alterations of myeloid neoplasms has expanded significantly in the past few years.[4,5] The diagnosis and classification of myeloid neoplasms is likely to evolve over the next few years when this new information is incorporated into the classification scheme. Although the impact of genetic alterations on diagnosis, treatment, and prognosis is still in its early stages, the themes discussed in the following sections have emerged from the vast amount of new information gained so far.

THE SAME GENETIC ALTERATIONS ARE SHARED AMONG DIVERSE TYPES OF MYELOID NEOPLASMS

For example, when it was first described, JAK2 V617F was once thought to be a specific marker for MPN.[6] Since then, JAK2 V617F has been detected in almost every type of myeloid neoplasm, including most patients with polycythemia vera (PCV), a significant portion of patients (30%–70%) with essential thrombocythemia (ET), myelofibrosis (PMF), and refractory anemia with ring sideroblasts associated with marked thrombocytosis (RARS-T), and a small percentage (<10%) in almost all types of myeloid neoplasms.[7–9] Moreover, a small percentage of patients with AML with no prior history of MPN have a JAK2 V617F mutation and JAK2 V617F is a frequent finding among older individuals with normal complete blood count, a phenomenon called "age-related clonal hematopoiesis" (ARCH).[10,11] Similar findings apply to almost all other frequently mutated genes in myeloid neoplasms. No one gene is specific to a diagnostic entity and no gene is mutually exclusive with a specific diagnosis. The strongest genotype-phenotype associations (Table 1) are seen between splicing factor 3b subunit 1 (SF3B1) mutations and the presence of ring sideroblasts and KIT D816V with mastocytosis,[12,13] but again, SF3B1 and KIT mutations are also seen in many other types of myeloid neoplasms (Fig. 1).

COMPLEX COMBINATION OF GENETIC ALTERATIONS IS THE RULE, NOT THE EXCEPTION

Early in the disease course, patients may present with one or a few mutations but frequently acquire more mutations as the disease progresses. Total number of mutations present in a patient may be indicative of the duration and severity of the disease. Specific co-mutation patterns, either in the form of content or temporal sequence, have not been consistently described across specific entities, with a few notable exceptions.

Table 1
Phenotype-genotype associations between genomic alterations and myeloid neoplasms

Gene	Disease/Phenotype
JAK2 V617F and exon 12 deletions	PV, ET, and PMF
CALR exon 9 frameshift	PV, ET, and PMF
MPL W515 missense	PV, ET, and PMF
JAK2 + SF3B1	RARS-T
SF3B1	Ring sideroblasts/RARS
RUNX1	Thrombocytopenia
TET2 + SRSF2	Monocytosis
U2AF1	Male predominance
TP53	5q-, 7q-, complex karyotype
CSF3R	Chronic neutrophilic leukemia
KIT D816V	Mastocytosis

Abbreviations: ET, essential thrombocytopenia; PMF, primary myelofibrosis; PV, polycythemia Vera; RARS, refractory anemia with ring sideroblast; RARS-T, refractory anemia with ring sideroblasts associated with thrombocytosis.

Co-mutation of TET2 and SRSF2, for example, has emerged as the most common pair of alterations for chronic myelomonocytic leukemia (CMML)[14,15] and JAK2 and SF3B1 combination is seen in more than 50% of patients with RARS-T.[8] Concurrent DNMT3A, NPM1 mutation and FLT3-ITD is a common finding in de novo AML.[16] However, even when these co-mutations are present, they are rarely the only changes in a patient. One notable exception is in patients with tumor protein p53 (TP53) mutations who generally have relatively few co-mutated genes. The genomic instability associated with TP53 loss may be sufficient to drive leukemogenesis (Fig. 2).

GENETIC ALTERATIONS IN THE SAME PATHWAY TEND TO BE MUTUALLY EXCLUSIVE

Patients with mutations in one splicing factor gene (SF3B1, SRSF2, ZRSR2, U2AF1) rarely acquire mutations in other splicing factors and patients with isocitrate dehydrogenase 1 (IDH1) mutation typically lack IDH2 mutation.[17–20] When mutations in the same pathways are observed in the same patient, it is more likely that they are present in different subclones rather in the same cells (see the following section). Presumably, tumoral evolution does not select for mutations in other genes within shared pathways, as they do not offer additional growth or survival advantage.

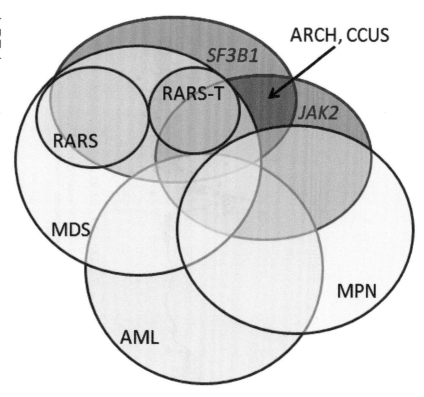

Fig. 1. Genetic alterations are shared among diverse types of myeloid neoplasms that by themselves are overlapping.

CLONAL HETEROGENEITY IS A WAY OF LIFE FOR MYELOID NEOPLASMS

Two forms of clonal evolution are well recognized: convergent and divergent evolution (Fig. 3).[21] In convergent evolution, non-neoplastic HSC first acquires a sentinel mutation to become neoplastic and proliferate to form a founding clone. Individual neoplastic cells of the founder clone then acquire different mutations in the same gene or in different genes of the same pathway. These neoplastic cells with different mutations compete with each other leading to the emergence of multiple subclones that share the same founding mutations. The different progression alleles occurring either in the same gene, or different genes in the same pathway, offer similar selective advantage. This convergent evolution frequently uses activating mutations within the RAS pathway[14,22] (ie, multiple subclonal NRAS, KRAS, CBL, and PTPN11 mutations in the same individual) but also occurs in other genes such as WT1, STAG2, and TP53. These are later events and generally take place in a preleukemic background.

Divergent evolution is thought to occur as the bone marrow niche becomes pro-mutagenic. Multiple different HSCs harboring different founding mutations in different pathways may emerge in this environment. The rule of survival of the fittest eventually determines which clone(s) dominate this evolution process. Because multiple minor subclones exist, some undetectable, different subclones may overtake the original clone during chemotherapy or other therapeutic intervention. When acute leukemia develops in patients with JAK2 V617F mutations, the leukemic clones sometimes have wild-type JAK2, indicating that they may not be derived from a previously dominant clone.[23,24] Patients with aplastic anemia[25] or paroxysmal nocturnal hemoglobinuria[26] also present good examples of divergent evolution, where an injured and diminished HSC pool within an altered bone marrow niche is prone to clonal hematopoiesis.

WHEN MULTIPLE ALTERATIONS ARE PRESENT, TEMPORAL SEQUENCE OF ACQUISITION OF GENETIC ALTERATIONS VARIES AMONG DIFFERENT INDIVIDUALS AND MAY HAVE PROGNOSTIC SIGNIFICANCE

Multistep tumorigenesis was first proposed for colon cancer more than 15 years ago.[27] Since then, it is clear that the same process applies to most, if not all organs/tissues, as normal cells

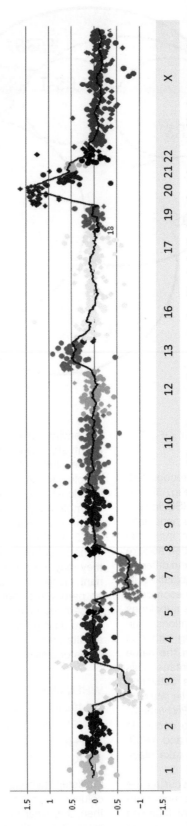

Fig. 2. An AML with *TP53* mutation with a complex karyotype as assessed by next-generation-sequencing. Y-axis: log2 ratio of read count of leukemic sample versus normal control. A log2 ratio of 0 denotes normal number of chromosomes and log2 ratio of −1 indicates loss of 1 chromosome, whereas log2 ratio of 0.53 is gain of 1 chromosome. X-axis: chromosome. This AML has loss of chromosomes 3q, 7q, and gain of chromosomes 13, 20, and 21.

Fig. 3. During convergent evolution (shown on the right side of the figure), an HSC first acquires a mutation (*blue circle*) to form a founding clone. Individual daughter cells of the founder then acquire different mutations either in the same gene or different genes in the same pathway (*yellow circles with different colored-borders*) and gain similar growth advantage and undergo clonal expansion. During divergent evolution, the hostile bone marrow niche selects against wild-type HSC and favors HSCs with various mutations in diverse pathways (*variously colored circles*). Competition among these mutant HSCs eventually lead to outgrowth of the "fittest clone."

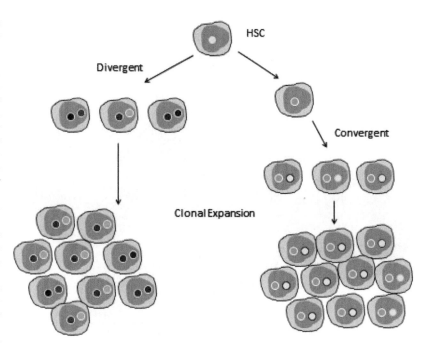

have to disrupt several key pathways on their way to full malignancy.[28] In myeloid neoplasms, epigenetic regulators (*DNMT3A*, *TET2*, and *ASXL1*, for example) are frequently mutated early.[18,29–31] This is followed by alterations of gene expression (splicing factors,[19,20,32] cohesion[33,34] and transcription regulators[35]) and/or activation of growth signals (activating mutations in RAS signaling molecules and growth factor receptors). The earlier steps affect the maturation/differentiation programs, whereas the later steps promote proliferation. In de novo AML in which a precursor myeloid neoplasm is either absent or clinically not detectable, they are more likely to be driven by a strong growth signals generated by a fusion protein resulting from a chromosomal translocation, such as *PML-RARA* in t(15;17) or runt-related transcription factor 1 (*RUNX1*)-*RUNX1T1* in t(8;21).

FOCAL COPY NUMBER ALTERATIONS AND LOSS OF HETEROZYGOSITY (UNIPARENTAL DISOMY) INVISIBLE TO CONVENTIONAL CYTOGENETICS ARE COMMON IN MYELOID NEOPLASM AND CONTRIBUTE TO ITS PATHOGENESIS

JAK2, *TET2*, *EZH2*, *TP53*, and *DNMT3A* are loci frequently affected by these types of alterations, leading to bi-allelic alterations.[36,37] Although fluorescence in-situ hybridization

(FISH) can detect small deletions, it is expensive and can assess only 1 locus at a time. The prevalence and the clinical significance of this type of change are thus probably underappreciated because array or NGS-based testing is needed to identify it.[38,39]

DIAGNOSTIC UTILITY OF NEXT-GENERATION SEQUENCING–BASED MOLECULAR TESTING IN MYELOID NEOPLASM

Morphologic evaluation of bone marrow biopsy and aspirate along with flow cytometry and karyotypic analysis has long been the gold standard in diagnostic workup of myeloid disorders. Single gene molecular testing for *JAK2*, *KIT*, *FLT3*, and *CEBPA* were added to the diagnostic algorithm following the 2008 World Health Organization (WHO) classification. With the rapidly expanding list of frequently mutated genes in myeloid disorders, NGS for a panel of frequently mutated genes has quickly become the test of choice. Currently, it is offered as a clinical test by more than 20 commercial and academic laboratories and has gained wide clinical acceptance (**Fig. 4**). For technical discussion and design of an NGS panel test, the reader is referred to other recent reviews.[40,41] Currently, the NGS panel test is most useful in the settings described in the sections that follow.

Fig. 4. Genes frequently mutated in myeloid neoplasms belong to several key functional groups.

PATIENTS HAVE CYTOPENIA AND CLINICAL SUSPICION OF MYELODYSPLASIA BUT NO OVERT MORPHOLOGIC ABNORMALITY/DYSPLASIA IS SEEN IN MARROW EXAMINATION AND THE KARYOTYPE IS NORMAL

A common clinical scenario is for an otherwise healthy patient to present with mild to moderate degree of cytopenia. After excluding other non-neoplastic causes of cytopenia, such as alcohol, drug, or virus, the patient is labeled as having the so-called idiopathic cytopenia of uncertain significance (ICUS).[42–44] The decision to perform bone marrow biopsy and aspirate under this situation is not always straightforward, especially with older individuals. NGS-based testing on peripheral blood specimens has become a reasonable option before a bone marrow procedure. The presence of one or more modest allele-fraction (>10%) pathogenic mutations can establish the presence of clonal hematopoiesis; that is, the patient now has clonal cytopenia of uncertain significance (CCUS).[44] There is

growing evidence and consensus that there is higher risk for patients with CCUS to progress to overt myeloid neoplasms[45] and bone marrow examination is recommended in this situation. CCUS is not to be confused with ARCH[10,44] in that individuals with ARCH do not have cytopenia and they usually have 1 or at most 2 mutations with low (<10%) allele fractions.

Morphologic dysplasia is notoriously difficult to ascertain or standardize among even experienced hematopathologists.[46,47] It is possible that a fraction of patients with CCUS have sufficient dysplasia to meet the current diagnostic criteria for myelodysplasia but the dysplasia is underappreciated. The identification of high allele-fraction (>10%) pathogenic mutations warrants more careful review of morphologically borderline cases for presence of subtle dysplasia. The concept of CCUS has only come into existence in the past few years and with more data on the natural history and prognosis of CCUS, it is likely that the classification of myelodysplastic syndrome may evolve to include a subset of these patients.

UNLIKE CYTOPENIA, MUTATIONS CAN BE DETECTED IN MOST PATIENTS WITH CYTOSIS BY NEXT-GENERATION SEQUENCING–BASED PANEL TESTING

JAK2, *MPL*, *CALR*, and *TET2* are the more commonly mutated genes in patients with cytosis and are seen in patients with pure MPN (PCV, ET, or PMF).[48–52] Only a small (<10%) fraction of patients have other co-mutated genes. Patients with *CALR* mutation had a lower risk of developing anemia, thrombocytopenia, and marked leukocytosis compared with other subtypes. They also had a lower risk of thrombosis compared with patients carrying *JAK2* (V617F). In some studies, CALR-mutant patients also had better overall survival than JAK2-mutant patients,[53,54] especially for patients with smaller insertions.

In the WHO MPN/MDS, unclassifiable category or CMML, additional mutations in splicing factors, cohesins, and epigenetic regulators are found in addition to *JAK2*, *MPL*, *CALR*, and *TET2*. The presence of complex combinations of genetic alterations may contribute to the varied clinical presentations commonly seen in this group of patients. Aside from the association of mutations in *SF3B1* with ring sideroblast and anemia, and *RUNX1* with propensity for thrombocytopenia,[55] there is no other strong genotype-phenotype association discovered thus far. However, with the widespread use of NGS-based testing, it is anticipated that further correlations may become apparent.

PROGNOSTICATION IN PATIENTS WITH NEWLY DIAGNOSED ACUTE MYELOID LEUKEMIA

Because panel-based testing not only can identify established prognostic markers and therapeutic targets, such as *FLT3*, *NPM1*, *CEBPA*, and *KIT*, but also other genes of increasing importance, such as *TP53*, *RUNX1*, *SF3B1*, *WT1*, and *IDH1/IDH2*, it has become more economical to perform panel testing than to run multiple single-gene tests. NGS panel testing has a higher initial setup cost, but it is relatively inexpensive to introduce additional genes to the panel. As new targeted therapeutics become available, it is more economical and time-saving to screen patients for all potential targets and prognostic markers up front. *RUNX1* mutation, for example, is present in 5% to 15% of AML and is associated with a poor outcome.[56,57] *IDH1* and *IDH2* mutations are usually driver events in leukemogenesis and clinical

trials of specific IDH1/IDH2 inhibitors have shown promising results in early-stage trials.[58,59] Because few targeted therapies have entered the realm of standard of care for AML, the vast majority of patients are treated with conventional cytarabine-based chemotherapy. The impact on treatment decision therefore focuses on identification of patients with high likelihood of relapse and therefore greater benefit from bone marrow transplantation. The current risk stratification is based on a set of clinical parameters, including age, performance status, and a defined set of cytogenetic and molecular findings. As NGS-based testing becomes routine, one can expect more genetic alterations to be incorporated into future risk stratification schemes.

The determination of therapy-related MDS/AML (t-MDS/t-AML) and secondary AML (sAML) in the current WHO classification is based on clinical history of prior treatment and/or myeloid neoplasm.[60] Because a patient can develop a de novo AML unrelated to the prior chemotherapy and prior existence of a myeloid neoplasm is not always clinically apparent, the designation of t-MDS/AML and sAML is often not definitive. Characterization of the genetic alterations may provide a more accurate classification of these patients. For example, the identification of a gene mutation in 1 of 8 genes (*SRSF2*, *SF3B1*, *U2AF1*, *ZRSR2*, *ASXL1*, *EZH2*, *BCOR*, or *STAG2*) was highly specific for a diagnosis of secondary AML and a poor clinical outcome in one study.[61] *TP53* mutations and complex karyotypes are commonly seen in t-AML and both of these features are associated with a poor prognosis.[61–63] Further studies are needed to determine whether the poor prognosis in t-AML can be attributed to these molecular features.

MONITORING DISEASE PROGRESSION FOR PATIENTS WITH EXISTING DIAGNOSIS OF MYELOID NEOPLASM

Clinical change of status or progression for patients with myeloid neoplasms may be associated with acquisition of new variants or significant increase of allele fractions of preexisting variants (**Fig. 5**). Multiple subclonal activating mutations in genes in the RAS signaling pathway commonly emerge at the time of disease progression as the result of convergent co-evolution. Because the read depth (coverage) of an NGS test can reach tens of thousands, an NGS-based panel may be useful for detection of early, low-level variants that may herald the appearance of clinical relapse or progression.

Minimal residual disease (MRD) monitoring is another area that is under intense investigation.

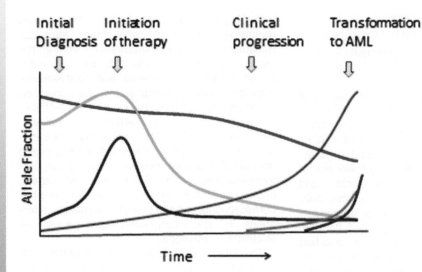

Fig. 5. Clonal evolution in myeloid neoplasm. Each color line represents one variant whose abundance changes during the course of treatment and progression. Some variants arise later in disease process, whereas some variants present early in the course may be supplanted by others.

Because of the multistage leukemogenesis process, leukemic clones and their precursors may share some genetic alterations. Thus, it is important to identify the appropriate leukemia-specific markers to monitor MRD for specific patients. Some variants, such as *DNMT3A* R882C, have been shown to persist in morphologic and clinical complete remission. Detection of such variants may not predict relapse or clinical outcome, whereas the *IDH1* R132 variant may.[64] Furthermore, the current NGS technology has an inherent error rate of 0.1% to 1.0% and therefore limits the sensitivity for MRD detection.

MONITORING BEFORE AND AFTER ALLOGENEIC BONE MARROW TRANSPLANTATION

A byproduct of NGS-based panel testing is information on naturally occurring germline single nucleotide polymorphisms (SNPs) carried by the individual. Comparing the pattern and allele fractions of germline SNPs between donor and recipient in the postallogeneic bone marrow transplantation (allo-BMT) setting can yield chimerism information similar to traditional chimerism testing, which has well-established prognostic implications.[65–67] Because many patients have low or undetectable leukemic burden before transplantation, detection of recipient-derived cells in the marrow by traditional chimerism testing does not necessarily mean these recipient cells are derived from the leukemic clone. Detection of pathogenic variants carried in leukemic clones may be a more predictive marker for relapse. In addition to SNPs, NGS-based testing can assess the leukemia-specific genetic alterations and provide an estimate of leukemic burden, which may better inform prognosis after transplantation.

PROGNOSTIC UTILITY OF NEXT-GENERATION SEQUENCING–BASED MOLECULAR TESTING IN MYELOID NEOPLASM

Because the response to treatment and overall survival of patients with myeloid neoplasms based on current classification is heterogeneous, there has been intense interest in identifying individual genes or scores from multiple genes as prognostic markers outside the clinical parameters, such as age, performance status, karyotype, and pathologic subtypes. Several modifications/enhancements to the exiting WHO risk stratification scheme that make use of additional genetic information have been proposed.[68–72] Randomized prospective clinical trials can now be designed based on these proposals and longer-term outcome data and tailored therapies will be needed to validate these models. The current trend points to a complex picture in which a single gene alteration is unlikely to be deterministic and both the combination and the temporal sequence of acquisition of alterations may be prognostic.[7] A gene may predict chance of progression in MDS/MPN but may not retain prognostic significance in the context of the acute leukemic stage and vice versa. Because of the clonal heterogeneity that can occur both temporally and spatially, as discussed earlier, the assignment of a prognostic group for an individual is not likely to be straightforward and patients may need to be reassessed periodically during the course of the disease.

Because the true clonal architecture[63,73] of neoplastic cells in an individual can be assessed only through single-cell analysis, it is also possible that models based on variant allele fractions of nonfractionated blood or marrow may never be adequate.

As targeted therapeutics begin to enter the treatment arsenals of myeloid neoplasms, it is likely that drug resistance will emerge similar to those seen in solid tumors. Unlike solid tumors in which detection of resistant mutations are difficult because access to tumors can be an issue (plasma-free DNA or "liquid biopsy"[74] notwithstanding), in myeloid neoplasms, monitoring of residual disease and detection of resistance mutations should be straightforward. Insight gained from study and monitoring of myeloid neoplasms may in fact inform us about the development of drug resistance to targeted therapy in general.

RECENT DEVELOPMENT IN PROGNOSTIC IMPACT OF GENE MUTATIONS

The prognostic significance of abnormalities in certain genes (mutations in *FLT3*, *NPM1*, *KIT*), as well as gene expression profiles, have long been recognized in adult patients with AML.[70] Newer information since the 2008 WHO classification is reviewed in the following sections.

CEBPA GENE

The favorable effect of *CEBPA* mutations is now thought to be limited to patients who carry 2 copies of the mutant allele and who are wild type for *FLT3*.[75–77] A large international study of more than 1000 patients with cytogenetically normal AML showed that although the presence of any *CEBPA* mutation was associated with a favorable outcome, only the presence of double mutations of *CEBPA* was an independent prognostic factor on multivariable analysis. The prevalence of *CEBPA* in MDS and MPN is low (<5%) and 2-allele loss is even lower (<1%) and thus no large outcome study is available.

ISOCITRATE DEHYDROGENASE GENES

IDH1/2 mutations are mutually exclusive and infrequently co-mutated with *TET2* and *WT1* mutations because they affect the same pathway in maintaining DNA methylation. Mutant *IDH1/2* can convert alpha-ketoglutarate to 2-hydroxyglutarate (2-HG), which acts as an "oncometabolite" that suppresses *TET2* and thus blocks differentiation.[17,78] *IDH1/2* mutations are found at much higher frequency in patients with AML than MDS/MPN, suggesting that they are secondary, driver-type mutations more often acquired by the neoplastic cells on their way to leukemic transformation. Data are conflicting regarding the prognostic impact of *IDH1/2* gene mutations on outcome and survival. Because *IDH1/2* inhibitors have shown promising results in early phases of clinical trials, the availability of these agents may improve the outcome for patients carrying these mutations.

KIT GENE

Mutations of the *KIT* gene can be detected in approximately 6% of patients with newly diagnosed AML and in 20% to 30% of patients with AML and either t(8;21) or inv(16).[79] Interest in screening for *KIT* mutations arises from the availability of tyrosine kinase inhibitors (TKIs), such as imatinib or dasatinib, which have in vitro activity against other tumors with *KIT* mutations. Clinical trials evaluating the addition of TKIs in selected patients with *KIT* mutations are under way.

Recent development around systemic mastocytosis (SM) and SM with associated hematologic clonal nonmast cell disease (SM-AHNMD) suggests that they may be 2 ends of a single disease spectrum caused by genetic alteration in the HSC.[13,80–82] Subclonal acquisition of *KIT* D816V mutation promotes the neoplastic cells toward mast cell differentiation, whereas other mutations, such as *ASXL1* and *TET2*, may manifest themselves in other clonal hematologic disorders.

WILMS TUMOR 1 GENE

The Wilms tumor 1 gene (*WT1*) is mutated in 10% to 15% of patients with AML and 5% to 10% of patients with MDS/MPN. Although a couple of studies did not find prognostic significance, most studies showed *WT1* mutation to be an unfavorable prognostic factor with adverse impact of event-free survival for cytogenetically normal AML.[70,83,84] *WT1* mutations are more common in patients younger than 60 years and less likely to coexist with *DNMT3A*, *ASXL1*, *IDH1*, and *IDH2* mutations. Multiple subclonal *WT1* mutations are common. Because individual clonal allele fractions are low and may be below the detection limit for Sanger sequencing, the prevalence of *WT1* mutations in myeloid neoplasms may increase as more patients are analyzed by more sensitive NGS-based testing.

ADDITIONAL SEX COMBS GENE

The additional sex combs gene (*ASXL1*) is located in chromosome 20q11 and is involved in regulation of gene expression through maintenance of histone methylation. Mutations in the *ASXL1* gene are frameshift and nonsense truncating mutations located in the last 2 exons (11 and 12) and are present in 6% to 30% of cytogenetically normal AML and about the same frequency in MDS/MPN and denote a poor prognosis.[85–87] Patients with *ASXL1* mutations in AML are older and more likely to be sAML.[61,86]

DNA (CYTOSINE-5)-METHYLTRANSFERASE 3A GENE

The DNA (cytosine-5)-methyltransferase 3A (*DNMT3A*) gene is located in 2p23.3. Eighty percent of the mutations in the *DNMT3A* gene are missense, loss-of-function mutations affecting the amino acid arginine 882 (R882) and the remainder are truncating mutations. Mutations are found in 20% to 22% of cytogenetically normal AML and at a similar frequency in MDS. Mutations in *DNMT3A* lead to hypomethylation and thereby disrupt normal gene regulation. The prognostic impact of *DNMT3A* is likely to be complex, as it is commonly co-mutated with *NPM1* and *FLT3-ITD*.[18,88,89]

SPLICING FACTOR 3B SUBUNIT 1 GENE

SF3B1 encodes a protein that is a component of the spliceosome. Missense mutations in a few hotspots have been identified in a subset of patients with MDS and there is a strong association with the presence of ring sideroblasts.[12,15,20,90] In MDS, *SF3B1* mutation is associated with a favorable prognosis. When co-mutated with *JAK2*, patients frequently have clinical features of refractory anemia with ring sideroblast and thrombocytosis (RARS-T). In AML, the presence of *SF3B1* is associated with sAML and is thus less favorable. In addition to myeloid neoplasm, *SF3B1* mutations have been reported in chronic lymphocytic leukemia, breast carcinomas and rarely in lung cancers.

SPLICING FACTORS OTHER THAN *SF3B1* (*SRSF2*, *U2AF1*, AND *ZRSR2*)

SRSF2 mutations are almost exclusively found at codon 95, either as missense or in-frame deletion. *U2AF1* mutations are missense mutations at codon 34 (S34F) or 157 (Q157P), whereas *ZRSR2* mutations are nonsense or frameshift loss-of-function mutations. Unlike *SF3B1*, mutations in these other splicing factors are unfavorable prognostic markers in most proposed models of MDS. In AML, they are indicators for prior MDS and therefore sAML. *SRSF2* and *ZRSR2* mutations appear to be myeloid-specific.

If *SRSF2* and *ZRSR2* mutations are seen in sequencing of solid tumors, it is more likely that they represent mutations present in the myeloid cells infiltrating the tumors rather than mutations in the tumors themselves. Similar statement also holds true for *JAK2* V617F. In contrast, *U2AF1* mutations are common in endometrial and lung cancers.

RUNT-RELATED TRANSCRIPTION FACTOR 1 GENE

RUNX1 encodes a transcription factor that regulates myeloid gene transcription. Loss of mutations are frequently biallelic, in 5% to 10% of AML, and are also associated with poor outcome.[56,91] Most of the mutations are truncating mutations, but a few hotspots also exist. Association of *RUNX1* mutation with thrombocytopenia has been noted.

TUMOR PROTEIN P53 GENE

TP53 is a prototypic tumor suppressor gene in this group. Loss-of-function mutations and deletions of *TP53* are among the most common genetic alterations in numerous tumor types and are associated with genomic instability, aneuploidy, and poor prognosis. In AML, it is associated with a complex karyotype and poor prognosis.[62] It can be detected in more than 50% of leukemic patients with a history of chemotherapy or radiation therapy and is thought to be a marker for therapy-related disease.

SUMMARY

Our knowledge of the complex, multiple genetic alterations implicated in leukemogenesis has increased significantly in recent years. The availability and ease of NGS-based testing of a panel of frequently mutated genes has improved sufficiently to allow routine clinical use and has impacted the way patients are diagnosed and managed. Some of these discoveries are likely to be incorporated into diagnostic and management scheme in the very near future. Morphologic examination, flow cytometry, FISH, and karyotype are still indispensable parts of workup for myeloid neoplasm, but NGS-based panel testing is likely to become an equally important modality. Demonstration of prognostic utility of genetic alterations through prospective randomized clinical trials has taken place and is likely to dominate in the next few years. Molecular classification of myeloid neoplasms should become a reality in the foreseeable future.

REFERENCES

1. Bonnet D, Dick JE. Human acute myeloid leukemia is organized as a hierarchy that originates from a primitive hematopoietic cell. Nat Med 1997;3: 730–7.

2. Shlush LI, Zandi S, Mitchell A, et al. Identification of pre-leukaemic haematopoietic stem cells in acute leukaemia. Nature 2014;506:328–33.

3. Metzker ML. Sequencing technologies—the next generation. Nat Rev Genet 2010;11(1):31–46.

4. Mardis ER, Ding L, Dooling DJ, et al. Recurring mutations found by sequencing an acute myeloid leukemia genome. N Engl J Med 2009;361: 1058–66.

5. Wong TN, Ramsingh G, Young AL, et al. Role of TP53 mutations in the origin and evolution of therapy-related acute myeloid leukaemia. Nature 2014. http://dx.doi.org/10.1038/nature13968.

6. Kralovics R, Passamonti F, Buser AS, et al. A gain-of-function mutation of JAK2 in myeloproliferative disorders. N Engl J Med 2005;352(17):1779–90.

7. Lundberg P, Karow A, Nienhold R, et al. Clonal evolution and clinical correlates of somatic mutations in myeloproliferative neoplasms. Blood 2014;123(14): 2220–8.

8. Zoi K, Cross NCP. Molecular pathogenesis of atypical CML, CMML and MDS/MPN-unclassifiable. Int J Hematol 2015;101(3):229–42.

9. Malcovati L, Della Porta MG, Pietra D, et al. Molecular and clinical features of refractory anemia with ringed sideroblasts associated with marked thrombocytosis. Blood 2009;114(17):3538–45.

10. Jaiswal S, Fontanillas P, Flannick J, et al. Age-related clonal hematopoiesis associated with adverse outcomes. N Engl J Med 2014;371(26): 2488–98.

11. Genovese G, Kähler AK, Handsaker RE, et al. Clonal hematopoiesis and blood-cancer risk inferred from blood DNA sequence. N Engl J Med 2014;371(26): 2477–87.

12. Papaemmanuil E, Cazzola M, Boultwood J, et al. Somatic SF3B1 mutation in myelodysplasia with ring sideroblasts. N Engl J Med 2011. http://dx.doi.org/10.1056/NEJMoa1103283.

13. Arock M, Sotlar K, Akin C, et al. KIT mutation analysis in mast cell neoplasms: recommendations of the European Competence Network on Mastocytosis. Leukemia 2015;29(6):1223–32.

14. Jankowska AM, Makishima H, Tiu RV, et al. Mutational spectrum analysis of chronic myelomonocytic leukemia includes genes associated with epigenetic regulation: UTX, EZH2, and DNMT3A. Blood 2011; 118:3932–41.

15. Malcovati L, Papaemmanuil E, Ambaglio I, et al. Driver somatic mutations identify distinct disease entities within myeloid neoplasms with myelodysplasia. Blood 2014;124(9):1513–21.

16. Loghavi S, Zuo Z, Ravandi F, et al. Clinical features of de novo acute myeloid leukemia with concurrent DNMT3A, FLT3 and NPM1 mutations. J Hematol Oncol 2014;7:74.

17. Figueroa ME, Abdel-Wahab O, Lu C, et al. Leukemic IDH1 and IDH2 mutations result in a hypermethylation phenotype, disrupt TET2 function, and impair hematopoietic differentiation. Cancer Cell 2010;18: 553–67.

18. Shih AH, Abdel-Wahab O, Patel JP, et al. The role of mutations in epigenetic regulators in myeloid malignancies. Nat Rev Cancer 2012;12: 599–612.

19. Hahn CN, Scott HS. Splice factor mutations and alternative splicing as drivers of hematopoietic malignancy. Immunol Rev 2015;263:257–78.

20. Yoshida K, Sanada M, Shiraishi Y, et al. Frequent pathway mutations of splicing machinery in myelodysplasia. Nature 2011;478:64–9.

21. Paguirigan AL, Smith J, Meshinchi S, et al. Single-cell genotyping demonstrates complex clonal diversity in acute myeloid leukemia. Sci Transl Med 2015; 7(281):281re2.

22. Bacher U, Haferlach T, Schoch C, et al. Implications of NRAS mutations in AML: a study of 2502 patients. Blood 2006;107(10):3847–53.

23. Campbell PJ, Baxter EJ, Beer PA, et al. Mutation of JAK2 in the myeloproliferative disorders: timing, clonality studies, cytogenetic associations, and role in leukemic transformation. Blood 2006; 108(10):3548–55.

24. Theocharides A, Boissinot M, Girodon F, et al. Leukemic blasts in transformed JAK2-V617F-positive myeloproliferative disorders are frequently negative for the JAK2-V617F mutation. Blood 2007; 110(1):375–9.

25. Yoshizato T, Dumitriu B, Hosokawa K, et al. Somatic mutations and clonal hematopoiesis in aplastic anemia. N Engl J Med 2015;373(1):35–47.

26. Brodsky RA. Paroxysmal nocturnal hemoglobinuria: stem cells and clonality. Hematology Am Soc Hematol Educ Program 2008;111–5. http://dx.doi.org/10.1182/asheducation-2008.1.111.

27. Vogelstein B, Fearon ER, Hamilton SR, et al. Genetic alterations during colorectal-tumor development. N Engl J Med 1988;319(9):525–32.

28. Vogelstein B, Kinzler KW. The path to cancer–three strikes and you're out. N Engl J Med 2015;373(20): 1895–8.

29. Ley TJ, Ding L, Walter MJ, et al. DNMT3A mutations in acute myeloid leukemia. N Engl J Med 2010; 363(25):2424–33.

30. Delhommeau F, Dupont S, Della Valle V, et al. Mutation in TET2 in myeloid cancers. N Engl J Med 2009; 360:2289–301.

31. Gelsi-Boyer V, Trouplin V, Adélaïde J, et al. Mutations of polycomb-associated gene ASXL1 in myelodysplastic syndromes and chronic myelomonocytic leukaemia. Br J Haematol 2009;145:788–800.

32. Graubert TA, Shen D, Ding L, et al. Recurrent mutations in the U2AF1 splicing factor in myelodysplastic syndromes. Nat Genet 2011;44:53–7.

33. Kon A, Shih L-Y, Minamino M, et al. Recurrent mutations in multiple components of the cohesin complex in myeloid neoplasms. Nat Genet 2013; 45:1232–7.

34. Thol F, Bollin R, Gehlhaar M, et al. Mutations in the cohesin complex in acute myeloid leukemia: clinical and prognostic implications. Blood 2014;123: 914–20.

35. Cazzola M, Della Porta MG, Malcovati L. The genetic basis of myelodysplasia and its clinical relevance. Blood 2013;122(25):4021–34.

36. Gondek LP, Tiu R, O'Keefe CL, et al. Chromosomal lesions and uniparental disomy detected by SNP arrays in MDS, MDS/MPD, and MDS-derived AML. Blood 2008;111(3):1534–42.

37. Volkert S, Haferlach T, Holzwarth J, et al. Array CGH identifies copy number changes in 11% of 520 MDS patients with normal karyotype and uncovers prognostically relevant deletions. Leukemia 2016;30(1): 259–61.

38. Bullinger L, Fröhling S. Array-based cytogenetic approaches in acute myeloid leukemia: clinical impact and biological insights. Semin Oncol 2012;39(1): 37–46.

39. Bullinger L, Krönke J, Schön C, et al. Identification of acquired copy number alterations and uniparental disomies in cytogenetically normal acute myeloid leukemia using high-resolution single-nucleotide polymorphism analysis. Leukemia 2010;24(2): 438–49.

40. Kuo FC, Dong F. Next-generation sequencing-based panel testing for myeloid neoplasms. Curr Hematol Malig Rep 2015;10(2):104–11.

41. Duncavage EJ, Tandon B. The utility of next-generation sequencing in diagnosis and monitoring of acute myeloid leukemia and myelodysplastic syndromes. Int J Lab Hematol 2015;37(Suppl 1): 115–21.

42. Wimazal F, Fonatsch C, Thalhammer R, et al. Idiopathic cytopenia of undetermined significance (ICUS) versus low risk MDS: the diagnostic interface. Leuk Res 2007;31:1461–8.

43. Steensma DP. Dysplasia has a differential diagnosis: distinguishing genuine myelodysplastic syndromes (MDS) from mimics, imitators, copycats and impostors. Curr Hematol Malig Rep 2012;7:310–20.

44. Kwok B, Hall JM, Witte JS, et al. MDS-associated somatic mutations and clonal hematopoiesis are common in idiopathic cytopenias of undetermined significance. Blood 2015;126(21):2355–61.

45. Cargo CA, Rowbotham N, Evans PA, et al. Targeted sequencing identifies patients with preclinical MDS at high risk of disease progression. Blood 2015; 126(21):2362–5.

46. Font P, Loscertales J, Benavente C, et al. Interobserver variance with the diagnosis of myelodysplastic syndromes (MDS) following the 2008 WHO classification. Ann Hematol 2013;92(1):19–24.

47. Font P, Loscertales J, Soto C, et al. Interobserver variance in myelodysplastic syndromes with less than 5 % bone marrow blasts: unilineage vs. multilineage dysplasia and reproducibility of the threshold of 2% blasts. Ann Hematol 2015;94(4):565–73.

48. **Vannucchi AM, Rotunno G, Bartalucci N, et al. Calreticulin mutation-specific immunostaining in myeloproliferative neoplasms: pathogenetic insight and diagnostic value. Leukemia 2014.** http://dx.doi.org/10.1038/leu.2014.100.

49. Tefferi A, Wassie EA, Lasho TL, et al. Calreticulin mutations and long-term survival in essential thrombocythemia. Leukemia 2014;28(12).

50. Klampfl T, Gisslinger H, Harutyunyan AS, et al. Somatic mutations of calreticulin in myeloproliferative neoplasms. N Engl J Med 2013;369:2379–90.

51. Pardanani AD, Levine RL, Lasho T, et al. MPL515 mutations in myeloproliferative and other myeloid disorders: a study of 1182 patients. Blood 2006; 108(10):3472–6.

52. Beer PA, Campbell PJ, Scott LM, et al. MPL mutations in myeloproliferative disorders: analysis of the PT-1 cohort. Blood 2008;112(1):141–9.

53. Tefferi A, Lasho TL, Finke CM, et al. CALR vs JAK2 vs MPL-mutated or triple-negative myelofibrosis: clinical, cytogenetic and molecular comparisons. Leukemia 2014;28(7):1472–7.

54. Rumi E, Pietra D, Pascutto C, et al. Clinical effect of driver mutations of JAK2, CALR, or MPL in primary myelofibrosis. Blood 2014;124(7):1062–9.

55. Ito Y, Bae S, Shyue L, et al. The RUNX family: developmental. Nat Rev Cancer 2015;15(2):81–95.

56. Gaidzik VI, Bullinger L, Schlenk RF, et al. RUNX1 mutations in acute myeloid leukemia: results from a comprehensive genetic and clinical analysis from the AML study group. J Clin Oncol 2011;29: 1364–72.

57. Mendler JH, Maharry K, Radmacher MD, et al. RUNX1 mutations are associated with poor outcome in younger and older patients with cytogenetically normal acute myeloid leukemia and with distinct gene and microRNA expression signatures. J Clin Oncol 2012;30(25):3109–18.

58. Yaqub F. Inhibition of mutant IDH1 in acute myeloid leukaemia. Lancet Oncol 2015;16(1):e9.

59. Wang F, Travins J, DeLaBarre B, et al. Targeted inhibition of mutant IDH2 in leukemia cells induces cellular differentiation. Science 2013;340(6132): 622–6.

60. Swerdlow S, Campo E, Harris N, editors. World Health Organization classification of tumours of haematopoietic and lymphoid tissues. Lyon (France): IARC press; 2008.

61. Lindsley RC, Mar BG, Mazzola E, et al. Acute myeloid leukemia ontogeny is defined by distinct somatic mutations. Blood 2014;125(9): 1367–76.

62. Haferlach C, Dicker F, Herholz H, et al. Mutations of the TP53 gene in acute myeloid leukemia are strongly associated with a complex aberrant karyotype. Leukemia 2008;22:1539–41.

63. Martincorena I, Roshan A, Gerstung M, et al. High burden and pervasive positive selection of somatic mutations in normal human skin. Science 2015; 348(6237):880–6.

64. Debarri H, Lebon D, Roumier C, et al. IDH1/2 but not DNMT3A mutations are suitable targets for minimal residual disease monitoring in acute myeloid leukemia patients: a study by the Acute Leukemia French Association. Oncotarget 2015;6(39):42345–53.

65. Lion T, Watzinger F, Preuner S, et al. The EuroChimerism concept for a standardized approach to chimerism analysis after allogeneic stem cell transplantation. Leukemia 2012;26(8):1821–8.

66. Lee HC, Saliba RM, Rondon G, et al. Mixed T lymphocyte chimerism after allogeneic hematopoietic transplantation is predictive for relapse of acute myeloid leukemia and myelodysplastic syndromes. Biol Blood Marrow Transplant 2015; 21(11):1948–54.

67. Koreth J, Kim HT, Nikiforow S, et al. Donor chimerism early after reduced-intensity conditioning hematopoietic stem cell transplantation predicts relapse and survival. Biol Blood Marrow Transplant 2014; 20(10):1516–21.

68. Sanchez M, Levine R, Rampal R. Integrating genomics into prognostic models for AML. Semin Hematol 2014;51(4):298–305.

69. Bejar R, Stevenson KE, Caughey BA, et al. Validation of a prognostic model and the impact of mutations in patients with lower-risk myelodysplastic syndromes. J Clin Oncol 2012;30(27):3376–82.

70. Santamaría CM, Chillón MC, García-Sanz R, et al. Molecular stratification model for prognosis in cytogenetically normal acute myeloid leukemia. Blood 2009;114(1):148–52.

71. Tefferi A, Guglielmelli P, Lasho TL, et al. CALR and ASXL1 mutations-based molecular prognostication in primary myelofibrosis: an international study of 570 patients. Leukemia 2014;28:1494–500.

72. Bejar R, Stevenson K, Abdel-Wahab O, et al. Clinical effect of point mutations in myelodysplastic syndromes. N Engl J Med 2011;364(26):2496–506.

73. Welch JS. Mutation position within evolutionary subclonal architecture in AML. Semin Hematol 2014; 51(4):273–81.

74. Buder A, Tomuta C, Filipits M. The potential of liquid biopsies. Curr Opin Oncol 2016. http://dx.doi.org/10.1097/CCO.0000000000000267.

75. Preudhomme C, Sagot C, Boissel N, et al. Favorable prognostic significance of CEBPA mutations in patients with de novo acute myeloid leukemia: a study from the Acute Leukemia French Association (ALFA). Blood 2002;100:2717–23.

76. Fröhling S, Schlenk RF, Stolze I, et al. CEBPA mutations in younger adults with acute myeloid leukemia and normal cytogenetics: prognostic relevance and analysis of cooperating mutations. J Clin Oncol 2004;22:624–33.

77. Dufour A, Schneider F, Metzeler KH, et al. Acute myeloid leukemia with biallelic CEBPA gene mutations and normal karyotype represents a distinct genetic entity associated with a favorable clinical outcome. J Clin Oncol 2010;28(4):570–7.

78. Khasawneh MK, Abdel-Wahab O. Recent discoveries in molecular characterization of acute myeloid leukemia. Curr Hematol Malig Rep 2014; 9:93–9.

79. Kihara R, Nagata Y, Kiyoi H, et al. Comprehensive analysis of genetic alterations and their prognostic impacts in adult acute myeloid leukemia patients. Leukemia 2014;28(8):1586–95.

80. Damaj G, Joris M, Chandesris O, et al. ASXL1 but not TET2 mutations adversely impact overall survival of patients suffering systemic mastocytosis with associated clonal hematologic non-mast-cell diseases. PLoS One 2014;9(1):e85362.

81. Alvarez-Twose I, Morgado JM, Sánchez-Muñoz L, et al. Current state of biology and diagnosis of clonal mast cell diseases in adults. Int J Lab Hematol 2012; 34(5):445–60.

82. Jawhar M, Schwaab J, Schnittger S, et al. Molecular profiling of myeloid progenitor cells in multi-mutated advanced systemic mastocytosis identifies KIT D816V as a distinct and late event. Leukemia 2015;29(5):1115–22.

83. Krauth M, Alpermann T, Bacher U, et al. WT1 mutations are secondary events in AML, show varying frequencies and impact on prognosis between genetic subgroups. Leukemia 2015. http://dx.doi.org/10.1038/leu.2014.243.

84. Virappane P, Gale R, Hills R, et al. Mutation of the Wilms' tumor 1 gene is a poor prognostic factor associated with chemotherapy resistance in normal karyotype acute myeloid leukemia: The United Kingdom Medical Research Council Adult Leukaemia Working Party. J Clin Oncol 2008;26: 5429–35.

85. Chou W-C, Huang H-H, Hou H-A, et al. Distinct clinical and biological features of de novo acute myeloid leukemia with additional sex comb-like 1 (ASXL1) mutations. Blood 2010;116(20): 4086–94.

86. Schnittger S, Eder C, Jeromin S, et al. ASXL1 exon 12 mutations are frequent in AML with intermediate risk karyotype and are independently associated with an adverse outcome. Leukemia 2013;27(1): 82–91.

87. Metzeler KH, Becker H, Maharry K, et al. ASXL1 mutations identify a high-risk subgroup of older patients with primary cytogenetically normal AML within the ELN Favorable genetic category. Blood 2011; 118(26):6920–9.

88. Thol F, Damm F, Lüdeking A, et al. Incidence and prognostic influence of DNMT3A mutations in acute myeloid leukemia. J Clin Oncol 2011;29: 2889–96.

89. Patel JP, Gönen M, Figueroa ME, et al. Prognostic relevance of integrated genetic profiling in acute myeloid leukemia. N Engl J Med 2012;366(12): 1079–89.

90. Wassie EA, Itzykson R, Lasho TL, et al. Molecular and prognostic correlates of cytogenetic abnormalities in chronic myelomonocytic leukemia: a Mayo Clinic-French Consortium Study. Am J Hematol 2014;89(12):1111–5.

91. Tang JL, Hou HA, Chen CY, et al. AML1/RUNX1 mutations in 470 adult patients with de novo acute myeloid leukemia: prognostic implication and interaction with other gene alterations. Blood 2009;114: 5352–61.

Molecular Pathology
Predictive, Prognostic, and Diagnostic Markers in Lymphoid Neoplasms

Caleb Ho, MD[a], Michael J. Kluk, MD, PhD[b],*

KEYWORDS

• Lymphoma • Mutation • Molecular marker • Genetics • B cell • T cell • Targeted therapy

ABSTRACT

Lymphoid neoplasms show great diversity in morphology, immunophenotypic profile, and postulated cells of origin, which also reflects the variety of genetic alterations within this group of tumors. This review discusses many of the currently known genetic alterations in selected mature B-cell and T-cell lymphoid neoplasms, and their significance as diagnostic, prognostic, and therapeutic markers. Given the rapidly increasing number of genetic alterations that have been described in this group of tumors, and that the clinical significance of many is still being studied, this is not an entirely exhaustive review of all of the genetic alterations that have been reported.

OVERVIEW

Lymphoid neoplasms can be broadly divided into those derived from B lymphocytes, T lymphocytes, and natural killer (NK) cells, with the latter 2 sharing some immunophenotypic and functional properties.[1] Besides differences in morphology, immunophenotypic profile, and postulated cells of origin, the lymphoid neoplasm subtypes also differ tremendously in their underlying genetic alterations. The genetic alterations can be at the chromosomal level or at the level of individual genes, through mutations in the gene's coding region, promoter region, intronic changes that affect splicing sites, as well as epigenetic modifications. Understanding the genetic changes associated with the different subtypes of lymphoid neoplasms can shed light on the pathogenesis of the tumors,

and ultimately leads to better prognostication and therapy development. The availability of next-generation sequencing (NGS) platforms has allowed investigators to conduct whole-genome/exome sequencing (WGS/WES) more effectively than before, leading to breakthrough discoveries of genetic mutations associated with different lymphoid neoplasms. These in turn have set the foundation for precision medicine and personalized care in this area. Due to the scope of the topic, this review highlights selected B-cell and T-cell lymphoid neoplasms, with an emphasis on findings from more recent studies, particularly those made possible through the NGS platforms.

MATURE B-CELL LYMPHOID NEOPLASMS

CHRONIC LYMPHOCYTIC LEUKEMIA/SMALL LYMPHOCYTIC LYMPHOMA

Clinical Features

Chronic lymphocytic leukemia/small lymphocytic lymphoma (CLL/SLL) is one of the most common mature B-cell neoplasms in the Western countries.[1] The incidence increases with age, with the mean age of diagnosis at approximately 65, and it frequently involves the peripheral blood, bone marrow, lymph node, spleen, and liver.

Histologic Features and Immunophenotypic Profile

The neoplastic lymphoid cells are usually small, monomorphic, and round, with variable amount of larger admixed prolymphocytes[1] (**Fig. 1**).

The authors have no commercial or financial conflicts of interest to disclose.

[a] Department of Pathology, Memorial Sloan Kettering Cancer Center, 1275 York Avenue, New York, NY 10065, USA; [b] Department of Pathology, Weill Cornell Medical College, 525 East 68th Street, Mailbox #79, F-540, New York, NY 10065, USA
* Corresponding author.
E-mail address: mik9095@med.cornell.edu

Surgical Pathology 9 (2016) 489–521
http://dx.doi.org/10.1016/j.path.2016.04.011

surgpath.theclinics.com

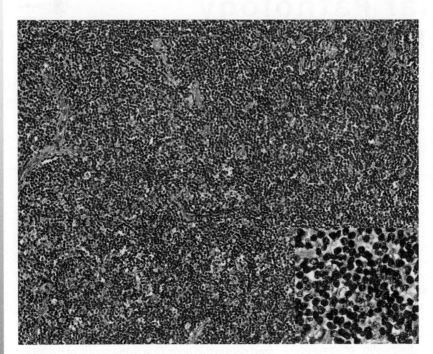

Fig. 1. Hematoxylin-eosin (H&E) image of small lymphocytic lymphoma (SLL) involving a lymph node at ×20 and ×60 (inset). Most neoplastic cells are small, monomorphic lymphoid cells with indistinct nucleoli and clumped chromatin pattern. There are scattered larger prolymphocytes.

Classically, CLL/SLL has a very distinctive immunophenotypic profile that allows its separation from other mature B-cell neoplasms: CD19+ CD20dim+ CD5+ CD10− CD23+ CD43+ CD11c weak+ FMC7− cyclin D1− surface heavy chains immunoglobulin (Ig)M/IgD and surface light chain dim+.[1]

Molecular Diagnostic and Prognostic Markers

Cytogenetic abnormalities are found in up to 80% of CLL/SLL cases, although none is unique to this disease.[1] Isolated deletion of 13q14.3, the most common cytogenetic alteration, is associated with a favorable prognosis, whereas other abnormalities, such as trisomy 12, deletion 11q22 to 23 (ataxia telangiectasia mutated [ATM] locus), deletion 17p13 (TP53 locus), and deletion 6q21, have been associated with an intermediate and/or adverse prognosis.[1] CLL/SLL is thought to arise from antigen-stimulated, post–germinal center B cells, and more than half of the cases have undergone somatic hypermutation (SHM) in the variable regions of the immunoglobulin heavy chain (IGVH) and light chain (IGVL) genes.[2,3] CLL/SLL cases with unmutated/germline IGVH, as well as cases with high ZAP-70 or CD38 expression, are associated with worse prognoses and higher chance of chemotherapy requirements.[1–6] A subsequent study showed that loss of methylation at a specific single CpG dinucleotide in the

5′ regulatory region of ZAP-70 transcription start site 1 (TSS1) predicted variably increased ZAP-70 mRNA and protein expression and poor prognosis.[7]

In addition to cytogenetic abnormalities, somatic mutations are also common in CLL/SLL (Table 1). The most frequent recurrent mutations occur in NOTCH1, SF3B1, TP53, ATM, and MYD88.[8–12] Among these, NOTCH1, SF3B1, and TP53 have been associated with adverse prognostic implications among untreated patients in retrospective studies,[13–15] as well as among patients with relapsed/refractory CLL in whom mutations in multiple genes (TP53, ATM, SF3B1) were frequent and associated with poor outcome.[16] A proposed model integrating cytogenetic abnormalities and gene mutations has been shown to improve prognostic accuracy compared with using karyotype alone.[15]

NOTCH1 is one of the most frequently mutated genes in CLL/SLL. It encodes for a ligand-activated transmembrane heterodimer that plays a key role in cell differentiation and lineage determination.[10,17–19] After ligand binding, a series of proteolytic cleavages leads to the liberation of the functionally active Notch intracellular domain (NICD), allowing its translocation to the nucleus and transcriptional activation of target genes, including MYC and NF-κB.[10,19–21] NOTCH1 mutations were described earlier in up to 50% to 60% of T-lymphoblastic leukemia/lymphoma (T-ALL)

Table 1
Selected recurrently mutated genes in chronic lymphocytic leukemia/small lymphocytic lymphoma (CLL/SLL)

	Normal Function	Mutational Frequency (%)	Mutation Hotspot	Effect of Mutations	Prognostic Marker
NOTCH1	Ligand-activated transmembrane protein, transcriptional activation	Up to 15	c.7541_7542delCT (p.P2514fs*4)	Truncation of PEST domain → Activation of NOTCH1 pathways	Adverse
FBXW7	Ubiquitin ligase	4	Mostly within substrate-binding domain	Decreased degradation of NICD	—
SF3B1	Catalytic core of spliceosome	15–20	Clustered within C-terminal PP2A-repeat Regions (eg, p.K700)	May alter splicing of gene transcripts	Adverse
TP53	Activates DNA repair, arrests cell cycle, initiates apoptosis	7–15	Mostly within the DNA-binding domain	Loss of gene function	Adverse
ATM	Coordinates response to dsDNA damage	9–12	None	Loss of gene function	Adverse
MYD88	Immune response via TLR signaling	3–10	c.794T>C (p.L265P)	Constitutive activation of the NF-κB and JAK-STAT pathways	—

Abbreviations: dsDNA, double-stranded DNA; NICD, Notch intracellular domain (Functionally active portion of Notch protein); TLR, Toll-like receptor.

cases, more commonly affecting the heterodimerization (HD) domain, and less commonly the PEST domain.[17,22–24] Mutations in the HD domain enhance cleavage of Notch 1 protein to liberate NICD, whereas mutations in the PEST domain reduce degradation of NICD, both of which lead to constitutive activation of Notch 1 signaling.[19,22] NOTCH1 mutations have been described in up to 15% of patients with CLL/SLL, most frequently as a recurrent 2 base pair frameshift mutation in exon 34 (c.7541_7542delCT [p.P2514fs*4])[8–10,25] (Fig. 2). The NOTCH1 p.P2514fs*4 results in a premature stop codon that truncates the PEST domain, impairing degradation of the active isoform NICD, and is associated with unmutated IGVH, higher chance of large cell transformation, and poor prognosis in patients with CLL/SLL.[8–10,25]

Besides mutations in NOTCH1, mutations in FBXW7 provide an alternative mechanism to deregulate the Notch 1 pathway. FXBW7 codes for a ubiquitin ligase that is critical in the proteasomal degradation of proteins such as N-myc, C-myc, and NICD. Mutations affecting the substrate-binding domain of FBXW7, usually in a heterozygous manner, have been identified in T-ALL[20,24] and CLL/SLL cases.[8] In general, heterozygous FBXW7 mutations were thought to induce a dominant negative effect on the binding and targeting of NICD for degradation, eventually resulting in the activation of Notch1 downstream targets in some cell types.[20,24]

Splicing factor 3b, subunit 1 (SF3B1) encodes for the catalytic core of the spliceosome that interacts with RNA sequences, and is also frequently mutated in CLL/SLL (15%–20% of cases), especially in treatment-refractory cases.[8,11] The mutations are typically missense mutations that cluster within the C-terminal PP2A-repeat regions, which may alter the splicing of a narrow spectrum of transcripts important for pathogenesis.[8,26,27] The most frequently affected codon is Lys700;

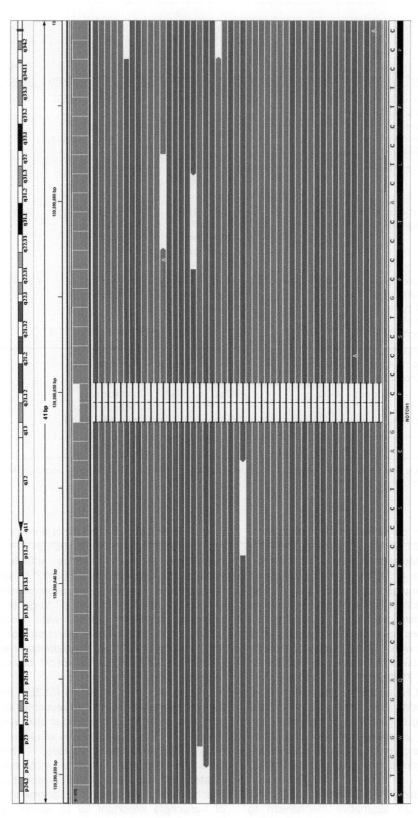

Fig. 2. NOTCH1 c.7541_7542delCT (p.P2514fs*4) as detected on an NGS platform and visualized using Integrated Genomics Viewer (Broad Institute, Cambridge, MA). Each horizontal bar represents a single sequencing read, with red and blue bars annotating reads from each of the 2 directions. The reference sequence is shown at the bottom. The white boxes represent the location of 2 base pair deletions within exon 34 of *NOTCH1*, resulting in frameshift and a premature stop codon, truncating the PEST domain.

other commonly affected codons include Glu622, Arg625, His662, Lys666, and Gly742.[11]

TP53 is a well-known tumor suppressor gene whose function can be lost through the deletion of 17p13 or via missense/nonsense/frameshift mutations, mostly within the DNA-binding domain (7%–15% of cases).[8,10,28,29] *TP53* mutation and deletion frequently coexist, resulting in functional loss of both alleles of the gene, especially in refractory/recurrent CLL/SLL cases.[16]

The *ATM* gene codes for a protein kinase that is important for maintaining genome stability through coordination of cellular response to double-stranded DNA (dsDNA) damage.[30,31] Loss of *ATM* function can be seen in CLL/SLL via the deletion of 11q22 to 23 (*ATM* locus) or mutations in the *ATM* gene itself (9%–12% of cases).[8,32] Similar to *TP53*, *ATM* mutation and deletion can coexist in refractory/recurrent CLL/SLL cases.[16]

Last, *MYD88* encodes for an adaptor protein that normally functions in immune responses leading to activation of the NF-κB and Janus kinase/signal transducers and activators of transcription (JAK-STAT) pathways.[33,34] The gain-of-function mutation, *MYD88* c.794T>C (p.L265P), is found in 3% to 10% of CLL/SLL,[8,9,35] and is discussed in more detail under the section Lymphoplasmacytic Lymphoma later in this article.

Targeted Therapies

Currently, targeted therapies approved by the Food and Drug Administration for the treatment of CLL/SLL include ibrutinib and idelalisib, which are available as oral medications that respectively inhibit Bruton tyrosine kinase (BTK) and the p110 δ isoform of phosphoinositide-3 kinase (PI3K).[36,37] Ibrutinib inhibits B-cell receptor signaling via the inhibition of BTK, whereas idelalisib inhibits the PI3K isoform specific for B cells, exerting the therapeutic effects with relative sparing of other hematopoietic elements. Idelalisib has been shown in clinical trials to improve treatment response and survival rates in relapsed/refractory CLL/SLL when administered in combination with the anti-CD20 agent rituximab,[36–39] whereas ibrutinib was superior to another anti-CD20 agent, ofatumumab, in improving response and survival rates for relapsed/refractory patients.[36,37,40,41] Furthermore, ibrutinib has shown some efficacy in studies specifically for patients with CLL/SLL with demonstrated 17p13.1 (*TP53*) aberrations, a group that tended to respond poorly to first-line therapy and associated with early relapse, higher chance of Richter transformation, and shorter survival.[42,43]

Additionally, venetoclax (ABT-199), an oral inhibitor of BCL2, has shown some promise in the treatment of CLL/SLL in a phase I study, and is currently being tested in late-stage clinical studies for CLL/SLL.[36,37,44] Another BCL2 inhibitor, navitoclax (ABT-263) also has showed some promise in CLL/SLL, but its use has been limited due to treatment-related thrombocytopenia.[37,44]

BURKITT LYMPHOMA

Clinical Features

Burkitt lymphoma (BL) is an aggressive B-cell lymphoma with very short doubling time, resulting in most patients presenting at advanced stages. BL most often involves extranodal sites, and can involve the central nervous system (CNS), lymph nodes, and bone marrow.[1] The 3 recognized clinical variants are endemic, sporadic, and immunodeficiency-associated BLs. BLs can be associated with Epstein-Barr virus (EBV) infection, especially in endemic BL cases.

Histologic Features and Immunophenotypic Profile

The neoplastic lymphoid cells are generally medium-sized, monomorphic in appearance, with fine to dispersed chromatin pattern, basophilic cytoplasm with lipid vacuoles, and somewhat resembling lymphoblasts[1] (**Fig. 3**).

The classic immunophenotypic profile is: CD20+ CD5– CD10+ BCL2– BCL6+ TdT– IgM+ MYC+ with a Ki67 proliferation index close to 100%.[1]

Molecular Diagnostic and Prognostic Markers

The most common translocation in BL, t(8;14)(q24;q32), involves the juxtaposition of *MYC* on 8q24 with the *IGH* enhancer on 14q32, leading to constitutive expression of the MYC protein in the neoplastic cells.[1] Less frequently, *MYC* is translocated to the immunoglobulin kappa *IGK* (2p12) or lambda light chain *IGL* (22q11) gene.[45] *MYC* translocations involving nonimmunoglobulin genes, such as *PAX5* (9p13) and *BCL6* (3q27), have also been described.[46–48]

Besides *MYC* gene translocation, there is growing evidence that DNA methylation and somatic mutations also contribute to the pathogenesis of BL (**Table 2**). In a whole-genome bisulfite sequencing (WGBS) study, BL showed global hypomethylation compared with non-neoplastic germinal center B cells.[49] The hypomethylated genes, which mostly correlated with upregulated gene expressions, are expressed in the so-called "dark-zone" of the germinal center, such as

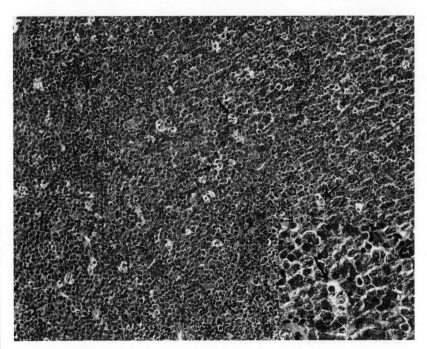

Fig. 3. H&E image of BL involving a lymph node at ×20 and ×60 (*inset*). The neoplastic cells are medium-sized, monomorphic lymphoid cells with fine to dispersed chromatin pattern. Admixed are scattered tingible-body macrophages (*arrows*), creating a "starry sky" appearance.

TCF3 and SMARCA4.[49,50] On the other hand, certain genes are hypermethylated, including Janus kinase 3 (*JAK3*) and signal transducer and activator of transcription 3 (*STAT3*), resulting in downregulation of the JAK-STAT pathway. DNA methylation alteration in BL also leads to upregulation of *IGF2BP1*, which encodes an RNA-binding protein that stabilizes MYC, and

upregulation of *MDM2*, which inhibits *TP53* transcription.

Highly recurrent somatic mutations in *TCF3*, or in its negative regulator *ID3*, are found in all subtypes of BL, although at different frequencies.[51–54] *TCF3* is a transcription factor that normally regulates the transcription of immunoglobulin and other B-cell–restricted genes, and promotes cell

Table 2
Selected recurrently mutated genes in Burkitt lymphoma (BL)

	Normal Function	Mutational Frequency (%)	Mutation Hotspot	Effect of Mutations	Prognostic Marker
TCF3	Regulates transcription of Immunoglobulin and B-cell related genes, promotes cell cycle progression	~30	Within the basic HLH DNA-binding and dimerization domain	Promotes survival of lymphoma cells	—
ID3	Negative Regulator of TCF3	60–70	Mostly within the HLH domain	Loss of inhibition of TCF3	—
S1PR2	Gα12/13 coupled receptor, regulates cell growth and local confinement of germinal center B cells	<5	None	Loss of/reduced inhibition on AKT pathway and migratory activity	—

Abbreviation: HLH, helix-loop-helix.

cycle progression by upregulating *CCND3* and *E2F2* and downregulating *RB1*.[53] Its overexpression is thought to be important for the survival of BL. Mutations in *TCF3* and *ID3* occur, respectively, in approximately 30% and 60% to 70% of BL overall, and are rare in other lymphoid neoplasms.[52,53] The identified *TCF3* mutations are gain-of-function mutations that specifically affect the E47 splice isoform, whereas the *ID3* mutations are mostly nonsense and frameshift mutations in the helix-loop-helix domain, which inactivate the ID3 protein.[51–54] The coexistence of *TCF3* hypomethylation[49] and activating mutation[53] further supports the importance of this gene in BL.

Other recurrent mutations reported in BL included those in *S1PR2*, which encodes for sphingosine-1-phosphate receptor 2, a Gα12-coupled and Gα13-coupled receptor that helps regulate growth and local confinement of germinal center B cells.[55] The reported *S1PR2* mutations (<5% of BL cases) result in the loss of inhibitory function on AKT and migratory activity. In a WGBS of BL, the sphingosine-1-phosphate (S1P) pathway signaling was further shown to be deregulated through a combination of DNA methylation and somatic mutations.[49] Several genes in this pathway, including *GNA11* (encoding stimulatory $G_q\alpha$), were hypomethylated and upregulated. At the same time, most BL harbored mutations in genes that were either part of the $G_{12/13}$ complex (*RHOA, P2RY8, GNA13,* and *ARHGEF1*) or the inhibitory $G_i\alpha$ complex (*GNAI1* and *GNAI2*). The net effect of the detected alterations was the activation of the AKT signaling pathway.

Other recurrently mutated genes in BL have been reported, including *SMARCA4*,[49,54] *CCND3*,[52,53] *TP53*,[52,53] and *PCBP1*.[56]

FOLLICULAR LYMPHOMA

Clinical Features

Follicular lymphoma (FL) is a B-cell neoplasm that accounts for approximately 20% of all lymphomas, involving predominantly lymph nodes, spleen, bone marrow, peripheral blood, and in widespread diseases, non-hematopoietic extranodal sites.[1]

Histologic Features and Immunophenotypic Profile

The neoplastic lymphoid cells are composed of a variable proportion of follicle center centrocytes and centroblasts, which is the basis for the World Health Organization (WHO) grading of 1 to 3.[1] Architecturally, FL usually forms at least a partially follicular pattern (Fig. 4).

The lesional cells are usually CD20+ CD5− CD10+ BCL2+ BCL6+ CD43− MUM1−.[1] In some cases, BCL2 can be negative by immunohistochemical (IHC) stain when the widely-used antibody clone 124 is used, but may be positive by using alternative antibodies raised against clones such as E17 or SP66, which target different epitopes of the BCL2.[1,57–59]

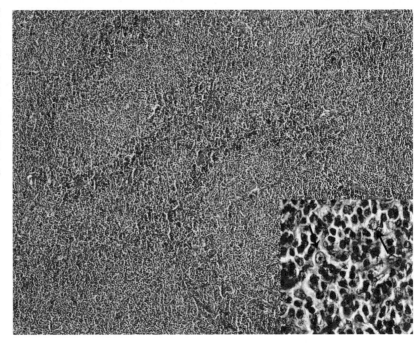

Fig. 4. H&E image of grade 1 to 2 FL involving a lymph node at ×10 and ×60 (*inset*). Most neoplastic lymphoid cells are centrocytes with irregular, cleaved nuclei and indistinct nucleoli. Admixed are occasional larger cells with more dispersed chromatin and prominent nucleoli, consistent with centroblasts (*arrows*).

Molecular Diagnostic and Prognostic Markers

Up to 90% of grade 1 to 2 FL, and a lower proportion of grade 3 FL, demonstrate the t(14;18) (q32;q21)/*IGH-BCL2* gene rearrangement, resulting in overexpression of the antiapoptotic protein BCL2.[1] Grade 3B FL can have t(14;18) or abnormalities involving 3q27/BCL6 in a mutually exclusive manner.[60]

FL, just as normal germinal center cells, undergo ongoing activation-induced cytidine deaminase–mediated somatic hypermutation (SHM) of the *IGH* variable gene region (*IGHV*), as well as atypical SHM (aSHM) in aberrant sites in nonimmunoglobulin oncogenes, such as *BCL2*, *BCL6*, *MYC*, and *PIM1*.[61] These sites of aSHM are therefore recurrently mutated in FL, and can occasionally result in hypermutated tumors.[61,62] Other recurrently mutated genes in FL include *CREBBP*, *KMT2D (MLL2)*, *EP300*, *EZH2*, and *MEF2B* (Table 3).

CREBBP is a transcription factor with intrinsic histone acetyltransferase (HAT) activity involved in epigenetic regulation, and *CREBBP* mutations are found in 30% to 70% of FL.[61–65] The prevalence of *CREBBP* mutations among FL and their presence in the early stage of disease development suggest that this gene could play an early role in pathogenesis.[61,63–65] The identified *CREBBP* mutations tend to be inactivating mutations within the HAT domain,[63] and are associated with higher degree of aSHM and DNA methylation changes.[61] In vitro studies have shown that *CREBBP* HAT domain mutations lead to a decrease in CREBBP-mediated activation of the p53 tumor suppressor as well as blunted inactivation of the BCL6 oncogene.[63]

Mutations in *KMT2D (MLL2)*, which encodes for a histone methyltransferase, has been found in up to 80% to 90% of FL,[61,62,64,66] whereas mutations in other histone modification–related genes (*EP300, EZH2, MEF2B*)[62–64,66–68] have been found in a large proportion of FL, further implicating epigenetic regulation in the pathogenesis of this tumor type. Of note, *EZH2* mutations are found in approximately 7% to 22% of FL, and p.Y646 (alternatively referred to as *EZH2* p.Y641) has emerged as a mutational hotspot.[62,66–68] This mutation appears to contribute to pathogenesis by increasing the levels of histone H3K27-specific trimethylation activity, resulting in transcriptional suppression of target genes.[69,70]

In a whole-genome bisulfite sequencing (WGBS) study, compared with BL, FL cases showed differences in methylation pattern, despite their common origin from follicle center B cells. For example, genes expressed in the so-called "dark-zone" of the germinal center,[49,50] such as *TCF3* and *SMARCA4*, were hypermethylated and downregulated in FL, opposite to the pattern seen in BL. On the other hand, *JAK3* and *STAT3* were hypomethylated and upregulated in FL, resulting in activation of the JAK-STAT pathway.[49] Interestingly, frequent mutations in *SOCS1* and *STAT6* were found in FL, serving as alternative mechanisms for constitutive activation of the JAK-STAT pathway.[64]

Based on retrospective analysis of mutation status in previously untreated FL cases, a clinicogenetic risk model (m7-FLIPI) was proposed, combining the mutation status of 7 genes (*EZH2, ARID1A, MEF2B, EP300, FOXO1, CREBBP,* and *CARD11*) with the Follicular Lymphoma International Prognostic Index (FLIPI) and Eastern Cooperative Oncology Group (ECOG) performance status.[62] The m7-FLIPI model stratifies patients into high-risk and low-risk groups, which improves prognostication and decision on treatment choices compared with using FLIPI alone.

Targeted Therapies

Most FL cases have high expression of BCL2, mostly due to t(14;18) (q32;q21). The BCL2 inhibitors venetoclax (ABT-199) and navitoclax (ABT-263) have shown responses in some patients with FL in phase I trials.[71,72] Phase I clinical trial of EZH2 inhibitor for FL is also in progress, and drug responses will be assessed in both EZH2-mutated and wild-type cases.[73]

Certain drugs targeting the NF-κB and B-cell receptor signaling pathways were also effective in treating FL in different clinical trials. For example, the combination of ibrutinib, a BTK inhibitor, with rituximab and bendamustine showed some degree of treatment responses in 90% of relapsed FL cases in one study.[74] The combination of bortezomib, a proteasome and NF-κB pathway inhibitor, with rituximab-based regimens showed good response rate in both treatment-naive and relapsed/refractory FL.[75,76]

DIFFUSE LARGE B-CELL LYMPHOMA, NOT OTHERWISE SPECIFIED

Clinical and Histologic Features

Diffuse Large B-Cell Lymphoma (DLBCL), not otherwise specified (NOS), is a biologically and morphologically heterogeneous group of B-cell neoplasms, grouped together based on the large cell morphology of the neoplastic cells[1] (Fig. 5). DLBCL can arise de novo, or from transformation from a more indolent lymphoma. Some specific types of large-cell lymphomas are recognized as distinct entities from DLBCL, NOS by the WHO

Table 3
Selected recurrently mutated genes in follicular lymphoma (FL)

	Normal Function	Mutational Frequency (%)	Mutation Hotspot	Effect of Mutations	Prognostic Marker	Targeted Therapy
CREBBP	Intrinsic HAT activity, epigenetic regulation	30–70	Within HAT domain	Loss of/reduced gene function	A component of m7-FLIPI	—
KMT2D (MLL2)	H3K4-specific histone methyltransferase	80–90	None	Loss of/reduced gene function	—	—
EP300	Acetyltransferase, epigenetic regulation	9–14	Frequently within HAT domain	Loss of/reduced gene function	A component of m7-FLIPI	—
EZH2	Regulate histone H3 methylation, early B-cell development and IgH gene rearrangement	7–22	p.Y646 (SET domain)	Increased Histone H3K27-specific trimethylation → Transcriptional suppression of target genes	A component of m7-FLIPI	EZH2 Inhibitors - Phase I Clinical Trial
MEF2B	Calcium-regulated gene involved in histone-modification	10–15	Mostly within exon 2 and 3 - MADS box and MEF2 domains (eg, p.Y69, p.N81, p.D83)	Uncertain, may alter histone modification and gene expression	A component of m7-FLIPI	—

Abbreviations: HAT, histone acetyltransferase; Ig, immunoglobulin; m7-FLIPI, Clinicogenetic risk model combining nonsilent mutation status of 7 genes (EZH2, ARID1A, MEF2B, EP300, FOXO1, CREBBP, and CARD11) with the Follicular Lymphoma International Prognostic Index (FLIPI) and Eastern Cooperative Oncology Group (ECOG) performance status.

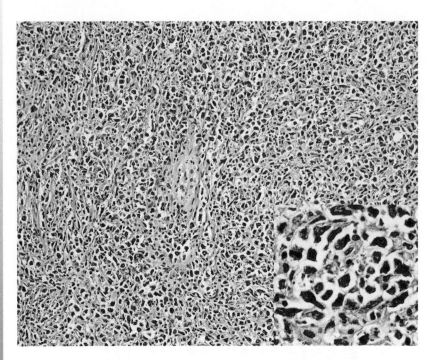

Fig. 5. H&E image of DLBCL involving the testicle at ×20 and ×60 (*inset*). The morphology of DLBCL is heterogeneous, and the neoplastic cells can show a range of cell sizes and shapes, with variably distinct nucleoli and variable amount of cytoplasm.

classification (eg, primary mediastinal large B-cell lymphoma, primary DLBCL of the CNS). The cell-of-origin (COO) classification remains important in DLBCL, dividing the cases into germinal center B-cell–like (GCB) type and activated B-cell–like (ABC) type, based on gene expression profile.[77,78] DLBCL, NOS is more common in the elderly (median age older than 70), usually with an aggressive clinical course, and is one of the most common types of lymphoma in the Western hemisphere.[1]

Molecular Diagnostic and Prognostic Markers

Clonal immunoglobulin gene rearrangements (eg, *IGH, IGK*) are detectable in most cases of DLBCL depending on the combination of immunoglobulin gene primer sets used (**Fig. 6**). In some cases with a bona fide morphologic diagnosis of DLBCL, molecular studies for immunoglobulin gene rearrangements may yield false-negative results due to somatic hypermutation, which can affect primer binding. The variety of chromosomal abnormalities reported in DLBCL underscores the biological heterogeneity of this group of tumors. ABC DLBCL more frequently has trisomy 3, gains of 3q, 18q21-q22, as well as loss of 6q21, whereas GCB DLBCL more frequently has gains of 12q12.[79] Furthermore, GCB DLBCLs are more likely to have translocations involving *BCL2* at 18q21 (including t[14;18]/*IGH-BCL2*, the hallmark of follicular lymphomas),[78,80,81] *REL* amplification,[78] and *IGH* gene somatic hypermutations.[82]

On the other hand, ABC DLBCLs are more likely to have translocations involving *BCL6* at 3q27.[83] In general, ABC DLBCLs are associated with more aggressive clinical course.[1,77,78]

High-grade B-cell lymphomas with concurrent translocations involving *BCL2* at 18q21 and *MYC* at 8q24, also known as "genetic double-hit lymphomas," constitute a distinct type of lymphoma associated with inferior prognosis and poorer response to treatment with conventional chemotherapy regimens used for DLBCL.[84,85] Even in DLBCLs without demonstrated *BCL2* or *MYC* translocation, high BCL-2 and MYC coexpression as detected by IHC ("double-expressers") is independently associated with poor prognosis and treatment response.[86–90] The poor prognostic value persists even when adjusted for ABC versus GCB gene signatures and International Prognostic Index (IPI) scores.[86–90] Nevertheless, a recent study found that COO classification remains prognostically important, especially for DLBCLs that are non–double-expressors.[91] Given these findings, efforts have been made to standardize the approach to the detection of elevated MYC protein using IHC[92–95] or molecular methods, such as gene expression profiling.[96]

GCB DLBCL and FL are both thought to be derived from germinal center B cells, a belief reinforced by their similar molecular alterations (**Table 4**). For example, both exhibit frequent mutations in *EZH2*, a gene encoding the catalytic unit of the histone methyltransferase PRC2.

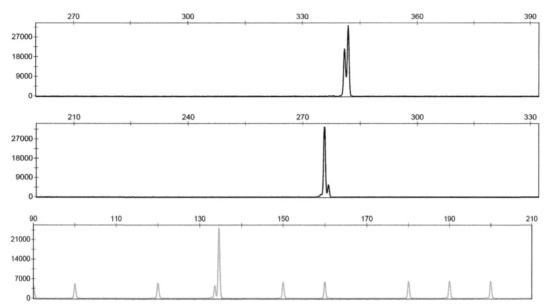

Fig. 6. IGH clonal gene rearrangement in a case of DLBCL using PCR primers validated by a European collaborative study (BIOMED-2 Concerted Action), with analysis of amplified products by capillary electrophoresis. (*Top*) Clonal peak detected by the VH-FR1 and JH consensus primers. (*Middle*) Clonal peak detected by the VH-FR2 and JH consensus primers. (*Bottom*) Clonal peak detected by the VH-FR3 and JH consensus primers.

EZH2 expression is usually high in sites of lymphopoiesis, such as normal proliferative germinal center B cells.[97] Through regulation of histone H3 methylation, *EZH2* is important for early B-cell development and IgH gene rearrangement. *EZH2* mutations are usually heterozygous, found in 7% to 22% of FL and more than 20% of GCB DLBCLs, but are rarely found in ABC DLBCL.[62,66–68] Of all the detected mutations, *EZH2 p.Y646* (*EZH2* p.Y641, see FL section, earlier in this article), located within the SET domain, has emerged as a mutational hotspot for GCB DLBCL, just as in FL. Through coordinated activities of the wild-type and Y646-mutated enzymes, heterozygous mutants show increased histone H3K27-specific trimethylation, leading to transcriptional repression of target genes.[69,70]

Other recurrent mutations seen commonly in both FL and GCB DLBCL also involve genes important for chromatin and histone modification. These include point mutations in *MEF2B*, a calcium-regulated gene involved in histone modification[66,98]; inactivating mutations in *KMT2D* (*MLL2*), a H3K4-specific histone methyltransferase[66,98]; and inactivating mutations or deletions in *CREBBP* or *EP300*, 2 closely related acetyltransferase.[63,98]

Other mutations shared in GCB DLBCL and BL, another lymphoma of germinal center B-cell origin, include inactivating mutations in *GNA13* (encoding Gα13) and *S1PR2* (encoding a Gα12/13 coupled

receptor), resulting in the loss of inhibitory function on AKT and migratory activity.[55,99]

Disinhibition of the oncogenic PI3K/AKT pathway can also occur through the loss of genomic material at 10q23, including *PTEN*, a negative regulator of the pathway. *PTEN* loss is much more common in GCB DLBCL (~55% of cases) compared with ABC DLBCL (~14% of cases).[100]

Although ABC DLBCLs share some similar genetic alterations to GCB DLBCLs, including mutations in *MLL2*,[66,98] in general the 2 groups exhibit different molecular signatures (see **Table 4**). Activation of NF-κB pathway is known to contribute to lymphomagenesis in ABC DLBCL,[101] which can be achieved through different mechanisms.[102–107] MyD88 is an adaptor protein that normally functions in immune responses via Toll-like receptor (TLR) signaling, and eventually leads to activation of the NF-κB and JAK-STAT3 pathways to promote cell survival.[33,34] Approximately 30% of ABC DLBCLs harbor the hotspot, gain-of-function mutation c.794T>C (p.L265P),[98,102] similar to the hotspot mutation found in lymphoplasmacytic lymphomas. *MYD88 p.L265P* has been shown through cell line knockdown experiments to be crucial for tumor pathogenesis and survival.[102] Other *MYD88* gain-of-function mutations have been reported in ABC DLBCLs, mostly in the Toll/IL-1 receptor domain (TIR) domain.[98,102]

Table 4
Selected recurrently mutated genes in diffuse large B-cell lymphoma (DLBCL)

		Normal Function	Mutational Frequency (%)	Mutation Hotspot	Effect of Mutations	Prognostic Marker	Targeted Therapy
Mutations more common in GCB DLBCL	*EZH2*	Regulates histone H3 methylation, early B cell development and IgH gene rearrangement	>20 GCB DLBCL	*p.Y646 (SET Domain)*	Increased Histone H3K27-specific trimethylation → Transcriptional suppression of target genes	—	EZH2 Inhibitors - Phase I Clinical Trial
	CREBBP	Intrinsic HAT activity, epigenetic regulation	~18–22 of DLBCL	Within HAT domain	Loss of/reduced gene function	—	—
	EP300	Acetyltransferase, epigenetic regulation	~5–10 of DLBCL	Frequently within HAT domain	Loss of/reduced gene function	—	—
	MEF2B	Calcium-regulated gene involved in histone-modification	~8–13 DLBCL	Mostly within exon 2 and 3 - MADS box and MEF2 domains (eg, p.Y69, p.N81, p.D83)	Uncertain, may alter histone modification and gene expression	—	—
	KMT2D (MLL2)	H3K4-specific histone methyltransferase	>20–30 DLBCL	None	Loss of/reduced gene function	—	—
Mutations more common in ABC DLBCL	*MYD88*	Immune response via TLR signaling	30–40 of ABC DLBCL	*c.794T>C (p.L265P)*	Constitutive activation of the NF-κB and JAK-STAT pathways	—	Drugs targeting the NF-κB and/or JAK-STAT pathways (eg, Bortezomib, Ibrutinib, JAK and IKKβ inhibitors)
	CARD11	Signaling scaffold protein required for NF-κB activation	~8–10 of ABC DLBCL	Mostly within exon of coiled-coil domain	Constitutive activation of the NF-κB and JAK-STAT pathways	—	
	CD79A/B	B-cell receptor-associated genes	~21–23 of ABC DLBCL	Mostly within ITAM domain	Constitutive activation of the NF-κB and JAK-STAT pathways	—	
	A20 (TNFAIP3)	Ubiquitin-modifying enzyme, terminates NF-κB signaling	23–54 of ABC DLBCL	Frequently within OTU domain	Loss of/reduced termination of NF-κB signaling	—	
	PRDM1/ BLIMP1	Transcriptional repressor, promotes B cell differentiation into plasma cells	~20–30 of ABC DLBCL	None	Loss of/reduced gene function	—	—

Abbreviations: ABC, activated B-cell–like; DLBCL, diffuse large B-cell lymphoma; GCB, germinal center B-cell–like; HAT, histone acetyltransferase; ITAM domain, immunoreceptor tyrosine-based activation motif domain; OTU domain, ovarian tumor domain.

Other recurrent mutations in ABC DLBCL also constitutively activate or synergize with the NF-κB pathway. These include mutations that enhance activity of *CARD11*, a signaling scaffold protein required for NF-κB activation[98,99,103,104]; mutations affecting the immunoreceptor tyrosine-based activation motif (ITAM) of the B-cell receptor-associated protein gene *CD79B* (and less commonly, *CD79A*)[98,107]; activating mutations in *TRAF2* and *TRAF5*, which activate NF-κB through IKK kinase[104]; and inactivating mutations or promoter methylation of *A20 (TNFAIP3)*, a ubiquitin-modifying enzyme involved in terminating NF-κB signaling.[104–106]

Other common genetic alterations in ABC DLBCL included inactivating mutations or deletion of *PRDM1/BLIMP1*, a transcriptional repressor required for terminal B-cell differentiation into and maintenance of plasma cell phenotype, via suppression of B-cell receptor signaling.[108–110] Interestingly, *A20* is located in the 6q23.3 locus and *PRDM1/BLIMP1* is located in the 6q21 to 22.1 locus, both within a region frequently deleted in aggressive B-cell lymphomas.[105,108,109]

Targeted Therapies

The BCL2 inhibitor venetoclax (ABT-199) has shown responses in some patients with GCB DLBCL in a phase I trial.[71] Other potential targeted drug therapies for GCB DLBCLs include inhibitor of EZH2, which is in a phase I clinical trial and is predicted to be more effective against GCB than ABC DLBCL.[73] Mammalian target of rapamycin inhibitors, such as everolimus, which affect the PTEN signaling pathway, have shown response in relapsed/refractory DLBCL cases when used in combination with rituximab.[111,112] Finally, inhibitors of BET, readers of histone acetylation marks, have therapeutic potential via downregulation of MYC.[113,114]

As mentioned previously, constitutive activation of the NF-κB and JAK-STAT3 antiapoptotic pathways are crucial to the tumor cell survival of ABC DLBCL. Interleukin (IL)-6 and IL-10 are key downstream targets of NF-κB signaling; high expression of both can be seen in a subset of ABC DLBCL, leading to JAK kinase activation and STAT3 phosphorylation (termed "STAT3-high").[115,116] In some cell line studies, small molecular inhibitors specific for JAK kinases were toxic to ABC DLBCL, most noticeably in STAT3-high populations.[115,116] Synergistic treatment effect was seen when the JAK inhibitor was used in combination with an IKKβ inhibitor that targeted the NF-κB pathway.[115]

Several drugs targeting the NF-κB pathway have been tested or are currently in clinical trials

for DLBCLs. Bortezomib, a proteasome and NF-κB pathway inhibitor, enhanced treatment responses of ABC DLBCL in conjunction with chemotherapy regimens.[117] Ibrutinib, a BTK inhibitor that blocks B-cell receptor signaling and, eventually, NF-κB activity, also showed efficacy in ABC DLBCL, and has entered phase III clinical trials.[73,118] However, some ABC DLBCLs do not respond to ibrutinib, thought to be due to mutations in *BTK* or in its downstream target genes, such as *CARD11*; it is speculated that these cases might respond better to downstream target inhibitors, such as IKKβ and mucosa-associated lymphoid tissue 1 (MALT1) inhibitors.[107,119,120]

HAIRY CELL LEUKEMIA

Clinical Features

Hairy cell leukemia (HCL) is a rare and indolent small B-cell neoplasm that predominantly involves the bone marrow and spleen, with low-level circulation in the peripheral blood in many cases, but usually spares the lymph node.[1] HCL affects patients with a median age of 50 years, who may present with splenomegaly, pancytopenia, and characteristically, monocytopenia.[1]

Histologic Features and Immunophenotypic Profile

The diagnosis of HCL is made by a combination of morphologic findings of lymphoid cells with pale blue cytoplasm, exhibiting circumferential hairylike villus projections, positivity for tartrate-resistant acid phosphatase, and characteristic immunophenotype (CD20+ CD5− CD10− CD11c+ CD25+ CD103+ CD123+ Annexin A1+ DBA.44+ FMC-7+ cyclin D1 weak+).[1,121,122]

Molecular Diagnostic and Prognostic Markers

Unlike many other mature B-cell neoplasms, recurrent cytogenetic abnormalities have not been reported in HCL.[1,122] However, in a WES study of 48 HCL cases using NGS technology, the recurrent somatic mutation *BRAF* c.1799T>A (p.V600E) was found in all HCL cases[123] (Fig. 7). Furthermore, this mutation has been rarely detected in other mature B-cell neoplasms, including entities that can be difficult to distinguish from HCL, such as splenic marginal-zone lymphomas and HCL-variant (HCL-v). Recurrent *BRAF* mutations in HCL were confirmed in several subsequent studies.[124–126] *BRAF* p.V600E results in its constitutive activation, leading to abnormal activation of the ERK/MAP kinase pathway and promoting tumor growth.[127] Besides molecular techniques, *BRAF* p.V600E can also be detected

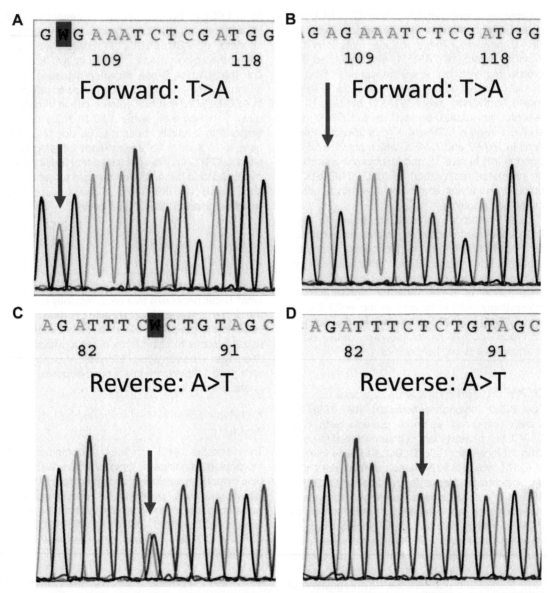

Fig. 7. *BRAF* c.1799T>A (p.V600E) as detected by Sanger sequencing using PCR primers flanking exon 15. (*A*) T>A nucleotide change detected with the forward primer set, with the wild-type allele (*red*) present at a slightly lower frequency than the mutant allele (*green*). (*B*) T>A nucleotide change detected with the forward primer set, in conjunction with a locked nuclei acid (LNA) oligonucleotide designed to suppress amplification of the wild-type allele. (*C*) A>T nucleotide change detected with the reverse primer set, with the wild-type allele (*green*) present at a similar frequency as the mutant allele (*red*). (*D*) A>T nucleotide change detected with the reverse primer set, in conjunction with an LNA oligonucleotide designed to suppress amplification of the wild-type allele. *Red arrows* represent location of nucleotide change detected in each Sanger sequencing reaction.

at the protein level by immunohistochemical staining of tissue sections with a mutant protein-specific antibody (Clone VE1)[128,129] (**Fig. 8**).

More recently, somatic mutations in *CDKN1B*, which codes for p27, have been reported in up to 16% of classic HCL cases with coexisting *BRAF* p.V600E mutations.[130] *CDKN1B* is a known tumor suppressor gene that acts by preventing the

activation of cyclin E-CDK2 and cyclin D-CDK4/6 complexes, controlling G0 to S phase cell-cycle transition.[131] Most *CDKN1B* mutations identified are inactivating splice site and nonsense mutations.[130] *CDKN1B* is now thought to be the second most commonly mutated gene in HCL.

A rare subgroup of HCL lacking *BRAF* mutations has been shown to harbor activating somatic

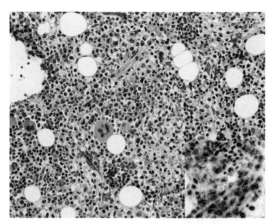

Fig. 8. H&E image of HCL involving the bone marrow at ×40. On marrow core biopsy, the neoplastic cells are small to medium-sized with moderate amount of pale cytoplasm. Due to HCL's tendency for interstitial pattern of marrow involvement, the extent of disease is better assessed with the aid of immunohistochemical stains, using an anti-CD20 antibody or a BRAF V600E-mutant protein-specific antibody, clone VE1 (*inset*).

mutations in *MAP2K1*, which codes for the kinase MEK1, a direct effector of BRAF.[132] This subgroup is more likely to express the IGHV4-34 immunoglobulin heavy chain rearrangement, which can also be found in HCL-v cases.[133] Just like the HCL-v cases, this subgroup in general has higher disease burden, worse response to purine analog therapies, and shorter survival compared with typical HCL.[132,133]

The finding of activating *BRAF* and *MAP2K1* mutations in HCLs suggests that the activation of the ERK/MAP kinase pathway is an important pathogenic mechanism in this disease. **Table 5** shows a summary of recurrently mutated genes in HCL.

Targeted Therapies

Vemurafenib, a BRAF inhibitor, has been shown to result in tumor regression and improved overall and progression-free survival rate in most patients with *BRAF* p.V600E-mutated metastatic melanomas.[134,135] Based on these findings, vemurafenib was used in a patient with *BRAF* p.V600E-mutated refractory HCL, with remarkable improvement in symptoms shortly after treatment initiation, and eventual complete remission.[136,137] Even when vemurafenib was discontinued to reduce the risk of secondary skin cancers that result from prolonged exposure, the patient remained in remission. Recently, a multicenter trial of short oral course of vemurafenib in patients with refractory HCL showed an overall response rate of 96%.[138]

Currently, the mechanism for drug resistance to vemurafenib among *BRAF* p.V600E-mutated HCL is not clear. In a WES study on 3 patients with HCL refractory to purine analog and vemurafenib treatments, *CDKN1B* mutations were identified in 2 of 3 patients.[130] However, they were also found in treatment-naïve patients, and did not influence response to standard treatments. Missense or frameshift mutations in *EZH2*, *ARID1A*, and *KDM6A* were also found in the treatment-refractory patients, but none was thought to affect the RAS/RAF/ERK/MAPK pathway.[130] In another study, *KRAS* p.G12D and *KRAS* p.K117N mutations were detected in a patient with vemurafenib-resistant HCL, suggesting one potential mechanism of drug resistance.[138] However, more studies are needed to clarify the role of any genetic alteration in predicting resistance to therapy.

In *BRAF* p.V600E-mutated metastatic melanomas resistant to vemurafenib, the combination of the BRAF inhibitor dabrafenib and the MEK inhibitor, trametinib, has been used with success to increase progression-free survival.[139] It is currently unclear whether such strategy will be beneficial in HCL cases refractory to vemurafenib.

LYMPHOPLASMACYTIC LYMPHOMA

Clinical, Histologic Features, and Immunophenotypic Profile

Lymphoplasmacytic lymphoma (LPL) is an indolent small B-cell neoplasm composed of a varying proportion of small lymphocytes, plasmacytoid lymphocytes, and plasma cells, that does not fulfill the diagnostic criteria for other recognized B-cell neoplasms[1] (**Fig. 9**). LPL most commonly involves the bone marrow, and less frequently the lymph node and spleen. It has a strong association with serum IgM paraprotein and Waldenström macroglobulinemia, and may be associated with hepatitis C virus infections in some cases.[1] The usual immunophenotypic profile is not specific: CD20+ CD5− CD10− CD23− CD103− cIgM+.

Molecular Diagnostic and Prognostic Markers

Recurrent cytogenetic abnormalities specific to LPL have not been reported, although deletion of 6q21-23 appears to be present in a significant portion of cases.[1,140] LPL has posed some diagnostic challenges due to overlapping clinical and/or pathologic features with plasma cell neoplasms, especially cyclin D1-positive cases that tend to have small plasma cells with relatively scant cytoplasm, as well as with other small B-cell lymphomas with plasmacytic differentiation, such as marginal zone lymphomas. The important

Table 5
Selected recurrently mutated genes in hairy cell leukemia (HCL)

	Normal Function	Mutational Frequency (%)	Mutation Hotspot	Effect of Mutations	Prognostic Marker	Targeted Therapy
BRAF	Regulates cell survival, proliferation, differentiation	80–100	*c.1799T>A (p.V600E)*	Constitutive Activation of the ERK/MAP kinase pathway	—	BRAF Inhibitors (eg, Vemurafenib, Dabrafenib)
CDKN1B	Tumor suppressor, control G0 to S phase cell cycle transition	up to 16	None	Inactivation of gene function	—	—
MAP2K1	Dual-specificity kinase, direct effector of BRAF	—	Mostly within exon 2 and 3 - N-terminal autoregulatory domain	Constitutive Activation of the ERK/MAP kinase pathway	Adverse prognosis compared to BRAF p.V600E-mutated cases	MEK1/2 inhibitors (eg, Trametinib)[a]

[a] Potential therapeutic target.

Fig. 9. H&E image of LPL involving the bone marrow at ×20 and ×60 (*inset*). The top left corner shows uninvolved marrow with normal hematopoietic elements, whereas the rest of the marrow shows involvement by neoplastic cells composed of a mixture of small lymphocytes, plasmacytoid lymphocytes, and plasma cells (*thin arrows*). Scattered mast cells are also seen (*thick arrow*), a common finding in LPL.

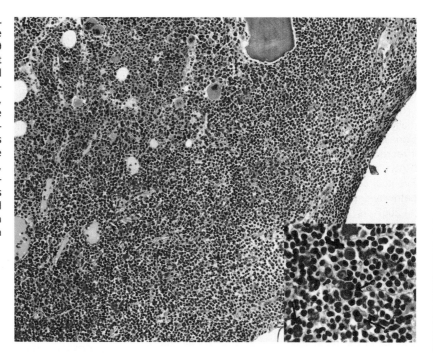

discovery of somatic *MYD88* c.794T>C (p.L265P) mutation in 80% to 90% of cases of LPL has improved the diagnostic accuracy of this entity[140–142] (**Fig. 10**). However, *MYD88* p.L265P is not unique to LPL, and has been reported in IgM monoclonal gammopathy of undetermined significance (MGUS),[35,143] ABC-type diffuse large B-cell lymphomas,[98,102,143] a small number of marginal

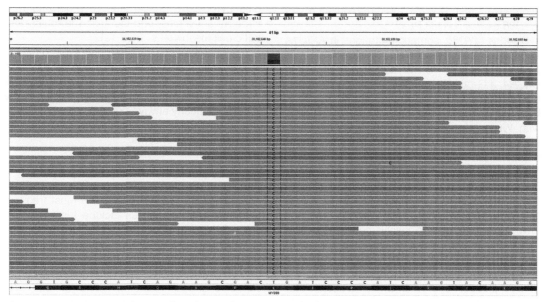

Fig. 10. MYD88 c.794T>C (p.L265P) as detected on an NGS platform and visualized using Integrated Genomics Viewer (Broad Institute). Each horizontal bar represents a single sequencing read, with red and blue bars annotating reads from each of the 2 directions. The reference sequence is shown at the bottom, and letters on the sequencing reads indicate changes in nucleotides. A single T>C nucleotide change results in an amino acid change from leucine (CTG) to proline (CCG).

zone lymphomas (including splenic marginal zone and MALT lymphomas),[35,102,141–144] CLL/SLL,[8,9,35] and FLs.[141] Therefore, the presence of the mutation alone cannot be used to render a diagnosis of LPL. **Table 6** shows a summary of recurrently mutated genes in LPL.

MyD88 is 1 of the 5 adaptor proteins that normally functions in immune responses via TLR signaling. After stimulation by TLR, IL-1, or IL-18, it recruits various downstream kinases, including IRAK1 and IRAK4, to form a signaling complex. Eventual activation of the NF-κB and JAK-STAT pathways promote cell survival.[33,34] *MYD88* p.L265P is a gain-of-function driver mutation affecting the key signaling domain, the Toll/IL-1 receptor domain (TIR) within the protein's hydrophobic core, resulting in the constitutive activation of the NF-κB and JAK-STAT pathways. In some *MYD88* p.L265P–mutated cases, acquired uniparental disomy in 3p (including the *MYD88* locus) resulted in the loss of heterozygosity.[140] *MYD88* p.L265P has been shown to be important for the survival of the neoplastic cells in in vitro cell line studies,[102,141] and has been hypothesized as an early oncogenic event in some patients who evolved from IgM MGUS to LPL.[35,140] More recently, BTK has been established as another downstream target of MYD88 L265P–expressing cells, which leads to B-cell receptor signaling activation.[145] Besides gain-of-function mutation, some LPL cases showed gain in copy number at 3p22 (including *MYD88*), which may suggest an alternative mechanisms in upregulation of *MYD88* expression in LPLs.[141]

Besides *MYD88*, somatic mutations in *ARID1A* and *CXCR4* were also found in a significant proportion of LPLs.[140,146] The mutations identified in *CXCR4*, a G-protein coupled chemokine receptor that promotes cell survival and migration, resembled the germline mutations found in patients with Warts, Hypogammaglobulinemia, Infection, and Myelokathexis-syndrome (WHIMS). These mutations all result in the truncation of the C-terminal cytoplasmic tail containing the regulatory phosphoserine region, leading to prolonged activation of CXCR4.[140,147] Patients with LPL with wild-type *MYD88* were more likely to have the lowest marrow disease involvement, regardless of *CXCR4* mutation status.[146] However, for uncertain reasons, this group also showed the worst survival rate, even adjusted for age. Concurrent *MYD88* p.L265P and WHIMS-like/frameshift mutations or wild-type *CXCR4* predicted intermediate disease burden. Patients with concurrent *MYD88* p.L265P and WHIMS-like/nonsense mutations in *CXCR4* were shown to have the highest marrow disease involvement, clinical symptoms, and disease severity, as well as the highest serum IgM levels, but lower chance of developing lymphadenopathy.

Targeted Therapies

BTK inhibitors, such as ibrutinib, which can inhibit MYD88 binding to BTK,[145] have been used successfully in treating patients with relapsed and refractory LPL, especially in patients with *MYD88* p.L265P and wild-type CXCR4.[148] CXCR4 mutations promote resistance to ibrutinib, but CXCR4 inhibitors, such as plerixafor, may be an option for these cases.[146]

MATURE T-CELL LYMPHOID NEOPLASMS

ANGIOIMMUNOBLASTIC T-CELL LYMPHOMA

Clinical and Histologic Features

Angioimmunoblastic T-cell lymphoma (AITL) is believed to originate from CD4+ follicular helper T cells (T$_{FH}$), and occurs more frequently in middle-aged and elderly patients.[1] It is almost always an EBV-associated systemic disease that involves the lymph nodes, with other frequent sites of involvement that include spleen, liver, skin, and bone marrow.

Histologic Features and Immunophenotypic Profile

AITL is characterized by a polymorphic infiltrate of small to medium-sized T lymphocytes with clear or pale cytoplasm, proliferation of arborizing high endothelial venules, and increased follicular dendritic cell meshworks[1] (**Fig. 11**).

The characteristic immunophenotypic profile of the lesional cells reflects its postulated origin from T$_{FH}$ cells: CD3+ CD4+ CD10+ BCL6+ CXCL13+ PD-1+.[1] Additionally, EBV-positive B cells are nearly always present, whereas CD21, CD23, and CD35 can show expanded follicular dendritic cell meshworks.

Molecular Diagnostic and Prognostic Markers

Recurrent cytogenetic abnormalities specific to AITL have not been reported. The most frequent cytogenetic abnormalities include trisomy 3, trisomy 5, and an additional X chromosome, none of which is specific to AITL.[1] Additionally, array-based comparative genomic hybridization (CGH) on AITL has identified several recurrent chromosomal changes, including gains of 22q, 19, 11p11-q14 (particularly 11q13), 20q13, and loss of 13q (particularly 13q22-q32).[149] Notably, these abnormalities are not specific, and many are also found in cases of peripheral T-cell lymphoma, NOS (PTCL-NOS).

Table 6
Selected recurrently mutated genes in lymphoplasmacytic lymphoma (LPL)

	Normal Function	Mutational Frequency (%)	Mutation Hotspot	Effect of Mutations	Prognostic Marker	Targeted Therapy
MYD88	Immune response via TLR signaling	80–90	c.794T>C (p.L265P)	Constitutive activation of the NF-κB and JAK-STAT pathways	Wild type MYD88 + Any CXCR4 status: Lowest disease burden but worst survival rate.	BTK Inhibitor Ibrutinib
CXCR4	G-protein coupled chemokine receptor, promotes cell survival and migration	27	p.S338 (C-terminal cytoplasmic tail)	Truncation of C-terminal regulatory region → Prolonged activation of receptor	Concurrent MYD88 p.L265P and WHIMS-like/frameshift mutation or WT CXCR4: Intermediate disease burden. Concurrent MYD88 p.L265P and WHIMS-like/nonsense CXCR4 mutation: Highest disease burden.	CXCR4 inhibitor Plerixafor[a]
ARID1A	Member of SWI/SNF, regulates transcriptions by chromatin alteration	17	None	Loss of/Reduced gene function	—	—

Abbreviations: SWI/SNF, switch/sucrose non-fermentable family; WHIMS, Warts, Hypogammaglobulinemia, Infection, and Myelokathexis syndrome.
[a] Potential therapeutic target.

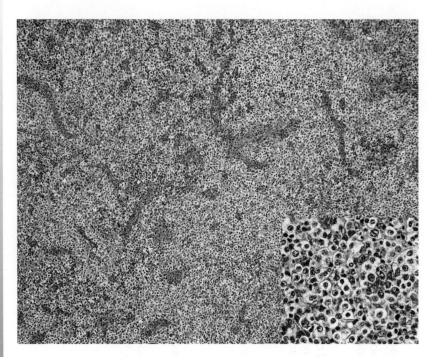

Fig. 11. H&E image of AITL involving a lymph node at ×10 and ×60 (*inset*). Some features characteristic of AITL include neoplastic lymphoid cells with clear or pale cytoplasm and proliferation of arborizing high endothelial venules.

The most common recurrent somatic mutations identified in AITL include *RHOA, IDH2, TET2,* and *DNMT3A*[150–155] (**Table 7**). Of these, *RHOA* and *IDH2* mutations are more specific for AITL among lymphoid malignancies, whereas *IDH2*-mutated tumors represent a subgroup with some promise for effective targeted therapies, and IDH2 inhibitors have already been used in clinical trials for *IDH2*-mutated acute myeloid leukemia (AML) and glioma cases.[156,157]

RHOA encodes for a small GTPase involved in the regulation of a wide variety of biological processes, such as actin stress fiber formation and cell adhesion,[158–161] and the serum response factor (SRF) pathway, a mediator of T-cell development in the thymus.[150,162,163] It is activated by specific guanine-exchange factors that catalyze the dissociation of bound GDP (inactive state) and rebinding of GTP (active state). Hydrolysis of GTP to GDP results in signal termination. The recurrent *RHOA* hotspot mutation c.50G>T (p.G17V), which affects the GTP-binding domain, is found in approximately 67% to 71% of AITL, and 17% to 18% of PTCL-NOS, but not in other T-cell neoplasms or any B-cell or myeloid neoplasms.[150,151] Not only does this mutation severely reduce GTP binding, the mutant GTPase also acts in a dominant-negative, dose-dependent manner to reduce GTP-binding by wild-type RHOA. Interestingly, the PTCL-NOS cases harboring the *RHOA* p.G17V mutation also tend to be cases that show a T_FH-like immunohistochemical

staining profile (eg, positivity for PD-1, BCL6, or CXCL13) or other features with T_FH-like phenotype (eg, presence of CD20+ large B cells, EBER+ cells, expanded CD21/CD23+ follicular dendritic cell network, and CD10 positivity), similar to AITL.[150,153] Furthermore, *RHOA* mutations tend to coexist with *TET2* mutations in AITL; *IDH2* mutations have been identified mostly in cases that also showed mutations in both *RHOA* and *TET2*.[150,151]

IDH2 encodes for the mitochondrial form of isocitrate dehydrogenase, which normally catalyzes the oxidative decarboxylation of isocitrate to α-ketoglutarate in the tricarboxylic acid (TCA) cycle.[164,165] Mutations affecting specific arginine residues in *IDH1* (eg, R132) and *IDH2* (eg, R140 and R172) have been identified in many gliomas,[166–168] and a subset of AML,[169–171] altering the enzymatic function of IDH and leading to the conversion of α-ketoglutarate (2-oxoglutarate) to D-2-hydroxyglutarate (D-2-HG).[164,165,171–173] D-2-HG is thought to be an oncometabolite that affects the hypoxia signaling pathway, histone and DNA methylation, and can inhibit TET protein function. *IDH1/2*-mutant AML shows a global DNA hypermethylation pattern and impaired hematopoietic differentiation.[164] The *IDH2* p.R140 (less frequent) or p.R172 (more frequent) mutations are identified in 13% to 45% of AITL, but not in any other T or B lymphoid neoplasms, including PTCL-NOS[150–152] (**Fig. 12**). On the contrary, *IDH1* mutation has not been found in AITL,

Table 7
Selected recurrently mutated genes in angioimmunoblastic T-cell lymphoma (AITL)

	Normal Function	Mutational Frequency (%)	Mutation Hotspot	Effect of Mutations	Prognostic Marker	Targeted Therapy
RHOA	GTPase, regulates actin stress fiber formation, cell adhesion, SRF pathway	67–71	c.50G>T (p.G17V)	Reduced GTP binding (active state)	—	—
IDH2	Oxidative decarboxylation of isocitrate to α-ketoglutarate in the TCA cycle	13–45	p.R140 and p.R172	Enzymatic alteration → Conversion of α-ketoglutarate to D-2-HG (oncometabolite)	—	IDH2 inhibitors (AG-221 in phase I Clinical Trial)
TET2	Epigenetic control of transcription	33–80	None	Loss of/reduced gene function	Adverse	Demethylating agent 5-azacytidine[a]
DNMT3A	DNA methyltransferase	up to 26	Mostly within methyltransferase domain	Loss of/reduced gene function	—	—

Abbreviations: D-2-HG, D-2-hydroxyglutarate; SRF, serum response factor; TCA cycle, tricarboxylic acid cycle.
[a] Potential therapeutic target.

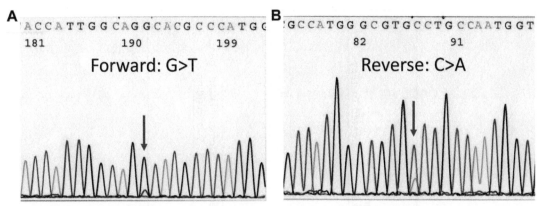

Fig. 12. *IDH2* c.516G>T (p.R172S) as detected by Sanger sequencing using PCR primers flanking exon 4. (*A*) G>T nucleotide change detected with the forward primer set, with the mutant allele (*red*) at a much lower frequency than the wild type allele (*black*). (*B*) C>A nucleotide change detected with the reverse primer set, with the mutant allele (*green*) at a much lower frequency than the wild type allele (*blue*).

although it has been reported in other T-cell neoplasms.[155] *IDH2* mutation status in AITL did not show any significant effect on progression-free or overall survival in one study, but that finding could be confounded by the heterogeneity of treatment regimens administered for AITL.[152]

TET2 encodes for a 2-oxoglutarate/Fe^{2+}-dependent oxygenase that is involved in epigenetic control of transcription, through oxidation of 5-methylcytosine to 5-hydroxymethylcytosine and DNA demethylation.[174–176] Somatic *TET2* mutations have been reported in various types of myeloid neoplasms, including myelodysplastic syndrome, myeloproliferative neoplasms (MPN), AML, and chronic myelomonocytic leukemia (CMML).[177,178] *TET2* is also one of the most frequently mutated genes in AITL, found in 33% to 80% of cases, with most mutations being nonsense mutations, or insertions and deletions leading to frameshift.[150,151,153–155] In comparison, *TET2* mutations are found in approximately 20% to 50% of PTCL-NOS.[150,151,153,154] PTCL-NOS cases harboring *TET2* mutations, just like those harboring *RHOA* mutations, also tend to be cases that exhibit T_{FH}-like immunohistochemical staining profile or T_{FH}-like phenotype, described earlier in this article.[153] *TET2* can coexist with other mutations, such as *RHOA* and *IDH2*,[150,151,153] in contrast to myeloid neoplasms, in which *TET2* and *IDH* mutations appear to be mutually exclusive.[164] Interestingly, *TET2* mutations have been found in both AITL tumor cells as well as non-neoplastic reactive T cells in tumor samples.[150] This finding, together with the significantly higher *TET2* mutant allele frequency compared with other mutations, suggest that *TET2* mutations are acquired earlier than other mutations in the pathogenesis of AITL.[150,154] In AITL and PTCL-NOS with T_{FH}-like features, the presence of *TET2* mutations is

associated with advanced-stage disease, presence of B symptoms, high IPI scores, and shorter progression-free survival.[153]

DNMT3A encodes for a DNA methyltransferase that can methylate specific repetitive regions of the genome, and can form dimers with another DNA methyltransferase encoded by *DNMT3B*.[179] *DNMT3A* can have gene-silencing activities through chromatin modification, such as interaction with histone H3K4. *DNMT3A* mutations are found in up to 26% of AITL, mostly within the methyltransferase domain, and are associated with the presence of *TET2* mutations.[150,151,155] Interestingly, *DNMT3A* mutations were found in both AITL tumor cells as well as non-neoplastic reactive T cells in the samples, which might suggest their early roles in pathogenesis.[150]

Targeted Therapies

Inhibitors targeting the IDH2 R140Q mutants have been developed and have been shown to induce differentiation of AML cells into more mature monocytes and granulocytes in vitro.[180] The reversible mutant IDH2 inhibitor, AG-221, is currently in phase I clinical trial for patients with AML and AITL with *IDH2* mutations.[156] For patients with AML, AG-221 led to a decrease of D-2-HG in up to 98% of patients with *IDH2* p.R140Q, and up to 88% of patients with *IDH2* p.R172K, with an the overall response rate at day 28 of treatment of 40% and complete remission rate of approximately 29% among the patients with relapsed/refractory AML. It remains to be seen if IDH2 inhibitors will have any effect on AITL, a neoplasm that has very different tumor biology from AML.

Brentuximab vedotin is a CD30-directed antibody-drug conjugate, which on binding to CD30 on the cell surface, internalizes and releases the active drug through proteolytic cleavage, eventually leading to tumor cell death.[181] It has shown objective response in up to 54% of patients with relapsed/refractory AITL (38% of patients with complete response) in a phase II multicenter study.[182] Notably, the response rate did not correlate with tumor cell CD30 expression as assessed by IHC, as many patients who responded had little to no CD30 expression in the tumor.

The demethylating agent 5-azacytidine (5-AZA) may hold promise for the treatment of *TET2*-mutated AITL. A patient with AITL associated with CMML, in which the identical frameshift *TET2* mutation was found in the bone marrow and involved lymph node, was treated with 5-AZA in combination with rituximab (due to high level of EBV-positive B cells) and showed complete response and remission of both AITL and CMML at 18-month follow-up.[183]

T-CELL LARGE GRANULAR LYMPHOCYTIC LEUKEMIA

Clinical Features

T-cell large granular lymphocytic leukemia (T-LGL) is a mostly indolent T-cell neoplasm characterized by a persistent (>6 months) increase (often >2 × 10^9/L) in large granular lymphocytes (LGLs) in the peripheral blood, usually clonal in nature, that cannot be attributed to a reaction to an identifiable etiology.[1] The underlying pathogenic mechanism of T-LGL may be related to sustained immune stimulation but is not well understood. T-LGL comprises 2% to 6% of chronic lymphoproliferative disorders, and mostly affects middle-aged to elderly patients, who can exhibit neutropenia, anemia, lymphocytosis, splenomegaly, and hypergammaglobulinemia.[1,184] The peripheral blood, bone marrow, liver, and spleen can be involved; however, lymph node involvement is uncommon.

Histologic Features and Immunophenotypic Profile

T-LGL cells often have moderate to abundant cytoplasm with azurophilic granules, and may not be easily distinguishable from non-neoplastic LGLs.[1]

The immunophenotypic profile can be very similar to normal cytotoxic T-cells: CD3+ CD8+ CD16+ CD57+ TCR α/β+; CD5 and CD7 may be intact or abnormally diminished or lost.[1]

Molecular Diagnostic and Prognostic Markers

No specific recurrent cytogenetic abnormalities for T-LGL have been reported.[1,185] The diagnosis of T-LGL can be challenging, as a variety of processes can give rise to an increase in LGLs that do not necessarily meet the criteria for T-LGL, including following allogeneic bone marrow transplantation, reactive processes secondary to viral infections, low-grade B-cell malignancies, and autoimmune disorders.[1,184,186,187] The detection of clonal T-cell receptor gene rearrangements (eg, TCR gamma, TCR beta) may be helpful in the diagnosis of T-LGL (**Fig. 13**). However, the demonstration of a clonal T-cell population(s) does not alone lead to an unequivocal diagnosis of T-LGL; correlation with clinical findings and chronicity of the LGL elevation is required because oligoclonal and even clonal T-cell populations may be seen in response to infection and in other clinical settings as mentioned previously. In T-LGL, the pattern seen on clonality studies may vary; for example, an oligoclonal pattern with multiple dominant peaks corresponding to 3 or more clones may be seen initially, and over time may evolve into a more typical clonal pattern seen in monoclonal populations as 1 dominant clone emerges. Documentation of the TCR family and base-pair size of the clonal polymerase chain reaction (PCR) product(s) facilitates tracking of the various clones over time with this testing modality.

Recently, mutations in the *STAT3*[185,188,189] and *STAT5b*[190] genes, as well as genes leading to activation of STAT pathway and T-cell activation,[191] have been associated with T-LGL. These molecular alterations provide promise as diagnostic markers and therapeutic targets for this disease (**Table 8**).

The *STAT3* gene plays a role in immune system regulation and determination of T-cell differentiation in response to certain cytokines.[192] Specifically, IL-6 and STAT3 are involved in the differentiation of Th17 cells, and in the phenotypic expression of the CD4+ follicular helper T cells. Up to 40% of patients with T-LGL with 1 major Vβ clone, and up to 73% of patients with T-LGL with any clonal TCR gene rearrangement, harbor somatic mutations in *STAT3*, detected only in the neoplastic CD8+ T cells, in mutational hotspots, including c.1919A>T (p.Y640F), c.1981G>T (p.D661Y), c.1982A>T (p.D661V), and c.1940A>T (p.N647I).[185,188,189] These mutations have not been found in other mature T-cell neoplasms, T-lymphoblastic leukemia, and cases with reactive, non-neoplastic increase in LGLs,[185,189] but are found in 30% to 40% of chronic lymphoproliferative disorders of NK cells.[188,189] Compared with *STAT3*–wild-type T-LGL, *STAT3* mutants are more commonly associated with neutropenia, symptomatic disease, requirement for therapies, rheumatoid arthritis, and autoimmune hemolytic

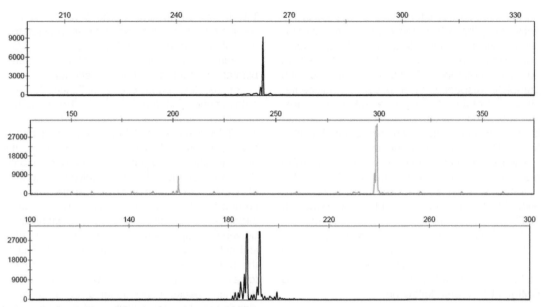

Fig. 13. *TCRB* and *TCRG* clonal gene rearrangements in a case of T-cell lymphoma using PCR primers validated by a European collaborative study (BIOMED-2 Concerted Action), with analysis of amplified products by capillary electrophoresis. (*Top*) Clonal peak detected by the Vβ and Jβ2 primers. (*Middle*) Clonal peak detected by the Dβ and Jβ1 primers. (*Bottom*) Clonal peaks detected by a mixture of primers that target various Vγ and Jγ regions.

anemia.[185,188] However, *STAT3* mutation status does not predict overall survival.[188] All the hotspot mutations are located in exon 21, within the SH2 domain, which mediates dimerization and activation of STAT protein. The mutations result in more hydrophobic protein structure, with increased nuclear localization and phosphorylation (activation) of STAT3, and upregulation of downstream target genes, including those involved in the anti-apoptotic pathways. In vitro experiments also have shown that compared with wild type, the *STAT3* p.Y640F and p.D661V mutants have enhanced basal and IL-6–stimulated transcriptional activity of STAT3-responsive cis-inducible element.[185] Besides somatic mutations, heterozygous germline *STAT3* p.K392R mutation, located within the DNA-binding domain, has been associated with the development of T-LGL in a 14-year-old patient with multiorgan autoimmunity, hypogammaglobulinemia, and lymphoproliferative disease, in an age group that is otherwise rarely affected by this disease.[193]

The *STAT5b* gene also plays a role in immune system regulation and determination of T-cell differentiation.[192] Specifically, IL-2 and STAT5B are critical for the differentiation of Treg cells, and can also negatively regulate Th17 cell differentiation. A small fraction of large granular lymphocytic leukemia (2%) harbors mutations in the SH2 domain of *STAT5b*, analogous to the recurrent mutations in *STAT3*.[190] The mutations identified

include c.1994A>T (p.Y665F) and c.1924A>C (p.N642H) in exon 16, in cases with T-cell phenotype or overlapping NK and T-cell phenotype (CD3+ CD56+). Of note, no mutations have been found in *STAT5a*, a closely related gene. The *STAT5b* mutants have increased STAT5b nuclear localization, phosphorylation, and transcriptional activity. *STAT5b* p.Y665F cases also have similar gene expression profile as *STAT3*-mutant T-LGL cases, with relatively indolent clinical courses and lower chance for requiring treatment. In contrast, the *STAT5b* p.N642H mutants are associated with aggressive, chemotherapy-refractory disease.[190]

Up to 50% of *STAT3*–wild-type and mutated T-LGLs share deregulation of STAT3 pathway-related genes.[188] T-LGL cases with wild-type *STAT3* and *STAT5b* can show mutations resulting in activation of the STAT3 pathway directly (inactivating mutation in *PTPRT*, a gene involved in STAT3 deactivation), or in alteration of T-cell activation and proliferation (*BCL11B*, *SLIT2*, and *NRP1*).[191] All these mutations are rare, but provide examples of alternative mechanisms of STAT3 pathway deregulation, which is found in most wild-type *STAT3* and *STAT5b* T-LGLs.[185,188,190,191]

Targeted Therapies

Interestingly, in a phase II clinical trial using methotrexate as a treatment for large granular

Table 8
Selected recurrently mutated genes in T-large granular lymphocytic leukemia (T-LGL)

	Normal Function	Mutational Frequency (%)	Mutation Hotspot	Effect of Mutations	Prognostic Marker	Targeted Therapy
STAT3	Immune system regulation and determination of T-cell differentiation	Up to 73 of all T-LGL	*Within SH2 domain (eg, p.Y640F, p.D661Y, p.D661V, and p.N647I)*	Increased nuclear localization and activation of STAT3 pathway	More severe symptoms and therapy requirement compared to WT	STAT3 p.Y640F shows better treatment response to methotrexate
STAT5b	Immune system regulation and determination of T-cell differentiation	2 of T-LGL	Within SH2 domain (eg, p.Y665F and p.N642H)	Increased nuclear localization and transcriptional activity	p.N642H associated with adverse prognosis compared to p.Y665F or WT cases	Inhibitors of STAT pathway (eg, STA-21, OPB-31121)[a]

Abbreviations: STAT, signal transducers and activators of transcription; WT, wild type.
[a] Potential therapeutic target.

lymphocytic leukemia, cases with the *STAT3* p.Y640F mutation showed improved treatment response.[194] In in vitro studies, STA-21, a synthetic inhibitor of STAT3 dimerization, DNA binding, and transcriptional activity, induced apoptosis of T-LGL tumor cells in a dose-dependent manner.[188] Apoptosis was observed in both *STAT3*–wild-type and mutant samples, but no significant effect was observed in normal lymphocytes. The STAT3 and STAT5-specific inhibitor, OPB-31121, has been evaluated in clinical trials for different hematologic malignancies, such as multiple myeloma, BL, and acute leukemia with *BCR-ABL* translocation, *FLT3* internal tandem duplication (*FLT3*-ITD) or *JAK2* p.V617F mutation.[195] It is possible that drugs, such as OPB-31121, also may show therapeutic benefit in T-LGL. Inhibition of the STAT pathway can be attained through other mechanisms as well. For example, the protein kinase CK2 is needed for activation of STAT1 and STAT3. Its inhibition can decrease phosphorylation of a constitutively active *STAT3*-mutant found in T-LGL.[196] Also, STAT3 SH2 domain inhibitors (eg, S3I-M2001, ISS 610) are under preclinical evaluation, whereas an antisense oligonucleotide that binds and degrades STAT3 mRNA is undergoing phase I clinical trials for certain refractory cancer types.[197]

SUMMARY

Lymphoid neoplasms have historically been classified based primarily on their morphologic features, immunophenotypic profiles, and in some instances, cytogenetic findings and clinical presentations. However, even within each type of lymphoid neoplasms, there is heterogeneity in clinical course and response to therapies. Recent discoveries in recurrent genetic mutations and epigenetic modifications among lymphoid neoplasms have not only provided valuable insight into pathogenesis and disease progression, but have also facilitated more precise tumor classification and stratification of patients into prognostic subgroups. Additionally, some of these genetic alterations may serve as biomarkers of targeted therapeutics. Molecular profiling of lymphoid neoplasms will likely gain more importance in routine clinical care in the future, as it provides clinicians with information important for personalized health care and targeted therapy selection. It is therefore crucial for pathologists, oncologists, and hematologists to have a good understanding of the molecular pathology of lymphoid neoplasms as it relates to their clinical practice.

REFERENCES

1. Swerdlow S, Campo E, Harris N, et al. WHO classification of tumours of haematopoietic and lymphoid tissues. 4th edition. Lyon (France): International Agency for Research on Cancer (IARC); 2008.
2. Damle RN, Wasil T, Fais F, et al. Ig V gene mutation status and CD38 expression as novel prognostic indicators in chronic lymphocytic leukemia. Blood 1999;94(6):1840–7.
3. Hamblin BTJ, Davis Z, Gardiner A, et al. Genes are associated with a more aggressive form of chronic lymphocytic leukemia. Blood 1999;94(6):1848–54.
4. Wiestner A, Rosenwald A, Barry TS, et al. ZAP-70 expression identifies a chronic lymphocytic leukemia subtype with unmutated immunoglobulin genes, inferior clinical outcome, and distinct gene expression profile. Blood 2003;101(12):4944–51.
5. Ghia P, Guida G, Stella S, et al. The pattern of CD38 expression defines a distinct subset of chronic lymphocytic leukemia (CLL) patients at risk of disease progression. Blood 2003;101(4):1262–9.
6. Del Poeta G, Del Principe MI, Consalvo MAI, et al. The addition of rituximab to fludarabine improves clinical outcome in untreated patients with ZAP-70-negative chronic lymphocytic leukemia. Cancer 2005;104(12):2743–52.
7. Claus R, Lucas DM, Stilgenbauer S, et al. Quantitative DNA methylation analysis identifies a single CpG dinucleotide important for ZAP-70 expression and predictive of prognosis in chronic lymphocytic leukemia. J Clin Oncol 2012;30(20):2483–91.
8. Wang L, Lawrence MS, Wan Y, et al. SF3B1 and other novel cancer genes in chronic lymphocytic leukemia. N Engl J Med 2011;365(26):2497–506.
9. Puente XS, Pinyol M, Quesada V, et al. Whole-genome sequencing identifies recurrent mutations in chronic lymphocytic leukaemia. Nature 2011; 475(7354):101–5.
10. Fabbri G, Rasi S, Rossi D, et al. Analysis of the chronic lymphocytic leukemia coding genome: role of NOTCH1 mutational activation. J Exp Med 2011;208(7):1389–401.
11. Quesada V, Conde L, Villamor N, et al. Exome sequencing identifies recurrent mutations of the splicing factor SF3B1 gene in chronic lymphocytic leukemia. Nat Genet 2012;44(1):47–52.
12. Landau DA, Carter SL, Stojanov P, et al. Evolution and impact of subclonal mutations in chronic lymphocytic leukemia. Cell 2013;152(4):714–26.
13. Baliakas P, Hadzidimitriou A, Sutton L, et al. Recurrent mutations refine prognosis in chronic lymphocytic leukemia. Leukemia 2015;29(2):329–36.
14. Jeromin S, Weissmann S, Haferlach C, et al. SF3B1 mutations correlated to cytogenetics and mutations in NOTCH1, FBXW7, MYD88, XPO1 and TP53 in

1160 untreated CLL patients. Leukemia 2014; 28(1):108–17.

15. Rossi D, Rasi S, Spina V, et al. Integrated mutational and cytogenetic analysis identifies new prognostic subgroups in chronic lymphocytic leukemia. Blood 2013;121(8):1403–12.

16. Guièze R, Robbe P, Clifford R, et al. Presence of multiple recurrent mutations confers poor trial outcome of relapsed/refractory CLL. Blood 2015; 126(18):2110–8.

17. Pear WS, Aster JC. T cell acute lymphoblastic leukemia/lymphoma: a human cancer commonly associated with aberrant NOTCH1 signaling. Curr Opin Hematol 2004;11(6):426–33.

18. Radtke F, Wilson A, Mancini SJC, et al. Notch regulation of lymphocyte development and function. Nat Immunol 2004;5(3):247–53.

19. Aster JC, Blacklow SC, Pear WS. Notch signalling in T-cell lymphoblastic leukaemia/lymphoma and other haematological malignancies. J Pathol 2011;223(2):262–73.

20. O'Neil J, Grim J, Strack P, et al. FBW7 mutations in leukemic cells mediate NOTCH pathway activation and resistance to -secretase inhibitors. J Exp Med 2007;204(8):1813–24.

21. Weng AP, Millholland JM, Yashiro-ohtani Y, et al. c-Myc is an important direct target of Notch1 in T-cell acute lymphoblastic leukemia/lymphoma. Genes Dev 2006;20(15):2096–109.

22. Weng AP, Ferrando AA, Lee W, et al. Activating mutations of NOTCH1 in human T cell acute lymphoblastic leukemia. Science 2004;306(5694):269–71.

23. Breit S, Stanulla M, Flohr T, et al. Activating NOTCH1 mutations predict favorable early treatment response and long-term outcome in childhood precursor T-cell lymphoblastic leukemia. Blood 2006;108(4):1151–7.

24. Asnafi V, Buzyn A, Le Noir S, et al. NOTCH1/FBXW7 mutation identifies a large subgroup with favorable outcome in adult T-cell acute lymphoblastic leukemia (T-ALL): a Group for Research on Adult Acute Lymphoblastic Leukemia (GRAALL) study. Blood 2009;113(17):3918–24.

25. Sportoletti P, Baldoni S, Cavalli L, et al. NOTCH1 PEST domain mutation is an adverse prognostic factor in B-CLL. Br J Haematol 2010;151(4):404–6.

26. Corrionero A, Miñana B, Valcárcel J. Reduced fidelity of branch point recognition and alternative splicing induced by the anti-tumor drug spliceostatin A. Genes Dev 2011;25(5):445–59.

27. Fan L, Lagisetti C, Edwards CC, et al. Sudemycins, novel small molecule analogues of FR901464, induce alternative gene splicing. ACS Chem Biol 2011;6(6):582–9.

28. Zenz T, Eichhorst B, Busch R, et al. TP53 mutation and survival in chronic lymphocytic leukemia. J Clin Oncol 2010;28(29):4473–9.

29. Trbusek M, Smardova J, Malcikova J, et al. Missense mutations located in structural p53 DNA-binding motifs are associated with extremely poor survival in chronic lymphocytic leukemia. J Clin Oncol 2011;29(19):2703–8.

30. Shiloh Y. ATM and related protein kinases: safeguarding genome integrity. Nat Rev Cancer 2003; 3(3):155–68.

31. Stankovic T, Hubank M, Cronin D, et al. Microarray analysis reveals that TP53- and ATM-mutant B-CLLs share a defect in activating proapoptotic responses after DNA damage but are distinguished by major differences in activating prosurvival responses. Blood 2004;103(1):291–300.

32. Austen B, Powell JE, Alvi A, et al. Mutations in the ATM gene lead to impaired overall and treatment-free survival that is independent of IGVH mutation status in patients with B-CLL. Blood 2005;106(9): 3175–82.

33. Loiarro M, Capolunghi F, Fanto N, et al. Pivotal advance: inhibition of MyD88 dimerization and recruitment of IRAK1 and IRAK4 by a novel poptidomimetic compound. J Leukoc Biol 2007;82(4):801–10.

34. O'Neill LAJ, Bowie AG. The family of five: TIR-domain-containing adaptors in Toll-like receptor signalling. Nat Rev Immunol 2007;7(5):353–64.

35. Xu L, Hunter ZR, Yang G, et al. MYD88 L265P in Waldenström macroglobulinemia, immunoglobulin M monoclonal gammopathy, and other B-cell lymphoproliferative disorders using conventional and quantitative allele-specific polymerase chain reaction. Blood 2013;121(11):2051–8.

36. Woyach JA, Johnson AJ. Targeted therapies in CLL: mechanisms of resistance and strategies for management. Blood 2015;126(4):471–8.

37. Tam CS, Stilgenbauer S. How best to manage patients with chronic lymphocytic leukemia with 17p deletion and/or TP53 mutation? Leuk Lymphoma 2015;56(3):587–93.

38. Brown JR, Byrd JC, Coutre SE, et al. Idelalisib, an inhibitor of phosphatidylinositol 3-kinase p110d, for relapsed/refractory chronic lymphocytic leukemia. Blood 2014;123(22):3390–7.

39. Furman RR, Sharman JP, Coutre SE, et al. Idelalisib and rituximab in relapsed chronic lymphocytic leukemia. N Engl J Med 2014;370(11):997–1007.

40. Byrd JC, Furman RR, Coutre SE, et al. Targeting BTK with ibrutinib in relapsed chronic lymphocytic leukemia. N Engl J Med 2013;369(1):32–42.

41. Byrd JC, Brown JR, O'Brien S, et al. Ibrutinib versus ofatumumab in previously treated chronic lymphoid leukemia. N Engl J Med 2014;371(3):213–23.

42. Farooqui MZH, Valdez J, Martyr S, et al. Ibrutinib for previously untreated and relapsed or refractory chronic lymphocytic leukaemia with TP53 aberrations: a phase 2, single-arm trial. Lancet Oncol 2015;16(12):169–76.

43. Stephens DM, Ruppert AS, Jones JA, et al. Impact of targeted therapy on outcome of chronic lymphocytic leukemia patients with relapsed del(17p13.1) karyotype at a single center. Leukemia 2014;28(6): 1365–8.

44. Souers AJ, Leverson JD, Boghaert ER, et al. ABT-199, a potent and selective BCL-2 inhibitor, achieves antitumor activity while sparing platelets. Nat Med 2013;19(2):202–8.

45. Hecht J, Aster J. Molecular biology of Burkitt's lymphoma. J Clin Oncol 2000;18(21):3707–21.

46. Bertrand P, Bastard C, Maingonnat C, et al. Mapping of MYC breakpoints in 8q24 rearrangements involving non-immunoglobulin partners in B-cell lymphomas. Leukemia 2007;21(3):515–23.

47. Bertrand P, Maingonnat C, Picquenot JM, et al. Characterization of three t(3;8)(q27;q24) translocations from diffuse large B-cell lymphomas. Leukemia 2008;22(5):1064–7.

48. Sonoki T, Tatetsu H, Nagasaki A, et al. Molecular cloning of translocation breakpoint from der(8) t(3;8)(q27;q24) defines juxtaposition of downstream of C-MYC and upstream of BCL6. Int J Hematol 2007;86(2):196–8.

49. Kretzmer H, Bernhart SH, Wang W, et al. DNA methylome analysis in Burkitt and follicular lymphomas identifies differentially methylated regions linked to somatic mutation and transcriptional control. Nat Genet 2015;47(11):1316–25.

50. Victora GD, Dominguez-sola D, Holmes AB, et al. Identification of human germinal center light and dark zone cells and their relationship to human B-cell lymphomas. Blood 2012;120(11):2240–8.

51. Dave S, Fu K, Wright G, et al. Molecular diagnosis of Burkitt's lymphoma. N Engl J Med 2006;354(23): 2431–42.

52. Richter J, Schlesner M, Hoffmann S, et al. Recurrent mutation of the ID3 gene in Burkitt lymphoma identified by integrated genome, exome and transcriptome sequencing. Nat Genet 2012;44(12): 1316–20.

53. Schmitz R, Young RM, Ceribelli M, et al. Burkitt lymphoma pathogenesis and therapeutic targets from structural and functional genomics. Nature 2012; 490(7418):116–20.

54. Love C, Sun Z, Jima D, et al. The genetic landscape of mutations in Burkitt lymphoma. Nat Genet 2012;44(12):1321–5.

55. Muppidi JR, Schmitz R, Green JA, et al. Loss of signalling via Gα13 in germinal centre B-cell-derived lymphoma. Nature 2014;516(7530):254–8.

56. Wagener R, Aukema S, Schlesner M, et al. The PCBP1 gene encoding poly(rc) binding protein I is recurrently mutated in Burkitt lymphoma. Genes Chromosomes Cancer 2015;54(9):555–64.

57. Schraders M, de Jong D, Kluin P, et al. Lack of Bcl-2 expression in follicular lymphoma may be caused by mutations in the BCL2 gene or by absence of the t(14;18) translocation. J Pathol 2005;205(3): 329–35.

58. Masir N, Campbell LJ, Goff LK, et al. BCL2 protein expression in follicular lymphomas with t(14;18) chromosomal translocations. Br J Haematol 2009; 144(5):716–25.

59. Adam P, Baumann R, Schmidt J, et al. The BCL2 E17 and SP66 antibodies discriminate 2 immunophenotypically and genetically distinct subgroups of conventionally BCL2-"negative" grade 1/2 follicular lymphomas. Hum Pathol 2013;44(9):1817–26.

60. Bosga-Bouwer AG, Van Imhoff GW, Boonstra R, et al. Follicular lymphoma grade 3B includes 3 cytogenetically defined subgroups with primary t(14; 18), 3q27, or other translocations: t(14;18) and 3q27 are mutually exclusive. Blood 2003; 101(3):1149–54.

61. Loeffler M, Kreuz M, Haake A, et al. Genomic and epigenomic co-evolution in follicular lymphomas. Leukemia 2015;29:456–63.

62. Pastore A, Jurinovic V, Kridel R, et al. Integration of gene mutations in risk prognostication for patients receiving first-line immunochemotherapy for follicular lymphoma: a retrospective analysis of a prospective clinical trial and validation in a population-based registry. Lancet Oncol 2015; 16(9):1111–22.

63. Pasqualucci L, Dominguez-Sola D, Chiarenza A, et al. Inactivating mutations of acetyltransferase genes in B-cell lymphoma. Nature 2011; 471(7337):189–95.

64. Okosun J, Bödör C, Wang J, et al. Integrated genomic analysis identifies recurrent mutations and evolution patterns driving the initiation and progression of follicular lymphoma. Nat Genet 2014; 46(2):176–81.

65. Green MR, Gentles AJ, Nair RV, et al. Hierarchy in somatic mutations arising during genomic evolution and progression of follicular lymphoma. Blood 2013;121(9):1604–11.

66. Morin RD, Mendez-Lago M, Mungall AJ, et al. Frequent mutation of histone-modifying genes in non-Hodgkin lymphoma. Nature 2011;476(7360): 298–303.

67. Morin RD, Johnson NA, Severson TM, et al. Somatic mutations altering EZH2 (Tyr641) in follicular and diffuse large B-cell lymphomas of germinal-center origin. Nat Genet 2010;42(2):181–5.

68. Bödör C, O'Riain C, Wrench D, et al. EZH2 Y641 mutations in follicular lymphoma. Leukemia 2011; 25(4):726–9.

69. Yap DB, Chu J, Berg T, et al. Somatic mutations at EZH2 Y641 act dominantly through a mechanism of selectively altered PRC2 catalytic activity, to increase H3K27 trimethylation. Blood 2011;117(8): 2451–9.

70. Sneeringer CJ, Scott MP, Kuntz KW, et al. Coordinated activities of wild-type plus mutant EZH2 drive tumor-associated hypertrimethylation of lysine 27 on histone H3 (H3K27) in human B-cell lymphomas. Proc Natl Acad Sci U S A 2010;107(49): 20980–5.

71. Phase I study of ABT-199 (GDC-0199) in patients with relapse/refractory non-Hodgkin lymphoma: responses observed in diffuse large B-cell (DLBCL) and follicular lymphoma (FL) at higher cohort doses. Clin Adv Hematol Oncol 2014;12(8 Suppl 16):18–9.

72. Roberts AW, Advani RH, Kahl BS, et al. Phase 1 study of the safety, pharmacokinetics, and antitumour activity of the BCL2 inhibitor navitoclax in combination with rituximab in patients with relapsed or refractory CD20 + lymphoid malignancies. Br J Haematol 2015;170(5):669–78.

73. Pon JR, Marra MA. Clinical correlates of molecular features in diffuse large B-cell lymphoma and follicular lymphoma. Blood 2016;127(2):181–6.

74. Maddocks K, Christian B, Jaglowski S, et al. A phase 1/1b study of rituximab, bendamustine, and ibrutinib in patients with untreated and relapsed/refractory non-Hodgkin lymphoma. Blood 2015;125(2):242–9.

75. Craig M, Hanna WT, Cabanillas F, et al. Phase II study of bortezomib in combination with rituximab, cyclophosphamide and prednisone with or without doxorubicin followed by rituximab maintenance in patients with relapsed or refractory follicular lymphoma. Br J Haematol 2014;16(6):920–8.

76. Evens AM, Smith MR, Lossos IS, et al. Frontline bortezomib and rituximab for the treatment of newly diagnosed high tumour burden indolent non-Hodgkin lymphoma: a multicentre phase II study. Br J Haematol 2014;166(4):514–20.

77. Alizadeh AA, Eisen MB, Davis RE, et al. Distinct types of diffuse large B-cell lymphoma identified by gene expression profiling. Nature 2000; 403(6769):503–11.

78. Rosenwald A, Wright G, Chan WC, et al. The use of molecular profiling to predict survival after chemotherapy for diffuse large-B-cell lymphoma. N Engl J Med 2002;346(25):1937–47.

79. Bea S, Zettl A, Wright G, et al. Diffuse large B-cell lymphoma subgroups have distinct genetic profiles that influence tumor biology and improve gene-expression-based survival prediction. Blood 2005; 106(9):3183–90.

80. Huang JZ, Sanger WG, Greiner TC, et al. Plenary paper. The t (14;18) defines a unique subset of diffuse large B-cell lymphoma with a germinal center B-cell gene expression profile. Blood 2002; 99(7):2285–90.

81. Iqbal J, Sanger WG, Horsman DE, et al. BCL2 translocation defines a unique tumor subset within the germinal center B-cell-like diffuse large B-cell lymphoma. Am J Pathol 2004;165(1):159–66.

82. Lossos IS, Alizadeh AA, Eisen MB, et al. Ongoing immunoglobulin somatic mutation in germinal center B cell-like but not in activated B cell-like diffuse large cell lymphomas. Proc Natl Acad Sci U S A 2000;97(18):10209–13.

83. Iqbal J, Greiner T, Patel K, et al. Distinctive patterns of BCL6 molecular alterations and their functional consequences in different subgroups of diffuse large B-cell lymphoma. Leukemia 2007;21(11): 2332–43.

84. Johnson NA, Savage KJ, Ludkovski O, et al. Lymphomas with concurrent BCL2 and MYC translocations: the critical factors associated with survival. Blood 2009;114(11):2273–9.

85. Snuderl M, Kolman OK, Chen Y-B, et al. B-cell lymphomas with concurrent IGH-BCL2 and MYC rearrangements are aggressive neoplasms with clinical and pathologic features distinct from Burkitt lymphoma and diffuse large B-cell lymphoma. Am J Surg Pathol 2010;34(3):327–40.

86. Johnson NA, Slack GW, Savage KJ, et al. Concurrent expression of MYC and BCL2 in diffuse large B-Cell lymphoma treated with rituximab plus cyclophosphamide, doxorubicin, vincristine, and prednisone. J Clin Oncol 2012;30(28):3452–9.

87. Green TM, Young KH, Visco C, et al. Immunohistochemical double-hit score is a strong predictor of outcome in patients with diffuse large B-Cell lymphoma treated with rituximab plus cyclophosphamide, doxorubicin, vincristine, and prednisone. J Clin Oncol 2012;30(28):3460–7.

88. Hu S, Xu-Monette ZY, Tzankov A, et al. MYC/BCL2 protein coexpression contributes to the inferior survival of activated B-cell subtype of diffuse large B-cell lymphoma and demonstrates high-risk gene expression signatures: a report from The International DLBCL Rituximab-CHOP Consortium Program. Blood 2013;121(20):4021–32.

89. Horn H, Ziepert M, Becher C, et al. MYC status in concert with BCL2 and BCL6 expression predicts outcome in diffuse large B-cell lymphoma. Blood 2013;121(12):2253–63.

90. Perry AM, Alvarado-Bernal Y, Laurini JA, et al. MYC and BCL2 protein expression predicts survival in patients with diffuse large B-cell lymphoma treated with rituximab. Br J Haematol 2014;165(3):382–91.

91. Scott DW, Mottok A, Ennishi D, et al. Prognostic significance of diffuse large B-Cell lymphoma cell of origin determined by digital gene expression in formalin-fixed paraffin-embedded tissue biopsies. J Clin Oncol 2015;33(26):2848–56.

92. Ruzinova MB, Caron T, Rodig SJ. Altered subcellular localization of c-Myc protein identifies aggressive B-cell lymphomas harboring a c-MYC translocation. Am J Surg Pathol 2010;34(6):882–91.

93. Kluk MJ, Chapuy B, Sinha P, et al. Immunohistochemical detection of MYC-driven diffuse large B-cell lymphomas. PLoS One 2012;7(4):e33813.

94. Schneider K, Banks P, Collie A, et al. Dual expression of MYC and BCL2 proteins predicts worse outcomes in diffuse large B-cell lymphoma. Leuk Lymphoma 2015;30:1–23.

95. Kluk MJ, Ho C, Yu H, et al. MYC Immunohistochemistry to identify MYC-driven B-Cell lymphomas in clinical practice. Am J Clin Pathol 2016;145(2): 166–79.

96. Carey CD, Gusenleitner D, Chapuy B, et al. Molecular classification of MYC-driven B-cell lymphomas by targeted gene expression profiling of fixed biopsy specimens. J Mol Diagn 2015;17(1):19–30.

97. Su I-H, Basavaraj A, Krutchinsky AN, et al. Ezh2 controls B cell development through histone H3 methylation and IgH rearrangement. Nat Immunol 2003;4(2):124–31.

98. Pasqualucci L, Trifonov V, Fabbri G, et al. Analysis of the coding genome of diffuse large B-cell lymphoma. Nat Genet 2011;43(9):830–7.

99. Lohr JG, Stojanov P, Lawrence MS, et al. Discovery and prioritization of somatic mutations in diffuse large B-cell lymphoma (DLBCL) by whole-exome sequencing. Proc Natl Acad Sci U S A 2012; 109(10):3879–84.

100. Pfeifer M, Grau M, Lenze D, et al. PTEN loss defines a PI3K/AKT pathway-dependent germinal center subtype of diffuse large B-cell lymphoma. Proc Natl Acad Sci U S A 2013;110(30):12420–5.

101. Davis RE, Brown KD, Siebenlist U, et al. Constitutive nuclear factor kappaB activity is required for survival of activated B cell-like diffuse large B cell lymphoma cells. J Exp Med 2001;194(12):1861–74.

102. Ngo VN, Young RM, Schmitz R, et al. Oncogenically active MYD88 mutations in human lymphoma. Nature 2011;470(7332):115–9.

103. Lenz G, Davis RE, Ngo VN, et al. Oncogenic CARD11 mutations in human diffuse large B cell lymphoma. Science 2008;319(5870):1676–9.

104. Compagno M, Lim WK, Grunn A, et al. Mutations of multiple genes cause deregulation of NF-kappaB in diffuse large B-cell lymphoma. Nature 2009; 459(7247):717–21.

105. Honma K, Tsuzuki S, Nakagawa M, et al. TNFAIP3/A20 functions as a novel tumor suppressor gene in several subtypes of non-Hodgkin lymphomas. Blood 2009;114(12):2467–75.

106. Kato M, Sanada M, Kato I, et al. Frequent inactivation of A20 in B-cell lymphomas. Nature 2009; 459(7247):712–6.

107. Davis RE, Ngo VN, Lenz G, et al. Chronic active B-cell-receptor signalling in diffuse large B-cell lymphoma. Nature 2010;463(7277):88–92.

108. Pasqualucci L, Compagno M, Houldsworth J, et al. Inactivation of the PRDM1/BLIMP1 gene in diffuse large B cell lymphoma. J Exp Med 2006;203(2): 311–7.

109. Tam W, Gomez M, Chadburn A, et al. Mutational analysis of PRDM1 indicates a tumor-suppressor role in diffuse large B-cell lymphomas. Blood 2006;107(10):4090–100.

110. Mandelbaum J, Bhagat G, Tang H, et al. BLIMP1 is a tumor suppressor gene frequently disrupted in activated B cell like diffuse large B cell lymphoma. Cancer Cell 2010;18(6):568–79.

111. Mehta-Shah N, Younes A. Novel targeted therapies in diffuse large B-cell lymphoma. Semin Hematol 2015;52(2):126–37.

112. Barnes JA, Jacobsen E, Feng Y, et al. Everolimus in combination with rituximab induces complete responses in heavily pretreated diffuse large B-cell lymphoma. Haematologica 2013;98(4):615–9.

113. Boi M, Gaudio E, Bonetti P, et al. The BET bromodomain inhibitor OTX015 affects pathogenetic pathways in preclinical B-cell tumor models and synergizes with targeted drugs. Clin Cancer Res 2015;21(7):1628–38.

114. Delmore JE, Issa GC, Lemieux ME, et al. BET bromodomain inhibition as a therapeutic strategy to target c-myc. Cell 2011;146(6):904–17.

115. Lam LT, Wright G, Davis RE, et al. Cooperative signaling through the signal transducer and activator of transcription 3 and nuclear factor-B pathways in subtypes of diffuse large B-cell lymphoma. Blood 2008;111(7):3701–13.

116. Ding BB, Yu JJ, Yu RY, et al. Constitutively activated STAT3 promotes cell proliferation and survival in the activated B-cell subtype of diffuse large B-cell lymphomas. Blood 2008;111(3): 1515–24.

117. Dunleavy K, Pittaluga S, Czuczman MS, et al. Differential efficacy of bortezomib plus chemotherapy within molecular subtypes of diffuse large B-cell lymphoma. Blood 2009;113(24):6069–76.

118. Younes A, Thieblemont C, Morschhauser F, et al. Combination of ibrutinib with rituximab, cyclophosphamide, doxorubicin, vincristine, and prednisone (R-CHOP) for treatment-naive patients with CD20-positive B-cell non-Hodgkin lymphoma: a non-randomised, phase 1b study. Lancet Oncol 2014; 15(9):1019–26.

119. Young RM, Staudt LM. Targeting pathological B cell receptor signalling in lymphoid malignancies. Nat Rev Drug Discov 2013;12(3):229–43.

120. Nagel D, Bognar M, Eitelhuber AC, et al. Combinatorial BTK and MALT1 inhibition augments killing of CD79 mutant diffuse large B cell lymphoma. Oncotarget 2015;6(39):42232–42.

121. Falini B, Tiacci E, Liso A, et al. Simple diagnostic assay for hairy cell leukaemia by immunocytochemical detection of annexin A1 (ANXA1). Lancet 2004;363(9424):1869–70.

122. Tiacci E, Liso A, Piris M, et al. Evolving concepts in the pathogenesis of hairy-cell leukaemia. Nat Rev Cancer 2006;6(6):437–48.

123. Tiacci E, Trifonov V, Schiavoni G, et al. BRAF mutations in hairy cell leukemia. N Engl J Med 2011; 364(24):2305–15.

124. Arcaini L, Zibellini S, Boveri E, et al. The BRAF V600E mutation in hairy cell leukemia and other mature B-cell neoplasms. Blood 2012;119(1): 188–91.

125. Blombery PA, Wong SQ, Hewitt CA, et al. Detection of BRAF mutations in patients with hairy cell leukemia and related lymphoproliferative disorders. Haematologica 2012;97(5):780–3.

126. Boyd EM, Bench AJ, van 't Veer MB, et al. High resolution melting analysis for detection of BRAF exon 15 mutations in hairy cell leukaemia and other lymphoid malignancies. Br J Haematol 2011;155: 609–12.

127. Davies H, Bignell GR, Cox C, et al. Mutations of the BRAF gene in human cancer. Nature 2002; 417(6892):949–54.

128. Capper D, Preusser M, Habel A, et al. Assessment of BRAF V600E mutation status by immunohistochemistry with a mutation-specific monoclonal antibody. Acta Neuropathol 2011;122(1):11–9.

129. Andrulis M, Penzel R, Weichert W, et al. Application of a BRAF V600E mutation-specific antibody for the diagnosis of hairy cell leukemia. Am J Surg Pathol 2012;36(12):1796–800.

130. Dietrich S, Jennifer H, Lee SC, et al. Recurrent CDKN1B (p27) mutations in hairy cell leukemia. Blood 2015;126(8):1005–9.

131. Chu IM, Hengst L, Slingerland JM. The Cdk inhibitor p27 in human cancer: prognostic potential and relevance to anticancer therapy. Nat Rev Cancer 2008;8(4):253–67.

132. Waterfall JJ, Arons E, Walker RL, et al. High prevalence of MAP2K1 mutations in variant and IGHV4-34-expressing hairy-cell leukemias. Nat Genet 2014;46:8–10.

133. Xi L, Arons E, Navarro W, et al. Both variant and IGHV4-34-expressing hairy cell leukemia lack the BRAF V600E mutation. Blood 2012;119(14): 3330–2.

134. Flaherty K, Puzanov I, Kim K, et al. Inhibition of mutated, activated BRAF in metastatic melanoma. N Engl J Med 2010;363(9):809–19.

135. Chapman P, Hauschild A, Robert C, et al. Improved survival with vemurafenib in melanoma with BRAF V600E mutation. N Engl J Med 2011;364(26): 2507–16.

136. Dietrich S, Glimm H, Andrulis M, et al. BRAF inhibition in refractory hairy-cell leukemia. N Engl J Med 2012;366:2038–40.

137. Dietrich S, Hüllein J, Hundemer M, et al. Continued response off treatment after BRAF inhibition in refractory hairy cell leukemia. J Clin Oncol 2013; 31(19):e300–3. Available at: http://www.ncbi.nlm.nih.gov/pubmed/23690412.

138. Tiacci E, Park JH, De Carolis L, et al. Targeting mutant BRAF in relapsed or refractory hairy-cell leukemia. N Engl J Med 2015;373(18):1733–47.

139. Flaherty K, Infante J, Daud A, et al. Combined BRAF and MEK inhibition in melanoma with BRAF V600 mutations. N Engl J Med 2013;367(18): 1694–703.

140. Hunter ZR, Xu L, Yang G, et al. The genomic landscape of Waldenström macroglobulinemia is characterized by highly recurring MYD88 and WHIM-like CXCR4 mutations, and small somatic deletions associated with B-cell lymphomagenesis. Blood 2014;123(11):1637–46.

141. Poulain S, Roumier C, Decambron A, et al. MYD88 L265P mutation in Waldenstrom macroglobulinemia. Blood 2013;121(22):4504–11.

142. Treon SP, Xu L, Yang G, et al. MYD88 L265P somatic mutation in Waldenström's macroglobulinemia. N Engl J Med 2012;367(9):826–33.

143. Jiménez C, Sebastián E, Chillón MC, et al. MYD88 L265P is a marker highly characteristic of, but not restricted to, Waldenström's macroglobulinemia. Leukemia 2013;27:1722–8.

144. Gachard N, Parrens M, Soubeyran I, et al. IGHV gene features and MYD88 L265P mutation separate the three marginal zone lymphoma entities and Waldenström macroglobulinemia/lymphoplasmacytic lymphomas. Leukemia 2013;27(1):183–9.

145. Yang G, Zhou Y, Liu X, et al. A mutation in MYD88 (L265P) supports the survival of lymphoplasmacytic cells by activation of Bruton tyrosine kinase in Waldenström macroglobulinemia. Blood 2013; 122(7):1222–32.

146. Treon SP, Cao Y, Xu L, et al. Somatic mutations in MYD88 and CXCR4 are determinants of clinical presentation and overall survival in Waldenström macroglobulinemia. Blood 2014;123(18):2791–6.

147. Lagane B, Chow KYC, Balabanian K, et al. CXCR4 dimerization and 2-arrestin mediated signaling account for the enhanced chemotaxis to CXCL12 in WHIM syndrome. Blood 2008;112(1):34–44.

148. Treon SP, Tripsas CK, Meid K, et al. Ibrutinib in previously treated Waldenström's macroglobulinemia. N Engl J Med 2015;372(15):1430–40.

149. Thorns C, Bastian B, Pinkel D, et al. Chromosomal aberrations in angioimmunoblastic T-cell lymphoma and peripheral T-cell lymphoma unspecified: a matrix-based CGH approach. Genes Chromosomes Cancer 2007;46(1):37–44.

150. Sakata-Yanagimoto M, Enami T, Yoshida K, et al. Somatic RHOA mutation in angioimmunoblastic T cell lymphoma. Nat Genet 2014;46(2):171–5.

151. Palomero T, Couronné L, Khiabanian H, et al. Recurrent mutations in epigenetic regulators, RHOA and

FYN kinase in peripheral T cell lymphomas. Nat Genet 2014;46(2):166–70.

152. Cairns RA, Iqbal J, Lemonnier F, et al. IDH2 mutations are frequent in angioimmunoblastic T-cell lymphoma. Blood 2012;119(8):1901–4.

153. Lemonnier F, Couronne L, Parrens M, et al. Recurrent TET2 mutations in peripheral T-cell lymphomas correlate with TFH -like features and adverse clinical parameters. Blood 2012;120(7):1466–70.

154. Quivoron C, Couronné L, Della Valle V, et al. TET2 inactivation results in pleiotropic hematopoietic abnormalities in mouse and is a recurrent event during human lymphomagenesis. Cancer Cell 2011; 20(1):25–38.

155. Couronné L, Bastard C, Bernard O. TET2 and DNMT3A mutations in human T-cell lymphoma. N Engl J Med 2012;366(1):95–6.

156. Stein EM. IDH2 inhibition in AML: finally progress? Best Pract Res Clin Haematol 2015;28(2–3):112–5.

157. Stein EM. Molecular pathways: IDH2 mutations, co-opting cellular metabolism for malignant transformation. Clin Cancer Res 2016;22(1):16–9.

158. Bustelo XR, Sauzeau V, Berenjeno IM. GTP-binding proteins of the Rho/Rac family: regulation, effectors and functions in vivo. Bioessays 2007;29(4):356–70.

159. Etienne-Manneville S, Hall A. Rho GTPases in cell biology. Nature 2002;420(6916):629–35.

160. Hanna S, El-Sibai M. Signaling networks of Rho GTPases in cell motility. Cell Signal 2013;25(10): 1955–61.

161. Hall A. Rho family GTPases. Biochem Soc Trans 2012;40(6):1378–82.

162. Mylona A, Nicolas R, Maurice D, et al. The essential function for serum response factor in T-cell development reflects its specific coupling to extracellular signal-regulated kinase signaling the essential function for serum response factor in T-cell development reflects its specific coupling. Society 2011;31(2):267–76.

163. Fleige A, Alberti S, Gröbe L, et al. Serum response factor contributes selectively to lymphocyte development. J Biol Chem 2007;282(33):24320–8.

164. Figueroa ME, Abdel-Wahab O, Lu C, et al. Leukemic IDH1 and IDH2 mutations result in a hypermethylation phenotype, disrupt TET2 function, and impair hematopoietic differentiation. Cancer Cell 2010;18(6):553–67.

165. Yen KE, Bittinger MA, Su SM, et al. Cancer-associated IDH mutations: biomarker and therapeutic opportunities. Oncogene 2010;29(49):6409–17.

166. Parsons W, Jones S, Zhang X, et al. An integrated genomic analysis of human glioblastoma multiforme. Science 2008;321(5897):1807–12.

167. Yan H, Parsons DW, Jin G, et al. Mutations in gliomas. N Engl J Med 2009;360(8):765–73.

168. Cancer Genome Atlas Research Network. Comprehensive, integrative genomic analysis of diffuse lower-grade gliomas. N Engl J Med 2015;372(26): 2481–98.

169. Mardis ER, Ding L, Dooling DJ, et al. Recurring mutations found by sequencing an acute myeloid leukemia genome. N Engl J Med 2009;361(11): 1058–66.

170. Marcucci G, Maharry K, Wu YZ, et al. IDH1 and IDH2 gene mutations identify novel molecular subsets within de novo cytogenetically normal acute myeloid leukemia: a cancer and leukemia group B study. J Clin Oncol 2010;28(14):2348–55.

171. Ward PS, Patel J, Wise DR, et al. The common feature of leukemia-associated IDH1 and IDH2 mutations is a neomorphic enzyme activity converting α-ketoglutarate to 2-hydroxyglutarate. Cancer Cell 2010;17(3):225–34.

172. Dang L, White DW, Gross S, et al. Cancer-associated IDH1 mutations produce 2-hydroxyglutarate. Nature 2009;462(7274):739–44.

173. Gross S, Cairns RA, Minden MD, et al. Cancer-associated metabolite 2-hydroxyglutarate accumulates in acute myelogenous leukemia with isocitrate dehydrogenase 1 and 2 mutations. J Exp Med 2010;207(2):339–44.

174. Cimmino L, Abdel-Wahab O, Levine RL, et al. TET family proteins and their role in stem cell differentiation and transformation. Cell Stem Cell 2011;9(3): 193–204.

175. Ito S, Shen L, Dai Q, et al. Tet proteins can convert 5-methylcytosine to 5-formylcytosine and 5-carboxylcytosine. Science 2011;333(6047):1300–3.

176. Tahiliani M, Koh KP, Shen Y, et al. Conversion of 5-methylcytosine to 5-hydroxymethylcytosine in mammalian DNA by MLL partner TET1. Science 2009;324(5929):930–5.

177. Delhommeau F, Dupont S, Della Valle V, et al. Mutation in TET2 in myeloid cancers. N Engl J Med 2009;360(22):2289–301.

178. Abdel-Wahab O, Mullally A, Hedvat C, et al. Genetic characterization of TET1, TET2, and TET3 alterations in myeloid malignancies. Blood 2009; 114(1):144–7.

179. Yang L, Rau R, Goodell MA. DNMT3A in haematological malignancies. Nat Rev Cancer 2015;15(3): 152–65.

180. Wang F, Travins J, Delabarre B, et al. Targeted inhibition of mutant IDH2 in leukemia cells induces cellular differentiation. Science 2013;340(6132): 622–6.

181. Francisco JA, Cerveny CG, Meyer DL, et al. cAC10-vcMMAE, an anti-CD30–monomethyl auristatin E conjugate with potent and selective antitumor activity. Blood 2003;102(4):1458–65.

182. Horwitz SM, Advani RH, Bartlett NL, et al. Objective responses in relapsed T-cell lymphomas with single-agent brentuximab vedotin. Blood 2014; 123(20):3095–100.

183. Cheminant M, Bruneau J, Kosmider O, et al. Efficacy of 5-azacytidine in a TET2 mutated angioimmunoblastic T cell lymphoma. Br J Haematol 2015;168(6):913–6.

184. Steinway SN, LeBlanc F, Loughran TP. The pathogenesis and treatment of large granular lymphocyte leukemia. Blood Rev 2014;28(3):87–94.

185. Koskela H, Eldfors S, Ellonen P, et al. Somatic STAT3 mutations in large granular lymphocytic leukemia. N Engl J Med 2012;366(20):1905–13.

186. Rossi D, Franceschetti S, Capello D, et al. Transient monoclonal expansion of CD8+/CD57+ T-cell large granular lymphocytes after primary cytomegalovirus infection. Am J Hematol 2007;82(12):1103–5.

187. Mohty M, Faucher C, Vey N, et al. Features of large granular lymphocytes (LGL) expansion following allogeneic stem cell transplantation: a long-term analysis. Leukemia 2002;16(10):2129–33.

188. Jerez A, Clemente MJ, Makishima H, et al. STAT3 mutations unify the pathogenesis of chronic lymphoproliferative disorders of NK cells and T cell large granular lymphocyte leukemia. Blood 2012; 120(15):3048–58.

189. Fasan A, Kern W, Grossmann V, et al. STAT3 mutations are highly specific for large granular lymphocytic leukemia. Leukemia 2013;27(7):1598–600.

190. Rajala H, Eldfors S, Kuusanmaki H, et al. Discovery of somatic STAT5b mutations in large granular lymphocytic leukemia. Blood 2013;121(22):4541–51.

191. Andersson EI, Rajala HLM, Eldfors S, et al. Novel somatic mutations in large granular lymphocytic leukemia affecting the STAT-pathway and T-cell activation. Blood Cancer J 2013;3(12):e168.

192. O'Shea JJ, Plenge R. JAK and STAT signaling molecules in immunoregulation and immune-mediated disease. Immunity 2012;36(4):542–50.

193. Haapaniemi EM, Kaustio M, Rajala HLM, et al. Autoimmunity, hypogammaglobulinemia, lymphoproliferation and mycobacterial disease in patients with activating mutations in STAT3. Blood 2015; 125(4):639–49.

194. Loughran TP, Zickl L, Olson TL, et al. Immunosuppressive therapy of LGL leukemia: prospective multicenter phase II study by the Eastern Cooperative Oncology Group (E5998). Leukemia 2015; 29(4):886–94.

195. Hayakawa F, Sugimoto K, Harada Y, et al. A novel STAT inhibitor, OPB-31121, has a significant antitumor effect on leukemia with STAT-addictive oncokinases. Blood Cancer J 2013;29(3):e166.

196. Aparicio-Siegmund S, Sommer J, Monhasery N, et al. Inhibition of protein kinase II (CK2) prevents induced signal transducer and activator of transcription (STAT) 1/3 and constitutive STAT3 activation. Oncotarget 2014;5(8):2131–48.

197. Munoz J, Dhillon N, Janku F, et al. STAT3 inhibitors: finding a home in lymphoma and leukemia. Oncologist 2014;19(5):536–44.

Printed and bound by CPI Group (UK) Ltd, Croydon, CR0 4YY

03/10/2024

01040304-0013